T0235979

HELIOPHYSICS: PLASMA PHYSICS
OF THE LOCAL COSMOS

Edited by Carolus J. Schrijver and George L. Siscoe

Heliophysics is a fast-developing scientific discipline that integrates studies of the Sun's variability, the surrounding heliosphere, and the environment and climate of the planets. Over the past few centuries, our understanding of how the Sun drives space weather and climate on the Earth and other planets has advanced at an ever increasing rate. The Sun is a magnetically variable star and, for planets with intrinsic magnetic fields, planets with atmospheres, or planets like Earth with both, there are profound consequences.

This volume, the first in a series of three heliophysics texts, integrates these diverse topics for the first time as a coherent intellectual discipline, providing a core resource for courses and seminars at the advanced undergraduate and graduate level. It emphasizes the physical processes that couple the realm of the Sun to that of our planet and provides insights into the interaction of the solar wind and radiation with the Earth's magnetic field, atmosphere, and climate system. In addition to its utility as a textbook, it also constitutes a foundational reference for researchers in the fields of heliophysics, astrophysics, plasma physics, space physics, solar physics, aeronomy, space weather, planetary science, and climate science. Additional online resources, including lecture presentations and other teaching materials, can be accessed at www.cambridge.org/9780521110617.

CAROLUS J. SCHRIJVER is an astrophysicist studying the causes and effects of magnetic activity of the Sun, and of stars like the Sun, and the coupling of the Sun's magnetic field into the surrounding heliosphere. He obtained his doctorate in physics and astronomy at the University of Utrecht in The Netherlands in 1986 and has since worked for the University of Colorado, the US National Solar Observatory, the European Space Agency, and the Royal Academy of Sciences of the Netherlands. Dr Schrijver is currently principal physicist at Lockheed Martin's Advanced Technology Center, where his work focuses primarily on the magnetic field in the solar atmosphere. He is an editor or editorial board member of several journals including *Solar Physics*, *Astronomical Notices*, and *Living Reviews in Solar Physics*, and has co-edited three other books.

GEORGE L. SISCOE received his Ph.D. in physics from the Massachusetts Institute of Technology (MIT) in 1964. He has since held positions at the California Institute of Technology, MIT, and the University of California, Los Angeles – where he was Professor and Chair of the Department of Atmospheric Sciences. He is currently a Research Professor in the Astronomy Department at Boston University. Professor Siscoe has been a member and chair of numerous international committees and panels and is on the editorial board of the *Journal of Atmospheric and Solar Terrestrial Physics*. He is a Fellow of the American Geophysical Union and the second Van Allen Lecturer of the AGU, 1991. He has authored or co-authored over 300 publications that cover most areas of heliophysics.

HELIOPHYSICS: PLASMA PHYSICS OF THE LOCAL COSMOS

Edited by

CAROLUS J. SCHRIJVER
Lockheed Martin Advanced Technology Center

GEORGE L. SISCOE
Boston University

CAMBRIDGE
UNIVERSITY PRESS

CAMBRIDGE UNIVERSITY PRESS
Cambridge, New York, Melbourne, Madrid, Cape Town,
Singapore, São Paulo, Delhi, Tokyo, Mexico City

Cambridge University Press
The Edinburgh Building, Cambridge CB2 8RU, UK

Published in the United States of America by Cambridge University Press, New York

www.cambridge.org
Information on this title: www.cambridge.org/9781107403222

© Cambridge University Press 2009

This publication is in copyright. Subject to statutory exception
and to the provisions of relevant collective licensing agreements,
no reproduction of any part may take place without the written
permission of Cambridge University Press.

First published 2009
First paperback edition 2011

A catalogue record for this publication is available from the British Library

ISBN 978-0-521-11061-7 Hardback
ISBN 978-1-107-40322-2 Paperback

Cambridge University Press has no responsibility for the persistence or accuracy of
URLs for external or third-party internet websites referred to in this publication, and
does not guarantee that any content on such websites is, or will remain, accurate or
appropriate.

Contents

The plates are to be found between pages 406 and 407*.
* These plates are available for download in colour from
www.cambridge.org/9780521110617

Preface

Over the past few centuries, our awareness of the coupling between the Sun's variability and the Earth's environment, and perhaps even its climate, has been advancing at an ever increasing rate. The Sun is a magnetically variable star and, for planets with intrinsic magnetic fields, planets with atmospheres, or planets like Earth with both, there are profound consequences and impacts. Today, the successful increase in knowledge of the workings of the Sun's magnetic activity, the recognition of the many physical processes that couple the realm of the Sun to our galaxy, and the insights into the interaction of the solar wind and radiation with the Earth's magnetic field, atmosphere and climate system have tended to differentiate and insularize the solar heliospheric and geo-space sub-disciplines of the physics of the local cosmos. In 2001, the NASA Living With a Star (LWS) program was initiated to reverse that trend.

The recognition that there are many connections within the Sun–Earth systems approach has led to the development of an integrated strategic mission plan and a comprehensive research program encompassing all branches of solar, heliospheric, and space physics and aeronomy. In doing so, we have developed an interdisciplinary community to address this program. This has raised awareness and appreciation of the research priorities and challenges among LWS scientists and has led to observational and modeling capabilities that span traditional discipline boundaries. The successful initial integration of the LWS sub-disciplines, under the newly coined term "heliophysics", needed to be expanded into the early education of scientists. This series of books is intended to do just that: aiming at the advanced undergraduate and starting graduate-level students, our aim is to teach heliophysics as a single intellectual discipline. Heliophysics is important both as a discipline that will deepen our understanding of how the Sun drives space weather and climate at Earth and other planets and also as a discipline that studies universal astrophysical processes with unrivaled resolution and insight possibilities. The goal of this series is to provide seed materials for the development of new researchers and new scientific discovery.

Richard Fisher, Director of NASA's Heliophysics Division
Madhulika Guhathakurta, NASA/LWS program scientist

Heliophysics

helio-, prefix, on the Sun and environs; from the Greek *helios*.
physics, n., the science of matter and energy and their interactions.

Heliophysics is the

- *comprehensive new term for the science of the Sun–solar-system connection.*
- *exploration, discovery, and understanding of our space environment.*
- *system science that unites all the linked phenomena in the region of the cosmos influenced by a star like our Sun.*

Heliophysics concentrates on the Sun and its effects on Earth, the other planets of the solar system, and the changing conditions in space. Heliophysics studies the magnetosphere, ionosphere, thermosphere, mesosphere, and upper atmosphere of the Earth and other planets. Heliophysics combines the science of the Sun, corona, heliosphere and geospace. Heliophysics encompasses cosmic rays and particle acceleration, space weather and radiation, dust and magnetic reconnection, solar activity and stellar cycles, aeronomy and space plasmas, magnetic fields and global change, and the interactions of the solar system with our galaxy.

From NASA's *Heliophysics. The New Science of the Sun–Solar-System Connection: Recommended Roadmap for Science and Technology 2005–2035.*

1

Prologue

CAROLUS J. SCHRIJVER AND GEORGE L. SISCOE

1.1 A voyage through the local cosmos

The place that we call home, the surface of the planet Earth, presents us with an environment in which temperatures range over perhaps 80 kelvins from the cool arctic regions or mountain tops to the hottest deserts or jungles. We are composed largely of liquid water with a density of 1 gram per cubic centimeter; we walk on solid rock with a density that is about five times higher than this and breathe a gas with a density that is 1000 times lower. These conditions are such that chemical reactions and phase transitions between solids, liquids, and gases are the processes that dominate our everyday experience.

When we move away from the Earth's surface, conditions change markedly. Deep in the Earth, for example, where densities are still only a few times higher than those at the surface, the pressure rapidly increases and temperatures reach up to some 20 times those characteristic of the range that is comfortable to mammals. In the Sun's core densities are larger still, almost a hundred times that of liquid water, at temperatures that exceed ten million kelvins. Those same temperatures may be found again in the hottest, flaring parts of the Sun's outermost atmosphere, called the corona, and furthermore are often characteristic of the ion energies high above the Earth around the altitudes where geosynchronous satellites orbit. The temperatures may be comparable in these three domains (at least to astrophysical standards), but the corresponding densities are 17 to 25 *orders of magnitude* different. Overall, from the Sun's center to the solar wind in the outer heliosphere, densities range over about 29 orders of magnitude while temperatures range over five orders of magnitude. This astonishing range in conditions (Fig. 1.1) is compounded by another phenomenon that prevails in most of the sphere of influence of the Sun, i.e. the domain of heliophysics: matter is generally so hot that many electrons are stripped off their atoms by collisions between these particles. This results in a electrically conducting gaseous medium that is known

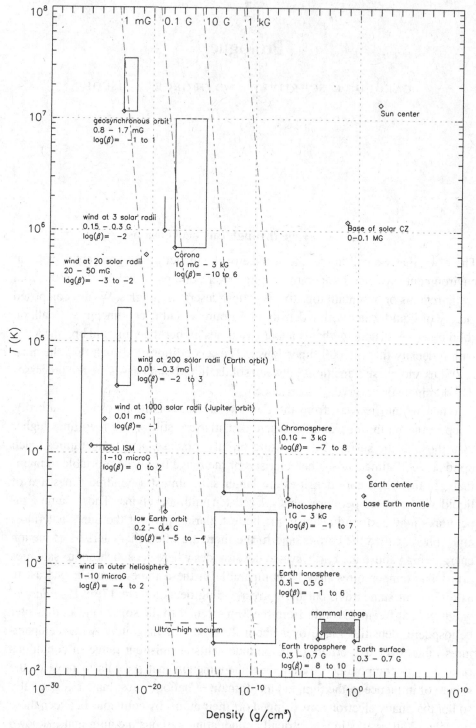

Fig. 1.1.

as plasma (often referred to as the fourth state of matter after solids, liquids, and gases).

Whereas for the largest scales of the universe physicists are puzzled by mysterious properties that have become known as "dark energy" and "dark matter", on the scale of the local cosmos it would appear that the classical forces of gravity, electromagnetism, and the weak and strong nuclear interactions are known so well that one may ask what it is that occupies solar, heliospheric, and space physicists. As it turns out, it is not in the nature of the forces that mysteries reside but rather in the complexity of the processes by which they interact within the ionized plasma.

In this series of three volumes on heliophysics we will discuss many of these processes. In this introductory chapter, we will take you on an exploratory journey from within the Sun to the edges of the solar system, to introduce the variety of environments and processes that would be encountered on that journey. In doing so, we also introduce some of the technical terms that form the jargon of the sub-disciplines of heliophysics. This jargon is often specific to a discipline: many terms are used actively only by part of the heliophysics community. These terms may not be familiar to colleagues in relatively distant sub-disciplines, while some terms may even be used by them for distinctly different processes.

1.1.1 Solar interior

Let us start in the deep interior of the Sun, some 200 000 km below the surface. Here we are just underneath the Sun's convective mantle in which hot upwellings alternate with cool downflows to effect a net transfer of heat through the mantle to the visible surface, the photosphere. Here, just below this convectively unstable envelope, we are in an environment that is stably stratified and in which the opacity remains low enough for the Sun's energy to be transported exclusively in the form of electromagnetic radiation. This radiation makes its way from the nuclear furnace

---

Fig. 1.1. Temperature vs. density for a variety of conditions within the local cosmos. Some typical ranges are indicated, and labeled with magnetic field strengths (in gauss) found in that domain and estimated ranges of the plasma β, i.e. the ratio of energy density in the plasma and that in the magnetic field (Eq. 3.11). This is but one of many diagrams that characterize the multidimensional world of heliophysics; in other figures collision rates, ionization states, scales of time, length, or velocity could be compared, but this diagram relates directly to our intuitive knowledge of the world around us. The reader is encouraged to experiment with such other diagrams using the data found throughout the three volumes of this series.

in the core to the convective envelope, in a random walk caused by frequent scattering off the ionized plasma particles. The rumbling of distant convective motions overhead is heard as sound waves traveling through this deep domain of the resonant cavity, which shapes the millions of pressure modes into standing wave patterns analyzed by helioseismologists.

Some large-scale plasma flows that occur here are imposed by convective down-draft plumes that overshoot beyond the base of the convective envelope, owing to their large momentum and inertia. As they gradually overturn into upwellings, they may deposit entrained magnetic field into this overshoot region or, conversely, they may dredge up magnetic field from there into the convective envelope. The only other large-scale motion in this region is a weak shearing flow, called the differential rotation, that extends from within the convective envelope to a thin layer, called the tachocline, just below it. The tachocline is sandwiched between the differentially rotating convective envelope, where the rate of rotation decreases from the equator to the pole, and the radiative interior, which rotates essentially as a solid body (except, maybe, for the deep core that may still possess some of the initially high angular momentum from when the Sun – and its attendant planetary system – formed 4.5 billion years ago).

Let us start an imaginary journey by selecting, in this region below the base of the convective envelope, a hydrogen ion embedded in a strand of magnetic field that is in the process of being scooped up by convective overshoot. Shortly after its rise begins into the convective envelope, the opacity increases when electrons begin to recombine with helium (and higher up with hydrogen) ions as the temperature drops. As that happens, the exchange of energy between a moving volume of plasma and its surroundings is impeded. Consequently, when the gas moves and responds to the changing pressure in the gravitationally stratified surroundings these changes will be largely adiabatic, leaving the rising gas hotter and lighter and the sinking gas cooler and heavier: convection becomes the avenue of choice for energy transport as radiative transport decreases in relative importance.

The hydrogen ion that we are tracking is not only driven by the buoyancy of the adiabatically rising gas but also by a magnetic effect. Our ion was, in fact, contained in a region of lower gas pressure because the pressure of the magnetic field around the ion adds to that of the gas, so that the volume of gas has expanded to establish pressure equilibrium with the non-magnetized surroundings. As the temperature between the interior and the surroundings is likely to be equalized by radiative exchange, the result is that field-carrying regions are buoyant relative to regions with weaker field. This makes it hard for field to stay deep within the convective envelope. In fact, many theorists argue that field is stored in the non-convecting overshoot layer or just above it where buoyancy is statistically countered by the downward drag from strong downflows. The overshoot region that we have just left behind may thus be one of the most important domains in heliophysics if indeed that is where much of the solar magnetic field is stored, caught in flows that strengthen it by pulling on it with the shearing winds that we know as the differential rotation (see Chapter 3 and Vol. III).

Most upward convective motions rapidly overturn into downflows again: as we gain altitude the gas pressure rapidly decreases and the density drops; upflows expand and cool adiabatically and return to the depths of the envelope. The small fractions of the gas and field that reach the surface take a few weeks for that journey. This time is comparable with the Sun's rotation period. The rising plasma consequently senses the Sun's rotation, which causes the field strands to have a preferential twist on emergence. This twist, with much scatter from case to case because of interaction with the turbulent convective motions, is opposite in the two hemispheres on either side of the solar equator, imposing some degree of order and resulting in a net dipole moment that is the ultimate cause of the heliospheric structure (Chapter 8 and Vol. III). The coupling of buoyant flows and rotation is an important ingredient in the dynamo theory that aims to describe the evolution of the Sun's magnetic field (Chapter 3 and Vol. III).

Only one in 100 000 ions that start their rise near the bottom of the convective envelope typically makes it to the topmost layers. There, convective flows move fast, at a large fraction of the sound speed, and consequently this is where the convection is, literally, loudest: here most of the power in the helioseismic pressure waves is generated. Here, too, is where upward propagating waves effectively hit a wall and are mostly reflected back down: the pressure gradient becomes so steep in the surface layers that long-period waves lift most of the mass in the overlying atmosphere without generating the elastic restoring forces that allow them to propagate. Above the photosphere it is consequently much quieter, and the deep rumbling with periods of several minutes is replaced by a higher pitch of lower intensity as lower frequencies are reflected and higher frequencies damped by radiative exchange between the compression and rarefaction regions in the waves.

Just below the surface, or photosphere, lies a mystery for solar physicists: the pressure gradient here is so large that it is hard to imagine how anything could avoid enormous expansion on rising. Yet much of the magnetic field that surfaces does so in a fibril state with structures of order 100 km, i.e. smaller than even the dominant scale of convection near the surface, the 1000 km scale of the granulation. As theorists try to understand this phenomenon, one scenario stands out: if the magnetic field is twisted, i.e. if it carries an electrical current, then the increased field tension resists expansion and convective shredding much more effectively. Let us therefore imagine that the ion we are tracking is caught in a magnetic environment that also supports a substantial electrical current.

1.1.2 Solar atmosphere

As the field breaches the solar surface, the plasma that it carries becomes transparent – this is, quite simply, the transition that defines the surface of this gaseous sphere. Once transparent, the photons emitted by the gas travel long distances. Most in fact are free to fly into interplanetary space, or – if they do not run into a planet or satellite – even into interstellar space and beyond. The efficient escape of its thermal energy causes most plasma within the magnetic field to cool down and slump back below the solar surface, sliding down along the emerging field.

As the magnetic field rises from the photosphere, it enters a domain where electromagnetic forces generally dominate over gas pressure and inertia and where the Alfvén speed becomes so large compared with the field's spatial and temporal scales that the field, for the first time, behave as is truly three-dimensional and continuous. Thus the field emerging into the solar atmosphere almost immediately begins to interact with any pre-existing field around it (Chapters 5 and 8, and Vol. II).

The magnetic field in the solar atmosphere (e.g. Figs. 1.2 and 8.1) has characteristics that are very different from those of the geomagnetic field in the observable domain above the Earth's surface (Fig. 1.3). The Sun's atmospheric magnetic field is the result of a multitude of flux ropes that emerge through the solar surface, ranging from millions of small ones that live for only minutes to a few very large ones that persist for weeks (at most a dozen of these exist on the solar surface at any one time even at cycle maximum); they extend over at least six orders of magnitude in total absolute flux, as far as we can currently observe. These fields move about, pushed by (sub-)surface flows ranging from small-scale granulation, with cells the size of Texas, Germany, or the Republic of the Congo, to the largest patterns, associated with the differential rotation and hemispheric meridional advection from the equator to the poles (Chapter 8 and Vol. III). The Earth's magnetic field, in contrast, appears to be dominantly dipolar, which is likely to be in part a consequence of the fact that the higher-order multipoles decay away between the outer layers of the dynamo region (i.e. the outer core) and the top of the Earth's slowly moving mantle: sensitive measurements and computer simulations suggest that the inner field in the Earth has a much more complex dynamic structure (see Vol. III; the same is true for the other magnetized planets, see Chapter 13).

As our hydrogen ion rises within the embedding field into the solar atmosphere, the environment changes rapidly. The density drops by seven orders of magnitude within a few thousand kilometers (which is less than $\sim 0.5\%$ of the solar radius) to values comparable with those in the much cooler ionosphere of the Earth. The only reason why the tenuous region above the solar photosphere, called the chromosphere, is visible to us at all from Earth is that its large volume compensates for the

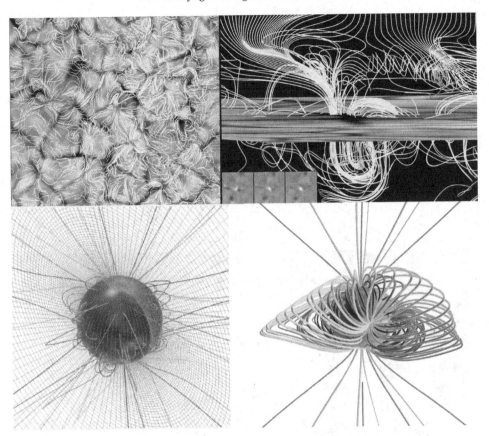

Fig. 1.2. The multitude of scales in the solar magnetic field. The upper panels show a model computation of the magnetic field (by Abbett, 2007; see also Fig. 8.11) on the scale of the dominant convective motions at the solar surface, the 1000 km scale of the granulation (see also Fig. 8.3); the upper left panel is a top view of the solar surface, with sample magnetic field lines (see Section 4.1) overplotted, while the upper right panel shows a vertical cut through one of the convection cells to illustrate how the field in this model can thread the surface multiple times, evolving on a time scale of a few minutes. The lower two panels (from the Center for Integrated Space Weather Modeling) show models of the global solar field, tracing field lines up to the cusps of the streamers that outline the topologically distinct regions of closed field and the field that is open to the heliosphere; this global-scale field evolves on time scales of months to a decade (see Chapters 4 and 9).

low density, resulting in an observable spectral signature. The Earth's ionosphere, in contrast, is virtually invisible from the Earth's surface at visible wavelengths under quiescent daytime conditions – it glows visibly at night and is readily seen looking past the edge of the Earth from space (Chapter 12); the "airglow" is a combination of fluorescence associated with ionization and dissociation processes, resonant

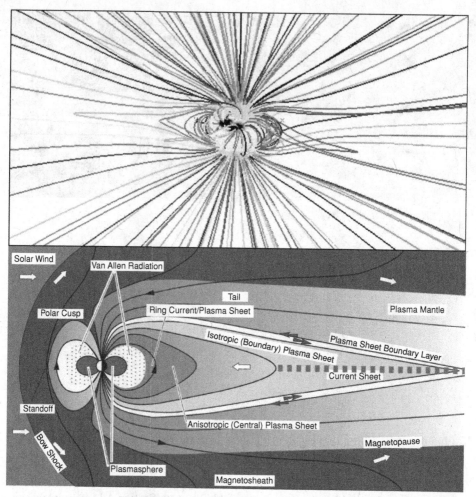

Fig. 1.3. (Upper panel) Magnetohydrodynamic model computation of a simplified solar magnetic field reaching into the heliosphere. The large-scale dipolar field of the Sun is blown open by the solar wind (Chapter 8), stretching out the relatively strong, closed, field of the activity belt into "streamers" with cusps that lie at the base of the heliospheric current sheet. (Courtesy of P. Riley, SAIC.) The stretching of magnetic field arcades into a cusp-and-sheet structure is a common features in space plasmas (compare, for example, Figs. 2.7, 5.1, 6.3, and 10.1 with the lower panel here). (Lower panel) Schematic of the Earth's magnetic field and magnetospheric features. Earth provides the prototype magnetosphere; it is produced by the interaction of the solar wind with the planet's internally generated magnetic field. The boundary of the region dominated by the planet's magnetic field is the magnetopause. The supersonic solar wind is sharply slowed and deflected at the bow shock, flowing around the planet in a magnetosheath boundary layer. It pulls the magnetosphere back behind the planet into a long magnetotail with a current sheet and cusp structures. (Courtesy of L. Weiss.)

scattering by select ions, and recombination emission ranging from infrared to visible spectral lines.

We are now in the low chromosphere, and the environment changes again. For one thing, the acoustic spectrum is different: not only is there less power in the waves but shock waves begin to form. Sound waves traveling upward into the ever more tenuous outer atmosphere steepen, and eventually form shocks that throw chromospheric plasma up to heights of some 10 000 km, before it falls down again to be caught by another shock front. The compression in these shocks can temporarily heat the plasma, but it cools again as it radiates much of that energy away during the passage of the shock. Other, magnetohydrodynamic, waves are also likely to dissipate energy, as do electrical currents in the dynamic magnetic field (Chapter 8). These heat the plasma, by processes that remain to be discovered, to temperatures of 10 000 to 20 000 K. Elsewhere in this environment, temperatures as low as 2500 K may exist because of adiabatic expansion of the now transparent plasma. These low temperatures allow some simple molecules, such as CO and CH, to form. But apart from these cool supra-photospheric regions and the cool sunspots (where e.g. TiO molecular lines can be used to measure the field strength) the Sun contains only monatomic gases.

As the rising field expands into the highest layers of the solar atmosphere, i.e. the corona, very little of the plasma can follow it against gravity. When our ion reaches this domain, the densities are so low that rocket engineers on Earth would effectively call it outer space; these are densities that only high-quality vacuum machines can achieve. But, high in this corona, waves and currents somehow deposit energy which finds its way down to the top of the chromosphere both in the form of energetic particles and as thermal energy conducted largely by electrons. That energy heats a small fraction of the chromospheric plasma (including our proton) to a few million kelvins. This gives the plasma the energy to fill the corona against gravity and to radiate in the extreme ultraviolet or X-rays region. These losses are balanced, at least statistically, by the dissipation of more non-radiative energy as currents or waves; eventually, it is converted to thermal energy.

The field is now so strong relative to the plasma, which has a mass density some 15 to 18 orders of magnitude below that of water, that plasma motions can only occur along the field, leading to the giant glowing arches called coronal loops. The plasma thus embedded by the field can carry electrical currents, and in our case the strong electrical current that emerges with the field leads somehow to an instability: the field and current are subjected to a poorly understood process called reconnection (Chapter 5), and some fraction of the field may temporarily be opened up into the heliosphere, again by processes that are far from being understood. The ion that we are tracking is now no longer on a field line that has two connections with the solar surface.

1.1.3 Inner heliosphere

The heating and acceleration of the solar wind low in the corona remain major problems (Chapter 9). A variety of waves and turbulence mechanisms have been proposed as playing a role: the dissipation of waves can obviously heat the plasma, and this is the primary requirement to start a wind that is slow near the base but supersonic in the heliosphere (as proposed by Parker around the middle of the twentieth century). The density in the nascent solar wind is so low that the constituent species in the plasma begin to decouple. A description of the physics requires that the electron, proton, and ionized helium populations, at the least, be differentially treated; the low mass of the electrons tends to give them a large scale height above the solar surface, which initiates an electric field that pulls the protons and helium ions up and in which helium, as the heavier element, is partly left behind. A similar process happens in open magnetospheres, where a so-called polar wind outflow is driven by the readily escaping electrons, which pull up protons and heavier atmospheric ions with them (Chapters 10 and 12).

Now our ion is on its way to the planets as part of the solar wind. It is interesting to note that the mass lost from the Sun by the solar wind is, in fact, only about half the mass lost by the outflow of light as a result of nuclear fusion in the Sun's core.

As our ion moves away from the Sun the expansion of the plasma causes its temperature to fall (speaking loosely, recognizing that temperature is a concept that applies to a collection of particles), even in this environment of infrequent collisions. The fast electrons conduct energy outward from the lower regions of the solar corona, but more is needed to keep the temperature high enough to match in situ observation in the inner heliosphere. The damping of Alfvénic waves, possibly after conversion into a magnetohydrodynamic (MHD) turbulent state, is a possible source of this heat deposition over much of the inner solar system (Chapter 7).

As the wind flows out, essentially radially, the Sun rotates underneath it (Chapter 9). Faster wind-flows from neighboring regions can thus move underneath and subsequently overtake slower streams that started earlier, and at their interfaces shocks may form if the velocity differential exceeds the Alfvén speed. A few times a day the solar corona is rocked by a massive explosion called a coronal mass ejection (CME). These explosions (Vol. II) are the result of the destabilization of large magnetic configurations associated with the substantial electrical currents in the solar corona. These current systems form either by the direct injection of prefabricated currents that emerge from below the surface or by induction as the atmospheric field is subjected to stress by surface flows. Destabilization occurs perhaps because a newly emerging flux system is at an odd angle with the pre-existing field, because the system is pushed over a limit beyond which the overall field configuration must change, or because of a gradual evolution of the surrounding and overlying magnetic field; in fact, each process appears to occur under certain circumstances (Vol. II).

Once the field destabilizes, the rising field plows through the overlying field, forcing it to reconnect as it moves around it. The stretched field lines anchoring the initial configuration also reconnect at some point, pinching off most of the field in the rising coronal mass ejection. These reconnection processes, as well as the shock that precedes the coronal mass ejection, result in the acceleration of a small fraction of the particle population to remarkable energies (Vol. II). Many of these particles associated with the reconnection behind the mass ejection propagate back toward the Sun, where they penetrate into the lower atmosphere. There they deposit their energy in collisions with the chromospheric plasma. That raises its temperature (with associated brightening of the flare ribbons), and this increases the pressure scale height so that hot matter "evaporates" into the corona. This fills the post-eruption (or post-flare) magnetic loop systems. The subsequent cooling of the plasma by conduction and radiation results in successive appearances of ever higher loops, as seen in soft X-ray and extreme ultraviolet images of the solar corona. Similar processes occur in the Earth's magnetosphere when plasmoids in the geotail are expelled in magnetospheric substorms associated with aurorae, the equivalent of flare ribbons, as energetic electrons deposit their energy in the highest layers of the Earth's atmosphere (Vol. II).

Some energetic particles formed during the reconnection processes or associated with the shock leading coronal mass ejection (see Vol. II) follow the heliospheric magnetic field. Heliospheric shocks send such particles both outward and inward, so that the particle beams passing an observer come from both shocks closer to the Sun than the observer and from shocks farther away from the Sun than the observer. Such particle storms traveling within the ecliptic plane can reach the planets and their (natural and artificial) satellites. In the absence of a magnetic shield (as in the cases of Venus and the Earth's Moon), the particles will hit the atmosphere or even the solid surface. Where the planet has a magnetic field, the particles find their way through the evolving yet persistent configuration of the interacting planetary and interplanetary magnetic fields (all these aspects are addressed in Vol. II).

1.1.4 Outer heliosphere

Our ion, once flowing out in the relatively quiet background solar wind, has a chance of very roughly one in 10 of being overtaken by the shock front of a coronal mass ejection before it reaches Earth's orbit. More frequently other, faster, wind streams overtake it and if the velocity difference is large enough then an interaction front forms, associated with a shock. More often than not, however, solar wind particles *en route* to Earth will find themselves first confronted with the shock wave that stands in front of the planetary magnetosphere (Chapters 10 and 11). The plasma is decelerated by this shock (and heated in the process) and then forced to flow subsonically sideways around the magnetopause, the interface between the interplanetary magnetic field and the Earth's magnetic field. It will gradually pick up speed again as it flows around the Earth and along the geomagnetic tail.

The small fraction of the solar wind that approaches the Earth very near to the nose of the magnetospheric shock will flow close by the Earth's magnetic field, and the interplanetary field can reconnect to it. This happens at the magnetopause, which in the case of Earth lies somewhere between a geosynchronous orbit and a position slightly beyond the Moon's orbit, depending on the solar wind speed (or, more precisely, the associated bulk kinetic momentum density or ram pressure; see Chapters 10 and 13).

As the fields of Sun and Earth connect on the sunward edge of the magnetopause, an "open" field line is formed connecting the interplanetary magnetic field to the Earth's polar caps. Energetic particles (electrons or protons) formed in the reconnection process or originating far away in the solar wind or corona at the shock fronts of coronal mass ejections find their way into the upper atmosphere, contributing to aurorae over the polar caps (Chapter 12 and Vol. II).

The open field lines, with ends in the fast flowing solar wind, are swept across the polar cap and eventually into the tail of the magnetosphere. There, again, reconnection can happen (energizing both electrons and protons) as the field closes within the plasma sheet, the layer between the oppositely directed magnetic fields that are stretched into the geotail. The processes of opening, transfer by advection in the solar wind, closing in the geotail, and subsequent sunward advection of the closed field form a cycle with a time scale that, for the Earth, is much shorter than the Earth's day. For a planet with a very large magnetosphere, a long tail, and a fairly short rotation period, such as Jupiter, this cycle takes longer than a planetary day, which considerably complicates the picture of the field's evolving geometry (Chapters 10 and 13).

Our ion, trapped in the Earth's magnetosphere after reconnection in the tail behind the Earth of the magnetic field to which it is bound, now advects forward (the term "convection" is used here by magnetospheric physicists). The ion has been energized by traversing the bow shock and by two reconnection events. Many of its fellow travelers will impact the ionosphere, affecting radio traffic there or contributing to the luminescent aurorae (Chapter 12). But our ion is fortunate and, in fact, after completing the convection cycle finds itself again on a newly opened field line and escapes into the solar wind.

Here, it finds itself in a very similar process to that to which it was subjected as it first found itself in the solar wind: again the population of energetic electrons readily makes its way into the solar wind, the heavier ions (many of which are

Earth-atmospheric oxygen ions) lagging behind because of their inertia and the resulting electric field in fact accelerating the protons to follow the electrons into the newly accessible heliosphere (see Chapter 8).

The proton again sweeps over the polar cap. The field's motion induces a strong electrical current in the lower ionosphere, in turn leading to a strong Lorentz force on the ions and electrons there. This ionized population couples, through collisions, with the surrounding neutral-particle population (typically 10 000 times more prevalent) of the thermosphere, which drives into the Earth's atmosphere a thermospheric wind that can affect the highest weather patterns (Chapter 12). Such differential motions of charged- and neutral-particle populations are rarely considered in solar conditions, except in the chromosphere, where the neutral-particle population is comparable with that of the ionized population and where chemical fractionation that depends on the ionization energy of the atomic species occurs, or in the low solar wind, where differential drag forces appear to be important; see Chapter 9.

As our ion finds itself again in the solar wind, subject to waves and turbulence, it continues to be energized so that the temperature of the populations remains orders of magnitude above what it would be if the expansion were adiabatic. After it has traveled a few astronomical units, increasingly a new population of particles adds to the solar wind's pressure. These are called pickup ions, neutral particles in the interstellar medium that can penetrate deep into the heliosphere and, upon a charge-exchange collision with solar wind ions, become a high-energy part of the solar wind population.

As far as our ion is concerned, the environment becomes ever more quiescent: many shocks have merged with each other into so-called merged interaction regions. These shocks and their associated ejecta sweep up much of the solar wind in front of them, compressing both matter and field *en route* to the heliopause. These phenomena are in some respects similar to those in the magnetosphere of a planet subjected to the supersonic wind: if the merged interaction regions move fast enough relative to the wind ahead of them then here too standoff shocks will form and matter will attempt to flow around them. But differences also exist. For one thing, these global merged shocks are so large, and expand so rapidly, that plasma cannot readily flow around them. Thus, although some sideways flows are set up, much of the plasma and field remain compressed at the shock front simply because they cannot flow around the ever-expanding structures.

1.1.5 The edge of the solar system

More than a year after leaving the Sun, and assuming it did not encounter any solid bodies or magnetospheres *en route*, our ion encounters one final shock within the solar system. This time it is the solar-wind termination shock, which propagates

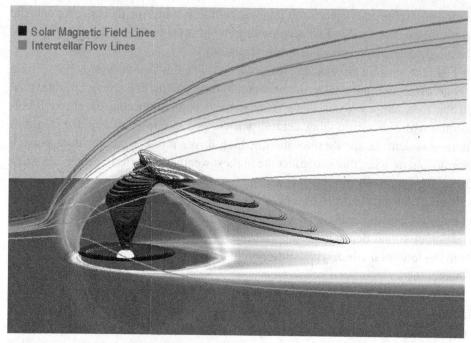

Fig. 1.4. The heliospheric cavity in the interstellar medium. The interstellar medium (ISM) flows around the heliopause, the interface between the interplanetary and interstellar plasmas, forming a cavity (or cell, see Section 6.7) similar to that around the planets in the solar wind. Here, however, the ram pressure of the ISM is balanced largely by that of the solar wind, which carries the heliospheric magnetic field with it. The sample field lines emanating from the Sun show that the solar rotation underneath the outflowing wind leads to a tightly wound spiral field. The polar outflows are blown into the heliospheric tail by the interaction with the ISM. (Courtesy of M. Opher, George Mason University.)

upstream from the heliopause, i.e. the interface between the interplanetary and interstellar plasmas. This shock slows the wind down and begins to deflect it into the tail region of the heliospheric volume, which is shaped by the interaction with the relative flow of the interstellar medium around the planetary system as the latter moves about the center of the Galaxy. Beyond this point, so far reached only by the two Voyager spacecraft launched in the 1970s, after more than three decades of fast travel, lie only the heliopause and its bow shock before the interstellar medium begins. These together form a system rather similar to that of a planetary magnetosphere or a coronal mass ejection plowing through the solar wind, but at a vastly larger scale (see Fig. 1.4).

Here, we are in an environment with a mass density that is over 10^{24} times lower than where we picked up our proton's travel. After traveling over $20\,000$ solar radii, or a million Earth diameters through many different environments, and being subject to shocks and reconnections, it is time for us to leave our proton as it flows into the interstellar medium and to go back to its origin to examine the many processes in some detail.

1.2 Magnetic field: a unifying force within heliophysics

In the preceding section, we followed the odyssey of a proton in its journey from within the Sun through the solar atmosphere, the solar wind, planetary magnetospheres, and ultimately into interstellar space. A complementary epic can be formulated to describe the life events of magnetic field within the extended heliosphere from solar interior to interstellar space. It is a story of birth, attempted escape, and (sometimes) violent death. Or, phrased only slightly less poetically, a story of creation, expansion, reconnection, and ultimately dissipation or annihilation.

The mother lode of magnetic flux in the solar system lies in the dynamo of the Sun, which, many solar physicists think, operates primarily between a quarter and a third of the way down from the surface to the center. There, near the base of the convection zone, a strong velocity shear stretches the field, strengthening it in the process. In a general sense, magnetic flux within the Sun is buoyant: it rises as flux tubes from the deep layers of the Sun and ultimately pokes through the photosphere, where it is shuffled around by the general motion of the convection cells. The details of what happens between the deep layers and the emergence of flux at the solar surface is a subject of ongoing research, but observations show that there is a near-balance between flux emergence and flux disappearance, so that any flux bundle in the photosphere survives for from a few minutes to at most a week (depending on its size and field strength), i.e. very much less than the full 22 years of the Sun's magnetic-reversal cycle.

Most flux through the solar surface is ground down into ever smaller bundles. Wherever oppositely directed flux tubes are pushed together by the vigorous convective plasma motions, the field reconnects there or in the overlying atmosphere. The process of reconnection, and the grinding down of the flux to smaller and smaller scales, converts magnetic energy into kinetic energy. On larger scales reconnection often occurs explosively, resulting in impulsive flares spanning many orders of magnitude in size and energy (Vol. II).

A small fraction of the solar magnetic flux connects to the heliosphere from the Sun's coronal holes, i.e. the long-lived dark patches in X-ray images of the Sun (Fig. 1.5). These coronal holes are fountain heads of solar wind streams, which expand until they merge together while flowing out to fill the heliosphere. A comparable amount of magnetic flux episodically bulges into the heliosphere from the solar corona in powerful eruptions called coronal mass ejections (CMEs), which

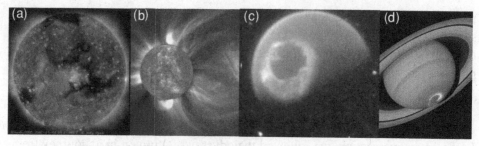

Fig. 1.5. (a) A Hinode X-Ray Telescope image of the Sun taken on 14 November 2007, showing coronal configuration resulting from the dynamic magnetic field, including large-scale dark regions called coronal holes. (b) A SOHO LASCO image of a coronal mass ejection on 8 January 2002. (c) Auroral storm on Earth on 15 July 2002 taken by the far-ultraviolet camera on the IMAGE spacecraft. (d) Auroral storm on Saturn taken by the Hubble spacecraft in October 1997. (Courtesy of nasaimages.org.)

propel huge missiles of plasma and magnetic field at high speed into interplanetary space.

When a CME impacts a planet, its coiled cargo of magnetic field binds with the planetary magnetic field and switches the normal rate of magnetospheric convection (a circulation of the planet's magnetic flux driven by the solar wind) into overdrive. The action converts some CME energy into internal magnetic storms marked by intense radiation belts and massive auroral displays. Ultimately, most of the escaped solar magnetic flux joins the galactic magnetic field.

One can say that the cycle of birth, life, and death of magnetic flux causes most of the interesting non-thermal processes and events that go on in the heliosphere (excluding intraplanetary processes). It is as if the magnetic field controls a distinct domain of cosmic structures and actions. To put this thought into a comparative context, we note that the structure and behavior of plasma on cosmic scales is governed by this equation of motion (which ignores viscosity; cf. Eq. 3.2):

$$\rho \frac{d\mathbf{v}}{dt} + \nabla p = \rho \mathbf{g} + \mathbf{j} \times \mathbf{B}, \tag{1.1}$$

in which ρ, $\mathbf{v}$, p, $\mathbf{g}$, $\mathbf{j}$, and $\mathbf{B}$ are the mass density, velocity, thermal pressure, gravitational field, electrical current, and magnetic field. The point is that the gravitational and magnetic forces on the right-hand side of the equation are the sources of motion and thermal energy on the left-hand side and that these sources are radically different in terms of the structures and actions of matter that they bring about.

The point is clarified by examples. Gravitationally organized matter tends to contract (e.g. star formation); magnetically organized matter tends to expand (e.g. coronal mass ejections). Systems of gravitationally organized matter tend to define planes or spheres (e.g. the galactic plane, the ecliptic plane, Oort cometary clouds, and galactic halos); magnetically organized matter tends to define shells, tubes, and sheets (e.g. radiation belts, flux ropes, and current sheets, Chapter 6). Gravitationally organized matter tends to dissipate energy through thermal radiation (e.g. stars); magnetically organized matter tends to dissipate energy through non-thermal radiation and particle acceleration (e.g. X-ray flares and cosmic rays). Gravitationally organized matter tends to convert gravitational energy into kinetic energy more or less continuously (e.g. at the inner edge of accretion disks); magnetically organized matter can convert magnetic energy into kinetic energy explosively (e.g. stellar flares and substorms). (We exempt collisions, which explosively convert gravitational energy into kinetic energy: collisions are accidental outcomes of independent factors whereas flares and substorms are necessary outcomes of interdependent processes.) Rings, satellite systems, planetary systems, and galaxies form a hierarchy of gravitationally organized matter; magnetic flux ropes in Venus' ionosphere, on the boundaries of planetary magnetospheres (flux transfer events, or FTEs), in Earth's magnetotail (plasmoids), on the Sun (sunspots), in the chromosphere (fibrils), in the corona (prominences and CMEs), at the core of the galaxy (radio filaments), and in the galactic disk (loops in the interstellar magnetic field) form one of a number of hierarchies of magnetically organized matter. Motions within gravitationally organized matter tend to be ordered (e.g. orbiting systems); motions within magnetically organized matter are often turbulent (e.g. in the solar wind, galactic radio jets, and the Earth's magnetotail, as exemplified by bursty flows of plasma).

The difference in the modes of organization brought about by gravitational and magnetic forces has its origin in the structure of the stress tensors for the two force fields. The term representing pressure in the gravitational stress tensor is negative definite whereas that for the magnetic field is positive definite. This is why the gravitational force causes matter to contract whereas the magnetic force causes matter to expand. A magnetic flux tube, however born, engages in a constant struggle to realize its natural state of equilibrium, namely, infinite expansion. The only way to hold a magnetic field down is to anchor it in a body held together by gravity, like a planet or a star. But, if it is born in a star then it will contrive to slip out, principally because magnetic flux is buoyant, leading to the story of perpetual creation, attempted escape, and annihilation already narrated. Another important difference between gravitational and magnetic fields relates to their sources. Gravitational fields are conserved in the sense that mass, the source of gravity (through $\nabla \cdot \mathbf{g} = -4\pi G\rho$), is conserved. However, magnetic fields have no material sources ($\nabla \cdot \mathbf{B} = 0$); they are generated by dynamos and destroyed by dissipation. Thus, whereas the gravitational field is fixed and limited by the total amount of mass present, the magnetic field is always in a non-steady equilibrium between creation and annihilation. It is the constant generation and dissipation

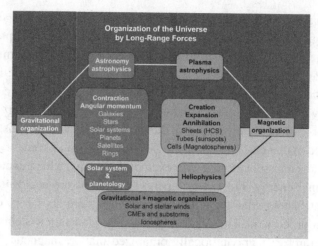

Fig. 1.6. The scientific disciplines concerned with structures and phenomena that result from gravity and magnetism, the two forces that operate at long range in the universe.

separated in time by the turbulent career of each magnetic flux tube that turns the otherwise homogeneous glowing surface of the Sun into a writhing, twisting, pulsing dynamic menagerie of fibrils, spicules, arcades, spots, flares, and CMEs, to name but a few of the zoo's magnetically engendered inhabitants and activities.

Figure 1.6 illustrates the contrasting ways in which gravitational forces and magnetic forces organize matter on cosmic scales. The diagram is divided horizontally into topics pertaining to the gravitational force (on the left) and topics pertaining to magnetic forces (on the right). Vertically the diagram is divided by relative proximity to the Sun: topics pertaining to outside the extended heliosphere are given above and topics pertaining to anywhere within the heliosphere are given below.

In the center of the diagram, the left-hand box lists structures that emerge under gravitational contraction restrained by conservation of angular momentum. The corresponding box on the right names structures that result naturally when magnetic fields are generated and interact in cosmic contexts. The box at the bottom of the diagram concerns mainly phenomena that result from the conflict between the two modes of organization (contraction and expansion). When magnetic fields are held down by gravity while being generated by dynamos, the magnetic energy builds up until some avenue of escape is found in gradual dissipation or explosive expulsion. Explosive release gives rise to cosmic rays. The final entry is planetary ionospheres. Ionospheres exist in the absence of planetary magnetic fields, for example, on Venus and Mars. Thus they are part of gravitationally organized matter just as are atmospheres. But they belong in the list because, for magnetized planets, they play an essential role in controlling the dynamics of magnetic interactions within the magnetosphere, and are themselves modified by magnetospheric processes (as in aurorae).

1.3 The three-volume series

This volume is the first of a three-part series of texts in which experts discuss many of the topics within the vast field of heliophysics. The texts reference the other volumes by number, as follows:

I   Plasma Physics of the Local Cosmos
II Space Storms and Radiation: Causes and Effects
III Evolving Solar Activity and the Climates of Space and Earth

The project is guided by the philosophy that the many science areas that together make up heliophysics are founded on common principles and universal processes, which offer complementary perspectives on the physics of our local cosmos. In these three volumes, the chapter authors point out and discuss commonalities and complementary perspectives in traditionally separate disciplines within heliophysics.

Many chapters in the volumes of this series have a pronounced focus on one or several of the traditional sub-disciplines within heliophysics, but we have tried to give each chapter a trans-disciplinary character that bridges gaps between these sub-disciplines. In some chapters stellar and planetary environments are compared, and in others the Sun is compared with its sister stars or planets are compared with one another; in yet other chapters general abstractions, such as magnetic field topology or magnetohydrodynamic principles, that are applicable to several areas are considered.

The vastness of the heliophysics discipline precludes completeness. We hope that our selection of topics helps to inform and educate students and researchers alike, thus stimulating mutual understanding and appreciation of the physics of the universe around us.

1.4 Additional resources

The texts were developed during summer schools for heliophysics held over three successive years at the facilities of the University Corporation for Atmospheric Research in Boulder, Colorado, and funded by the NASA Living With a Star program. Additional information, including text updates, lecture materials, (color) figures and movies, and teaching materials developed for the school can be found at http://www.vsp.ucar.edu/HeliophysicsSummerSchool.

1.5 Editors' note

The chapters in this volume were authored by the teachers of the heliophysics summer schools following the outlines provided by the editors. In the process of integrating these contributions into this volume, the editors have modified or added segments of the text, included cross references, pointed out related segments of text, introduced several figures and moved some others from one chapter to another, and attempted to create a uniform usage of terms and symbols, while allowing some differences to remain that are compatible with the discipline's literature usage. They bear the responsibility for any errors that have been introduced in that editing process.

2

Introduction to heliophysics

THOMAS J. BOGDAN

2.1 Preamble

Walk along an island beach on a clear, breezy, cloudless night, or stand on the spine of a barren mountain ridge after sunset, and behold the firmament of stars glittering against the coal black sky above. They fill the sky with their timeless, brilliant flickering. With binoculars or even a small telescope one finds that even the lacey dark matrix between the vast sea of stars is populated with still more stars that are simply too faint to be seen with the naked eye. Within the Milky Way Galaxy, which stretches from horizon to horizon, the density of stars against the background sky is even greater.

Each twinkling point of light is a star not too unlike our own Sun. The Sun is just an ordinary star but it features prominently on the pages of this book because of its proximity. The next closest star, α-Centauri (which is a triple system in which Proxima Centauri is currently the closest to Earth), is almost a million times farther away (at 4.22 light years), and the remainder are farther still. We may now say with some confidence that many stars are surrounded by planets of various sizes. Some of these orbital companions are so immense that they are stars in their own right: double-star systems are quite common.

With the same measure of confidence we may assert that most of these stars possess magnetic fields; that these magnetic fields create hot outer atmospheres, or coronae, that drive magnetized winds from their stars; and that these variable plasma winds blow past the orbiting planets, distorting their individual magnetospheres, and push outward against the surrounding interstellar medium. Where the ram pressure of the stellar wind becomes comparable with the surrounding pressure (gas, magnetic, and cosmic-ray) of the interstellar medium, a termination shock forms. This serves to mark the farthest extent of the mechanical impact of the star on its surrounding environment: a sphere of influence, so to speak.

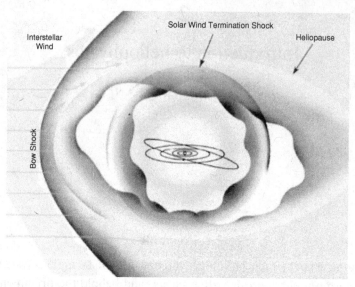

Fig. 2.1. The heliosphere is the sphere of influence carved out by the Sun in the surrounding interstellar medium. The solar wind flows out incessantly in all directions, past the orbiting planets, and is finally halted at the termination shock (the innermost nested surface). The shocked solar wind material diffuses outward into the surrounding material forming the heliopause (the prolate spheroid that encloses the spherical termination shock). The Sun and the heliosphere move with respect to the background interstellar medium; from right to left in this figure. This velocity is larger than the characteristic propagation speeds of waves supported by the interstellar medium and consequently a parabolic-shaped bow shock (the outer nested surface) forms around one side of the heliopause. (Courtesy of nasaimages.org.)

Our Sun has and does all these things, and we refer to the sphere of influence carved out by the solar wind as our heliosphere (see Fig. 2.1). It is not really spherical and it varies in extent with solar activity. But in broad terms we may safely think of it as extending about 100 times further from the Sun than the Earth's orbit. We have yet to agree on the name for such spheres of influence around the other stars (for which "astrospheres" has been proposed), but there can be little doubt that such environments are as commonplace as the many points of light we see strewn across the sky on a dark and cloudless light.

2.2 What is heliophysics?

Heliophysics encompasses the study of the various physical processes that take place within the sphere of influence of the Sun (i.e. the heliosphere) and, by analogy, those environments surrounding most other typical stars. But heliophysics also defines a specific method of study. This method embraces a holistic connected-system approach. It emphasizes a comparative context, in which one may come to understand a process by the many facets it presents in its various incarnations throughout the heliosphere. Taken together, these diverse facets serve to fill out a complete and physically satisfying picture of a given process or phenomenon.

The physical processes and phenomena we will encounter in these volumes are themselves especially diverse. They include the rapid and efficient energization of thermal particles to suprathermal energies, the generation and annihilation of magnetic fields, stellar variability and activity cycles, space weather, turbulent transport of energy and momentum, and the coupling between ionized and neutral atmospheres, to name just a few. Heliophysics fills a critical need to establish a unified science that connects these seemingly unrelated concepts in a manner that emphasizes complementarity over individuality, function over form, and generality over specificity.

Thus the first pillar on which heliophysics rests is unification. Coupling provides the second principal pillar. The heliosphere is a collection of coupled systems. It is fortunate that many linkages essentially operate in only one direction, that is to say, system A impacts system B but B has little influence on A. Under these circumstances it is expedient to treat system A as independent of the behavior of system B. This provides a certain economy of effort and scale and often reduces the (apparent!) complexity of a problem. For example, complex geomagnetic activity has no impact on solar flares, and the solar wind does not influence the Sun's cyclic variability.

Linkages, especially when several are present and working at cross purposes, can lead to confusion and spirited debate over what is a root cause and what is simply a resulting effect. The cause and effect relationship between solar flares and coronal mass ejections is a good case in point. Consider, for example, what the purported cause and effect relationship might be between a sore throat and a fever. Since a sore throat often starts before a fever develops one might be tempted to assign the effect to the fever and take the sore throat to be the cause. Fortunately, medical research informs us that both are effects and the root cause is the influenza virus. Heliophysics is needed to play this very same role in sorting out the appropriate relationships (or lack thereof) between the physical effects that often occur contemporaneously throughout the heliosphere.

Solar variability influences our climate here on Earth. This fact is certainly not negotiable in a purely scientific context and is arguably one of the most important linkages between the Sun and the Earth. Satellites have confirmed that the solar irradiance is variable on time scales from minutes to decades. The fluctuations are greatest on the shortest time scales. Day-to-day irradiance changes are on the order of one percent as against tenths of a percent over a solar cycle. The magnitude and sense of the irradiance trends over centuries and millennia are currently difficult to determine with any measure of certainty. Slow but steady progress on this question is being made through the studies of paleoclimate records. Over much longer time scales, stellar evolution theory provides assurances that significant changes in solar irradiance have taken, and will take, place with dramatic impacts on our climate and way of life.

What is debatable, however, is precisely what the direct relationship is between solar variability and climate change over any particular time scale, or epoch, of interest. For example, various opinions have been advanced that span the entire gamut from wholly inconsequential to complete solar responsibility for the gradual warming of the planet that has been observed since the middle of the twentieth century. Yet, it may not even make sense to speak of direct relationships between drivers and the behavior of systems which are as nonlocal, nonlinear, and plagued by various hystereses as is our climate here on Earth.

The third and final pillar upon which heliophysics rests is the exploration of Earth's neighborhood in space. As a space-faring civilization we have visited all the planets, two comets and numerous planetary satellites. We have ventured to the boundaries of the heliosphere and have flown through various parts of our magnetosphere. We have a spacecraft heading toward the Pluto–Charon system. Heliophysics enables successful exploration and, at the same time, it gains in knowledge and understanding from our exploration initiatives.

In summary, heliophysics is the systems-mediated study of the physical processes that take place within the Sun's sphere of influence. It is based upon the three pillars of the unification of physical processes and phenomena, the coupling of distinct physical systems, and the exploration of our neighborhood in space. Furthermore, it is broadly applicable to the environments around most average stars.

2.3 The language of heliophysics

The language of heliophysics is mathematics, and the body of literature from which heliophysics draws its substance and in turn records its accomplishments is the physics of magnetized plasmas. With only a rudimentary knowledge of a language, a literature is incomprehensible except, perhaps, in translation. And even in translation so much of the original meaning and the nuances that the author wished to convey are inevitably lost or, worse, misinterpreted by even the most conscientious translator.

The most precise, and intellectually demanding, literary prose of heliophysics assigns a phase-space distribution function to each individual species of particle (see Fig. 2.2). By a species one may simply mean free electrons, protons, oxygen molecules, or even photons. In some applications it might be necessary to distinguish between oxygen molecules in different excited (vibrational, rotational,

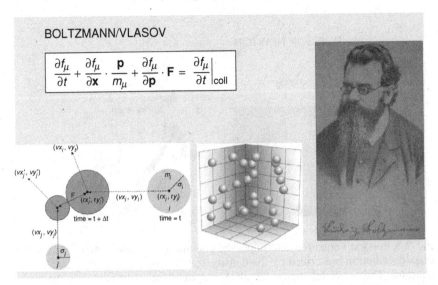

Fig. 2.2. The Boltzmann equation describes the evolution of the phase space density f_μ for particles of type μ and mass m_μ, in the presence of a force field **F**. The independent variables are the time t, the location **x**, and the momentum **p**. When binary collisions between particles can be ignored, the collision term on the right of the equals sign can be dropped. The collisionless Boltzmann equation is often referred to as the Vlasov equation. (Courtesy of K. Wayne, S.-H. Yu, and Lawrence Livermore National Laboratory.)

and electronic) states, or between iron atoms at different stages of ionization, or between different senses of photon polarization. In any case, the evolution of each distribution function is obtained by setting the total time derivative equal to the net production or loss of an individual species by various collisional or radiative processes. Such evolution equations are commonly referred to as Boltzmann, or Vlasov equations. When there is no net gain or loss, then Liouville's theorem asserts that the vanishing of the total time derivative of the distribution function conserves the phase space density for each species.

In specifying the total time derivative it is necessary to determine the forces acting upon a given particle species. For uncharged particles, gravitational attraction is the only important consideration. Accordingly, to the system of equations for the individual distribution functions one must add Poisson's equation in order to specify the gravitational field arising from the mass distribution provided by those particles with mass (see Fig. 2.3). Charged particles are also subject to electromagnetic interactions. Thus we must also include Maxwell's equations to deduce the electric and magnetic fields due to the distribution of charges and currents provided by the charged particle species (see Fig. 2.4).

In principle, this suffices to provide a complete description of the grammar and syntax of heliophysics at a very elegant, learned, and precise level. In practice

POISSON/NEWTON

$$\nabla^2 \Phi = -4\pi G \rho$$

Fig. 2.3. Poisson's equation determines the gravitational potential Φ in terms of the mass distribution prescribed by the density ρ; G denotes Newton's gravitational constant. The lower left figure simulates the distribution of matter in the early universe. (Courtesy of A. Kravtsov and A. Klypin.)

MAXWELL/FARADAY

MKS units	Gaussian units	
$\nabla \cdot \mathbf{D} = \rho$	$\nabla \cdot \mathbf{D} = 4\pi\rho$	
$\nabla \cdot \mathbf{B} = 0$	$\nabla \cdot \mathbf{B} = 0$	
$\nabla \times \mathbf{H} = \mathbf{J} + \frac{\partial \mathbf{D}}{\partial t}$	$\nabla \times \mathbf{H} = \frac{4\pi}{c}\mathbf{J} + \frac{1}{c}\frac{\partial \mathbf{D}}{\partial t}$	
$\nabla \times \mathbf{E} = -\frac{\partial \mathbf{B}}{\partial t}$	$\nabla \times \mathbf{E} = -\frac{1}{c}\frac{\partial \mathbf{B}}{\partial t}$	
$\mathbf{F} = q(\mathbf{E} + v \times \mathbf{B})$	$\mathbf{F} = q(\mathbf{E} + \frac{1}{c}v \times \mathbf{B})$	Lorentz force law
$\mathbf{D} = \varepsilon_0\mathbf{E} + \mathbf{P}$	$\mathbf{D} = \mathbf{E} + 4\pi\mathbf{P}$	(general)
$\mathbf{D} = \varepsilon_0\mathbf{E}$	$\mathbf{D} = \mathbf{E}$	(free space)
$\mathbf{D} = \varepsilon\mathbf{E}$	$\mathbf{D} = \chi\mathbf{E}$	(isotropic linear dielectric)
$\mathbf{B} = \mu_0(\mathbf{H} + \mathbf{M})$	$\mathbf{B} = \mathbf{H} + 4\pi\mathbf{M}$	(general)
$\mathbf{B} = \mu_0\mathbf{H}$	$\mathbf{B} = \mathbf{H}$	(free space)
$\mathbf{B} = \mu\mathbf{H}$	$\mathbf{B} = \mu\mathbf{H}$	(isotropic linear magnetic medium)

Fig. 2.4. Maxwell's equations determine the electromagnetic fields $\mathbf{E}$ and $\mathbf{B}$ in terms of the electric current density $\mathbf{J}$ and the charge density ρ. The two scalar equations were deduced by Gauss and Poisson. The induction equation (containing the curl of $\mathbf{E}$) is associated with the contributions of Faraday. Maxwell's breakthrough was to append the displacement current to the right side of Ampere's law, thereby accounting for electromagnetic radiation.

the task of following through with this program (i) is prohibitively difficult with or without the assistance of a computer, (ii) is subject to the problem that the initial conditions are not known with any degree of certainty, (iii) is complicated by the fact that many collisional and radiative transition probabilities are not even

IDEAL MAGNETOHYDRODYNAMICS

$$\frac{\partial \rho}{\partial t} + \nabla \cdot (\rho \mathbf{u}) = 0,$$

$$\frac{\partial S}{\partial t} + \mathbf{u} \cdot \nabla S = 0,$$

$$\frac{\partial \mathbf{B}}{\partial t} - \nabla \times (\mathbf{u} \times \mathbf{B}) = 0,$$

$$\frac{\partial \mathbf{u}}{\partial t} + (\mathbf{u} \cdot \nabla)\mathbf{u} + \frac{1}{\rho}\nabla p = \frac{1}{4\pi\rho}(\nabla \times \mathbf{B}) \times \mathbf{B} - g\hat{\mathbf{z}},$$

$$e = \frac{1}{\gamma - 1}\,\frac{p}{\rho}, \quad S = c_V \log (p/\rho^{\gamma}).$$

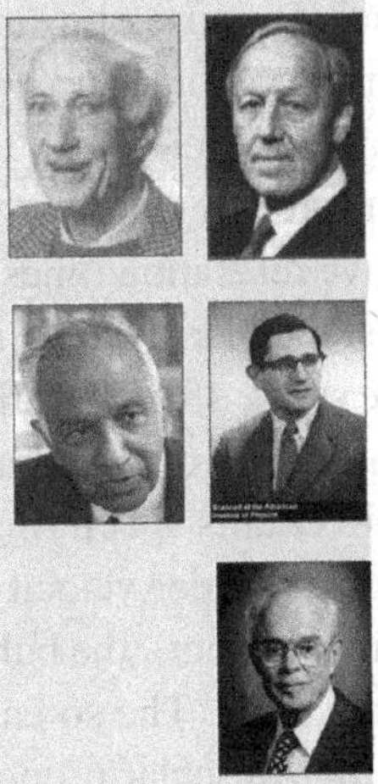

Fig. 2.5. Ideal magnetohydrodynamics (MHD) provides the simplest possible description of a heliophysical plasma. This closed system of equations may be derived directly from the kinetic equation of Boltzmann and the electromagnetic field equations of Maxwell, under certain limiting assumptions that are often realized in many situations of interest present in the Sun–Earth system. The principal architects and expositors of MHD have been Alfvén, Grad, Parker, Chandrasekhar, and Dungey. (Courtesy of Elsevier Publishing Company, AIP Emilio Segrè Visual Archives, and University of Chicago News Office.)

approximately known, and (iv) requires that certain conditions be fulfilled so that electromagnetic interactions can be separated into large-scale fields and small-scale collisions. Finally, this comprehensive description usually provides far more information than is usually necessary for comparing with observations or understanding the predictions of a theory over specified temporal and spatial scales.

At the opposite extreme from the scholarly literary prose is the common vernacular. For heliophysics, if the high literary prose centers on Poisson, Maxwell, Boltzmann, and Vlasov then the vernacular mode of expression is single-fluid, ideal, magnetohydrodynamics (MHD); see Fig. 2.5 and Chapter 3. Magnetohydrodynamics is a continuum fluid description that does not distinguish between particle species, averages in some sense over particle collisions, ignores radiative effects altogether, and is based on velocity moments of the underlying distribution functions. It retains Poisson without modification but takes certain liberties with Maxwell. Boltzmann and Vlasov drop out of the picture entirely.

Magnetohydrodynamics can be rigorously derived from the equations of Poisson, Maxwell, Boltzmann, and Vlasov under various conditions that are reasonable for many heliospheric applications. Usually this involves following the behavior of a physical process or phenomenon over coarse-grained spatial and temporal scales.

In other words it is a useful, and indeed often very accurate, description of the "big picture". Because of its relative simplicity, ideal MHD provides a useful context in which to interpret and understand the behavior of magnetized plasmas at a basic and often extremely intuitive level. Unfortunately, ideal MHD is often applied inappropriately to processes or phenomena. Generally speaking, if collisional and radiative relaxation times are short compared with the coarse-grained time scale of interest then ideal MHD is likely to be a reasonable option.

The successful derivation of the MHD equations requires a closure prescription, which may be regarded as a consequence of the familiar "no free lunch" maxim. Closure entails specifying a tractable procedure to determine the pressure tensor (the second-order velocity moment) in terms of the fluid density (the zeroth-order velocity moment), the bulk fluid velocity (the first-order velocity moment), and the magnetic field. The so-called polytropic approximation – in which the pressure is taken to be a scalar proportional to the density raised to a specified power – is the simplest option. A power law index equal to unity corresponds to an isothermal process (one at constant temperature). A power law index equal to the ratio of specific heats describes an isentropic (constant-specific-entropy) process that also manages to conserve energy. More complicated options are possible and are often tailored to accommodate specific situations. A successful and accurate closure scheme is inevitably based on some additional a priori knowledge of the behavior of the particle trajectories or of the general nature of the particle distribution functions.

In contrast with the Poisson–Maxwell–Boltzmann–Vlasov description, ideal MHD comprises a system of nine partial differential equations for nine dependent variables, the gravitational potential, fluid density, fluid pressure, fluid velocity (three components), and magnetic field (three components). These equations are (a) the Poisson equation, (b) the continuity equation expressing the conservation of mass, (c) the closure relation to specify the pressure tensor, (d) the equation for the conservation of momentum or force-balance equation (three components), and (e) the magnetic induction equation (three components).

Of course, between ideal MHD and the Poisson–Maxwell–Boltzmann–Vlasov description lies a plethora of compromise or hybrid descriptions. The number of such schemes is limited only by the imagination and ingenuity of the investigators. Multi-fluid treatments allow for individual densities, velocities, and pressures to be associated with different particle species or groupings of particle species but retain a single gravitational potential and magnetic field applicable to every fluid. This formulation is useful when the time scales of interest are short compared with characteristic inter-species collisional relaxation times but long compared with the analogous intra-species times.

Another intermediate scheme employs high-order moment closures. These schemes are necessary when the species distribution functions deviate significantly from a fully relaxed Maxwellian. Often this situation occurs when significant spatial gradients are imposed on the system. Additional partial differential equations are then used to describe the time evolution of the components of the pressure tensor. The closure is postponed to the next-higher level of the heat flux tensor (the third-order velocity moment) or, in extreme circumstances, to even higher-order moments.

Hybrid schemes treat some species as fluids and retain a Boltzmann–Vlasov – or kinetic – description for others. Indeed, even a single species of particle may be partitioned in such a fashion that some particles are treated kinetically (generally the high-energy suprathermal tail of the distribution function) while the remainder are described as a fluid (the thermal core of the distribution). Such schemes are particularly useful in describing the energization of charged particles.

In summary, there is a bewildering array of schemes that are presently invoked to describe the behavior of magnetized plasmas in the heliosphere. They encompass an extremely wide range of complexity. Each is specifically tailored to a given physical process and phenomenon. They are not simply interchangeable but have their own individual strengths and weaknesses. One should always choose the simplest description that will suffice for understanding the problem in hand. Use all the information and knowledge you have at your disposal about the nature and behavior of a physical system in selecting a scheme. If the heliophysics concepts can be adequately framed in the common vernacular then eschew the sophisticated flowery prose unless nothing less will do.

2.4 The creation and annihilation of magnetic field

The magnetic induction equation (3.4) provides two key insights about magnetized plasmas. First, if the magnetic field starts out with zero divergence then it cannot develop any divergence over time. In other words there are no magnetic monopoles, or at least too few to bother worrying about. An immediate consequence of this fact is that magnetic field lines do not end. Second, in an infinitely conducting fluid a small fluid element that is threaded by a given magnetic field line initially is always threaded by the same field line. That is, plasma cannot move across magnetic field lines but is free to slide along them as it pleases. Conversely, if some applied force or agent induces an element of plasma to move in a direction perpendicular to the embedded magnetic field line then the field line is obliged to move along with the plasma.

The uniform bulk motion of a magnetized plasma just advects the magnetic field without any further consequence. Differential motions, however, stretch, twist, and fold the magnetic field lines. Fluid dilations pull the magnetic field lines apart and spread a given magnetic flux over a larger cross sectional area, reducing the

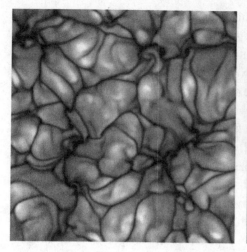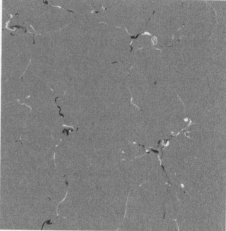

Fig. 2.6. Simulation of magnetoconvection. Magnetic fields are carried about by fluid motions in highly conducting plasmas. Accordingly they accumulate where flows converge and are swept out of regions of diverging flow. The two panels of this figure illustrate this behavior for a numerical simulation of magnetoconvection. In the left-hand panel we observe the vertical component of the flow close to the surface of the simulation. The lighter tones represent upwelling fluid and the darker tones correspond to downflows. The network of downflows is fed by horizontal diverging flows (not seen in this image) associated with the upwellings that converge abruptly at the dark downflow lanes. In the right-hand panel we see the vertical component of the magnetic field at the same instant of time. The light tones denote a positive-polarity magnetic field and the dark tones outline regions of negative polarity. We see clearly that the magnetic field predominantly lives in the downflow network. Some vertical magnetic flux that has just emerged can be seen in a few of the granular upwellings; it will subsequently be swept to the network to join the remainder of the magnetic field. (See also Fig. 1.2.)

intensity of the magnetic field. Contractions have the opposite effect. Magnetic fields accumulate and increase in strength where plasma flows converge and where they are dispersed from outflows (see Fig. 2.6). Cyclonic motions twist magnetic fields into tightly wound spirals and velocity shears (tachoclines) comb magnetic field out into unidirectional magnetic layers. One sees immediate heliospheric confirmation of these behaviors in the accumulation of magnetic fields along the intergranular lanes in the solar photosphere, in the large-scale spiral structure of the heliospheric magnetic field, and in the formation of magnetic flux ropes in the tachocline at the base of the solar convection zone.

Turbulent flows have the general property that neighboring fluid elements sepa-rate rapidly, often exponentially, with the passage of time. Consequently magnetic fields are quickly stretched, tangled, and intensified by turbulent flows. The mag-netic field develops structure on finer and finer spatial scales. At some point spatial scales are attained that are so fine that the coarse-grained spatial scale assumption of ideal MHD no longer holds. The immense but ultimately finite conductivity comes into play, either through charged particle collisions or collective wave–particle interactions. Physically, this limits the current density, adds thermal and kinetic energy back into the plasma, allows fluid elements to slip across magnetic field lines, and permits magnetic field lines to reconnect with each other. This annihilation of magnetic field serves to iron out small-scale structure, eliminate topological connectivity constraints, and anneal localized magnetic domains. Paradoxically, the annihilation of the magnetic field that is often the result of reconnection in these thin current sheets is absolutely necessary (but not in itself sufficient) for the generation of large-scale magnetic fields through induction.

One asserts that a magnetic dynamo is in operation if the rate at which energy is returned to the plasma through reconnection is, on average, less than the rate at which the plasma stretches and intensifies the magnetic field through induction. Under these circumstances the energy in the magnetic field tends to grow in time at the expense of the energy contained in the plasma or by means of work done on the system by an external agent. The growth in the magnetic energy does not occur without a cost; rather, the back reaction of the Lorentz force on the plasma through the force-balance equation increases as the field grows. If the external forcing is maintained, the Lorentz force tends to suppress most of those motions that stretch and twist the magnetic field in favor of those that incur no such growth in magnetic intensity. The induction flows that remain add energy to the magnetic field at the same rate at which the magnetic field returns that energy to the plasma through reconnection. A dynamic balance of sorts ensues.

However, the remarkable quality of dynamically balanced heliospheric dynamos is that they may operate with quasi-periodic, or even chaotic, exchanges of energy between the magnetic field and the background plasma. The solar dynamo, on the one hand, is now in a quasi-periodic phase with a periodicity of 11 to 12 years. The geodynamo, on the other hand, exhibits rather abrupt and aperiodic transitions. The dynamos operating within the other planets are different again in their behaviors. In any event, the key point is simply that the creation and annihilation of magnetic fields is an essential heliophysics activity.

2.5 Magnetic coupling

The motion of plasma along the magnetic field does not stress the field and incurs no dynamic back-reaction on the plasma through the action of the Lorentz force. Magnetic field lines therefore serve as conduits for moving energy, mass, momentum, and energetic particles from point to point in the heliosphere. Heliophysics accordingly focuses on the magnetic connectivity of the earth to the sun, of the magnetotail to the polar caps, of the Io plasma torus to the Jovian magnetic field, and so forth. Magnetic field lines are truly the interstate highway system, the Autobahn network, the autostrada web of the heliosphere.

Magnetic coupling ensures that heliospheric systems are not isolated. It is one of the key underlying reasons why there is a need for a systems approach to understanding processes in the heliosphere. Magnetic reconnection alters connectivities and tends to allow heliospheric systems some measure of individuality, isolation, and independence. For example, coronal mass ejections reconnect as they propagate away from the Sun and sever their ties with the Sun. This permits them to continue to move outward through the solar wind. Reconnection in the Earth's magnetotail allows the magnetosphere to expel plasmoids and relieve stresses built up over time.

Reconnection, such as in the nanoflaring observed throughout the solar corona, can be so rapid and pervasive that a given magnetic field line in the interplanetary medium is never connected to any photospheric footpoint for enough time to respond to the conditions in that region. Paradoxically, rapid reconnection can serve to isolate regions or, at the very least, permit them to respond to average conditions over a very large source region. A certain level of homogeneity is thus assured.

The concept of magnetic connectivity and reconnection raises the provocative question of how many closed magnetic field line systems exist at any one time within the heliosphere. In most theoretical models there are always infinitely many such closed magnetic field line circuits. This is due to the inordinate degree of symmetry that is usually imposed on such models, rightly or wrongly, to permit analytical progress. A popular philosophical answer is that there is but a single ergodic magnetic line of force that fills the whole of space, not just the heliosphere.

The ability of the magnetic field to connect regions with vastly different properties and characteristics creates the opportunity for many unique physical processes and phenomena. In a very real sense, it often forces heliophysical systems outside their comfort zones. In these situations, we usually learn the most about the underlying nature of the process or phenomenon.

2.6 Spontaneous formation of discontinuities

An outstanding characteristic of heliospheric magnetized plasmas is their propensity to develop sharp spatial transition zones where several physical properties change rapidly over very small spatial scales. These may in fact arise through several different avenues. A closely allied concept is the general lack of static equilibria in ideal MHD. Consequently, heliospheric plasmas are invariably in motion and can at best achieve dynamic equilibrium. And, as might be expected, ideal

MHD is often inapplicable for describing the physics in these thin, intermittent, transition zones.

To appreciate this point, it suffices to examine the ideal MHD force-balance equation. In order that a static equilibrium be attained, the Lorentz force must be balanced against the gravitational forces and the gradient of the plasma pressure. Now, in particular, in the two-dimensional manifold that is everywhere perpendicular to the gravitational force, the Lorentz force must be expressible as the gradient of a scalar field. That is to say, it must be irrotational, i.e. have zero curl, in this subspace. Now there is nothing intrinsic to the Lorentz force that requires it to have such a property, although one can certainly construct many Lorentz forces that do satisfy this constraint. Indeed, it is an unfortunate and misleading fact that most Lorentz forces which do satisfy this so-called compatibility constraint possess some degree of symmetry and are therefore also the special examples that have attracted the most attention. When compared with the Lorentz forces that are possible in reality, it is probably the case that these special examples are heliophysically irrelevant.

Among such "irrelevant" magnetostatic equilibria are potential magnetic fields, where the magnetic field is itself irrotational, and force-free magnetic fields, where the magnetic field is simply proportional to its own curl; the scalar function of proportionality is strictly constant along a given magnetic field line but may vary from one distinct magnetic field line to another. Obviously, for a space-filling ergodic force-free magnetic field the constant of proportionality is just a single number throughout all space.

Whenever the Lorentz force cannot be balanced by the plasma pressure gradient in a tangential two-dimensional manifold then flows develop, magnetic fields are stretched, twisted or folded, and the Lorentz force evolves. There is a fascinating tendency for the ensuing dynamics to create quasi-magnetostatic domains where the pressure, gravitational, and Lorentz forces very nearly balance, bounded by dynamic sheet-like discontinuities. One often refers to these discontinuities generically as current sheets, since the magnetic field can be quite different on either side of a discontinuity. But tangential flows can also vary across discontinuities (in tachoclines), as can the plasma density (in pycnoclines), the temperature (in thermoclines), and the composition and ionization states (which we will refer to in a general way as tangential discontinuities).

Other discontinuities do not depend much on the magnetic field for their creation. The low dissipation levels that are generally present in heliophysical plasmas permit large-amplitude disturbances to develop and dynamically steepen into shock waves as they move from regions of high plasma density to regions of low plasma density. Shock waves principally involve steep gradients in the normal fluid motion and invariably are surrounded by tenuous clouds of energetic particles. Shocks come in very many varieties because the magnetic field provides two additional wave modes for the conveyance of information from one point to another in the plasma. Again, the substructure within a shock is rarely amenable to an ideal MHD description, although the connection between the upstream and downstream states as well as the shock motion can often be determined without any detailed knowledge of the complex plasma physics present within the shock itself.

Shocks are an important way in which heliospheric systems react to impacts or forcing from other systems when these inputs occur rapidly in comparison with characteristic response times. For example, near the Earth the solar wind speed vastly exceeds the characteristic wave speeds available to the magnetized plasma. Thus there is no possibility for plasma encountering the Earth's magnetosphere to communicate back to the following plasma the need to divert to the left, right, top, or bottom, in order to slip past the Earth. Rather, this communication is achieved through the bow shock that develops around the Earth's magnetosphere (see Fig. 2.7). When the inflowing solar wind plasma passes through the shock, it is rapidly deflected in a direction that enables it to slip around the obstacle present in its way. Precisely the same thing happens at the edge of the heliosphere at the so-called termination shock, where the solar wind is integrated with the surrounding interstellar medium.

Not all quasi-discontinuities can be treated in the ideal MHD scheme. Double layers are small jumps in electric potential due to charge separation on the scale of a couple of Debye lengths, but they play a macroscopic role in planetary aurorae and in the closure of magnetospheric current systems. These layers can form when current densities and speeds exceed certain thresholds. Other discontinuities associated with the plasma skin depth or the ion gyroradius may develop around structures according to the constraints that are present.

2.7 Explosive energy conversion

The quasi-discontinuities that develop throughout the heliosphere are very often the sites of the explosive conversion of energy between magnetic field energy, plasma kinetic energy, and plasma thermal energy. These conversions are never very steady; then invariably display a bursty or intermittent character owing to the underlying non-linearity of the plasma process occurring in the transition layers. Accordingly, these events catch our attention.

Current sheets convert magnetic energy into plasma kinetic and thermal energy. The conversion is enhanced when the local plasma resistivity increases and is suppressed when greater conductivity and charged particle mobility prevail. Solar flares and magnetospheric substorms are thought to be ascribable to bursty magnetic reconnection in current sheets. The branching ratio between the energy provided

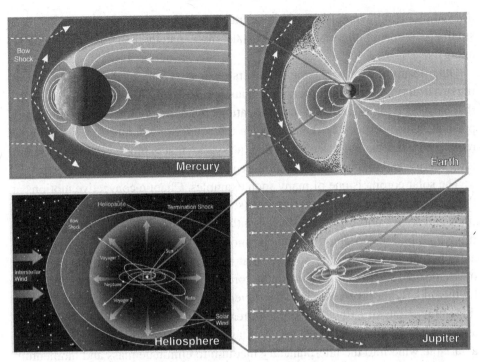

Fig. 2.7. Magnetospheres (the regions defined by the field lines) surround all planets with magnetic fields. They are constantly stressed and sheared by the incoming solar wind and by the rotation of their planet. In some cases, even the satellites that orbit a planet can impact magnetospheric dynamics. All magnetospheres are enveloped in bow shocks that serve to slow and deflect the solar wind around the magnetosphere. A particularly fascinating aspect of magnetospheres is how they adapt and change their connectivity to the magnetic fields entrained in the solar wind. The first three panels of this figure (clockwise) illustrate both the similarities and differences between three planetary magnetospheres of Mercury, Earth, and Jupiter. The final panel shows the entire heliosphere, a magnetosphere in its own right. The relative scales of the panels are indicated by the gray lines that connect one panel to an appropriately scaled outline of it in its neighboring panel. (Courtesy of F. Bagenal.)

to the plasma's thermal and kinetic energy is not well known, and some fraction of the thermal energy is unfairly distributed to a small subset of charged particles that achieve suprathermal, indeed often relativistic, energies. The detection signature of the explosive reconnection is provided by the photon byproducts (X-rays and γ-rays) created by these suprathermals when they collide with innocent bystanding atoms and nuclei or gyrate around the surrounding magnetic field lines (producing microwave and radio emission).

Shock fronts, however, generally convert plasma kinetic energy into plasma thermal energy, although magnetic fields can also gain or lose energy in the process depending upon the nature of the shock transition. There is again a tendency for a small group of suprathermal particles to take more than their share of the available energy.

Other forms of explosive energy conversion do not require the presence of existing quasi-discontinuities. Rather, rapid changes can occur in heliophysical systems that have been forced into untenable situations by their coupling with neighboring plasmas. This is especially the case when the coupling is in one direction and the recipient system cannot react back upon the system that is inputting mass, momentum, energy, and magnetic flux without regard for the potential consequences.

The solar corona is one of many such examples. Chaotic convective motions at the solar photosphere take the footpoints (see Section 4.2) of coronal magnetic field lines and weave and braid them incessantly. Jets of material, called spicules, are flung high into the corona from the chromosphere. Emerging magnetic flux buoyantly bubbles up through the visible solar surface and merges with the overlying magnetic field and material. In short, the corona is constantly jostled and aggravated by the underlying layers and has insufficient energy, mass, or momentum to alter this unfavorable situation. The corona creates myriad small-scale discontinuities and reconnects magnetic fields as much as it can to release energy, and it emits a variable wind into the interplanetary medium to eliminate mass and momentum and alleviate the stresses. But this, it seems, is still insufficient to fully compensate, on average, for the forcing from below. What all these actions fail to do is to relieve the twisting and knottedness, for lack of a better term, induced by the turbulent shuffling of the magnetic footpoints by the solar convection. However, the corona sporadically ejects clouds of twisted magnetized plasma into the solar wind and rids itself of the knottedness by simply ejecting the problem region and handing it off to the interplanetary medium (see Fig. 2.8). Precisely when and how this occurs remains unclear. Such coronal mass ejections occur maybe four or five times per day during periods of high solar activity, and perhaps once every four or five days during solar minimum. They are the space hurricanes or solar tsunamis of space weather.

2.8 Generation of penetrating radiation

Energy conversion in current sheets or shock fronts is neither democratic nor altruistic. A small subset of the charged particles takes away significantly more energy than it deserves and either directly contributes to, or indirectly generates through collisions, the penetrating radiations that are prevalent throughout the heliosphere. There is no intrinsic limit as to how much energy any single particle might obtain, but the number densities of these energetic particles usually decline very rapidly with energy, often as truncated power laws or exponentials.

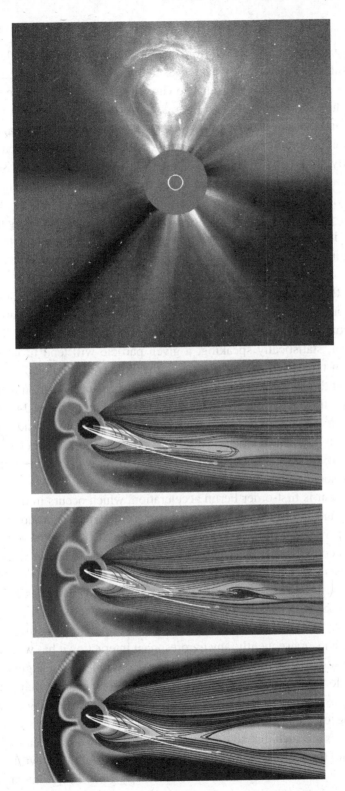

Fig. 2.8. The explosive release of energy, mass, momentum, and magnetic field from (top panel, SOHO-ESA and NASA) the solar corona and (lower three panels) the Earth's magnetosphere. (Slavín, 1998.)

Being rare and specially favored from the outset, the energetic particles continue to enjoy additional benefits and advantages that their thermal counterparts do not. They do not interact with each other, owing to their low number densities. On account of their elevated energy and momentum they do not interact much with the background thermal plasma either. They still gyrate around the magnetic field lines but their gyroradii can be comparable with the background spatial scales at sufficiently great particle rigidities. However, they are particularly susceptible to interactions with waves that travel along the magnetic field lines in lock-step with their gyro-motion. This gyroresonant interaction leads to a shift in the pitch angle of the particle with respect to the magnetic field and to a slight increase or decrease in the energy of the particle depending upon whether the wave is propagating in a direction anti-parallel or parallel to the particle along the magnetic field.

Provided that there are comparable numbers of resonant waves propagating in both directions along the magnetic field, it follows that energetic particles as a group will continue to gain energy through these resonant interactions. The reason is simply that, statistically speaking, a given particle will tend to have slightly more head-on than overtaking collisions, other things being equal. Thus it will gain energy at the expense of the wave field. In any given realization there will be some particles that were unlucky and had more overtaking than head-on collisions and vice versa. So in fact the energetic particles undergo a diffusion in energy. This process is known as stochastic, or second-order Fermi, acceleration. It is dependable, to be sure, but fairly slow because any given particle will execute many gyro-resonant interactions with the wave field.

More efficient is first-order Fermi acceleration, which occurs in the vicinity of shock fronts and is the reason why the latter are invariably shrouded in clouds of energetic particles. Energetic particles that live in the rest frame of the plasma behind the shock (i.e. they have just resonantly interacted with a wave) see the upstream plasma moving toward them, the shock wave moving away from them, and the plasma around them at rest. Since the upstream plasma is flowing into the shock at speeds in excess of any characteristic wave speed, it follows that the resonant waves on the upstream side of the shock are all propagating toward the energetic particle; consequently, any gyroresonant interaction it has with a wave on the upstream side will be of the head-on variety and will lead to a slight increase in energy. Now look at the situation as seen by an energetic particle living in the rest frame of the upstream plasma. It sees the downstream plasma moving toward it, the shock moving toward it, and the plasma around it at rest. Again, the flow speeds and the wave speeds are such that all the resonant waves in the downstream plasma are moving toward the particle, and any interaction it makes with a downstream wave will be of the head-on variety and lead to an energy gain. Therefore, an energetic

Jupiter Aurora
NASA and J. Clarke (University of Michigan) • STScI-PRC00-38
HST • STIS

Fig. 2.9. Aurorae occur near the magnetic polar caps of planets. They result from the precipitation of energetic particles, usually electrons, along the conduits formed by the dipolar magnetic field lines. The magnetic field lines that emanate from the polar caps often change their connectivity between the planetary field anchored at the opposite antipode and the magnetic fields carried by the solar wind. Rotation and the orientation of the solar wind magnetic field often determine when topological changes in connectivity will occur. Reconnection and current sheets are responsible for these changes and are also implicated in the acceleration of particles that gives rise to the aurora in the ionosphere and thermosphere. The figure gives for comparison two similar but distinct types of aurora around (left) Saturn and (right) Jupiter (see Chapter 13; the association of the features in Jupiter's aurora with its magnetic field and satellites is shown in Fig. 13.7). See also the color-plate section. (Courtesy of NASA.)

particle that bounces back and forth across a shock invariably gains energy with each collision, not simply on average.

This first-order Fermi process is fast and efficient and what limits its operation is that the shock eventually passes by and leaves the energetic particles behind in the downstream flow. The exception is a group of particles that are able to move great distances with the shock through a process known as "shock surfing". Eventually, even these particles stop gaining energy when they run out of waves with which to interact. Notice that these waves that scatter are a critical ingredient in this process and one may well wonder whether they are always present. The answer is yes, to some degree, because the energetic particles in fact create some of the waves by streaming at high speeds along the magnetic field lines! Indeed, during the overtaking resonant interactions the loss of energy by a particle leads to a gain by the wave field or the generation of a fresh wave quantum.

Electric fields, either in reconnection regions, along rotating magnetic field lines, or across double layers provide a much more obvious and direct way to energize particles to higher energies (see Fig. 2.9). Unlike the two Fermi processes, any charged particle can in principle take advantage of this energy boost quite independently of its energy. For this reason, large-scale electric fields are rare because they are effectively shorted-out by the mobility of the populous thermal plasma.

Not surprisingly, particle acceleration and energization take place throughout the heliosphere and are often closely tied to the presence of wave turbulence. The solar wind is full of waves that are mostly propagating away from the Sun. The magnetosphere is populated by standing waves that fit neatly onto the dipolar field lines. The solar atmosphere is criss-crossed by waves propagating in all different directions. The regions upstream from planetary bow shocks are veritable surf zones of wave activity.

The wave–particle connection is intimate and is based on equal rights. Waves scatter and energize particles, while particles create and damp waves. Energy is input to a wave–particle system through the generation of waves by external processes such as shock wave or shear-flow turbulence or through the direct acceleration of thermal particles by electric fields or adiabatic compression in magnetic traps. In the case of planetary bow shocks, the ultimate energy source is the kinetic energy of the inflowing solar wind plasma. For the solar atmosphere it is the underlying turbulent convection.

2.9 Concluding thoughts

In this very brief introduction to the concepts of heliophysics much has been left unsaid and very many fascinating processes and phenomena have not be touched upon in the limited space available. Fortunately, there are three volumes in this series, and over a thousand pages follow in which these lacunae will be addressed in depth and in breadth. This short chapter emphasizes the tremendous unity that pervades the diversity of processes within the heliosphere. This unity adds richness to the study of heliophysics and creates opportunities for penetrating insights, understandings, and most important, intuitions, that extend far beyond any specific realization.

A subordinate theme of the chapter has been the immense creative power of heliophysical magnetized plasmas. These plasmas carry out marvelous, surprising and often spectacular things on the basis of a rather limited number of fundamental processes and elementary building blocks. They make adroit use of a wide variety of spatial and temporal scales, structure and form, cause and effect. While all planetary magnetospheres ostensibly serve the same utilitarian purpose, each has its own individual character and appeal. While each dynamo dependably creates magnetic fields from the same raw ingredients, they all look and behave differently. And while every current sheet is busy severing and then reconnecting magnetic field lines, this is carried out with individual panache and explosive style.

The very language of heliophysics exhibits the same unity within diversity. The heliospheric plasma is on the one hand behaving as a self-gravitating magnetized fluid continuum if one blurs out the detail and leaves the camera shutter open for some time. Yet at the same time it is a myriad of individual particles traveling along a plethora of trajectories influenced by one another when viewed stroboscopically under the lens of a microscope. In reality it is both and it is neither. Just like the King in Antoine de Saint-Exupéry's *The Little Prince*, we need to ask the appropriate questions of each individual description of heliospheric plasmas. Only then may we rest assured of the veracity of the answers.

As we gaze upon the uncountable array of stars strewn across the vault of the heavens, we can remember that the remarkable phenomena of heliophysics are presently unfolding around those very stars and planetary systems that give light to the night sky. Heliophysics is truly a universal science.

3

Creation and destruction of magnetic field

MATTHIAS REMPEL

3.1 Introduction – magnetic fields in the universe

Objects in the universe from the scale of a planet to the size of a galaxy show evidence of large-scale magnetic fields. Despite the fact that the physical conditions in such objects are quite different, the creation and destruction of magnetic fields is closely linked to turbulent motions of a highly conducting fluid within these bodies. Dynamo theory focuses on the characterization of conditions under which a flow of highly conducting fluid can sustain a magnetic field against resistive decay. This chapter is an introduction to dynamo theory with a primary focus on general concepts rather than detailed applications; the latter are discussed in Vol. III of this series. We start this introduction with a brief overview of the properties of objects with large-scale magnetic fields in the universe.

3.1.1 The Earth and other planets

The magnetic field of the Earth has a strength of about 0.5 gauss and a mainly dipolar character. Currently the dipole axis is tilted by about 11° with respect to the axis of rotation. From studies of rock magnetism (when rocks cool below the Curie point they preserve the magnetic field that was present in them at that time) it is known that the Earth has had a magnetic field over the past 3.5×10^9 years and that the strength and orientation of the field has varied significantly on time scales of 10^3 to 10^4 years. A given polarity typically dominates for about 200 000 years, with quick reversals on a time scale of a few thousand years in between. While the orientation of the dipole axis changes significantly with time, the dipole moment is aligned with the axis of rotation when averaged over $\sim 10^4$ years.

Currently, the Moon does not have a large-scale magnetic field, but there are indications from lunar rocks that the Moon did have such a field more than 3×10^9 years ago. Of the other terrestrial planets only Mercury shows a large-scale field. All the gas giants (Jupiter, Saturn, Uranus and Neptune) show a magnetic field of significant strength (see Table 13.3 and, e.g. Fig. 13.2). Both Uranus and Neptune have a dipole axis significantly tilted to the rotation axis (indicating a different mode of the dynamo), while Saturn is the planet with the most axisymmetric field. Ganymede, a large satellite of Jupiter, also has a large-scale magnetic field.

While in terrestrial planets and the moons of the Jupiter system liquid iron cores are responsible for the dynamo action, the gas giants Jupiter and Saturn have inner regions of metallic hydrogen (formed under high pressure) with a conductivity similar to that of molten iron. For Uranus and Neptune, not much detail is known about the mixture of the conducting layer responsible for dynamo action.

3.1.2 The Sun and other stars

Magnetic fields on the Sun are measured primarily through the Zeeman effect. While it is in most cases difficult to observe the actual splitting of the spectral lines, most measurements make use of the linear and circular polarization resulting from the presence of a magnetic field. The Sun shows magnetic field on all observable scales with a significant range in field strength, from individual sunspots with magnetic field strengths of 2500 to 3000 gauss to the average field strength of the global field of only a few gauss.

The most prominent feature of solar magnetism is the 11 year sunspot cycle (if one considers field reversals the real period is 22 years), which is reflected in the changing number of sunspots appearing on the surface of the Sun. At the beginning of a cycle spots appear at latitudes of about 35°, while close to the end they appear almost at the equator. This property is commonly summarized in the so-called solar butterfly diagram. During the epoch of minimum, the large-scale field of the Sun is almost dipolar; the reversal of the poles takes place during solar maximum. On a longer time scale the magnetic activity changes significantly in amplitude and is interrupted by epochs, of 100–200 years' duration, when sunspots are completely absent. Over time scales beyond the direct observation records the magnetic behavior of the Sun can be derived from the concentration of cosmogenic isotopes in tree rings (^{14}C) and ice cores (^{14}C and ^{10}Be), allowing for a reconstruction of solar activity over the last 300 000 years. Over these time scales the Sun has exhibited a behavior similar to that in recent decades, namely a dominant 11 year sunspot cycle interrupted by epochs of sustained low activity. For more details on the solar magnetic field we refer to the textbook of Stix (2004).

Magnetic activity is common among other solar-like stars (stars with an outer convection zone and a radiative interior). Observations of the stellar luminosity or of chromospheric (UV and optical) and coronal (X-ray) emission show that a majority of solar-like stars are magnetically active and around a third to a half show cyclic activity with periods in the range from 3 to 30 years (see e.g. Schrijver and Zwaan, 2000).

3.1.3 Galaxies

Galactic magnetic fields are measured through the polarization of radio synchrotron emissions. The magnetic field strength of galactic fields is of the order of a few μG; galaxies of almost all structures and ages exhibit a magnetic field. The field typically shows features reflecting the structure of the matter distribution of a galaxy (e.g. spiral galaxies have spiral magnetic fields).

Galactic magnetic fields are of particular interest for dynamo theorists, since galaxies are the only objects in the universe in which the dynamo region is directly accessible through observation.

3.2 Magnetohydrodynamics

The mathematical basis for dynamo theory is magnetohydrodynamics (MHD). Since most derivations in this chapter are focused around the induction equation we will discuss only the latter in detail and refer to textbooks for a fuller discussion of MHD (Parker, 1979; Priest, 1982; Shu, 1992; Davidson, 2001; Parker, 2007). Here we summarize the most important equations for later reference.

3.2.1 MHD equations

The general assumptions underlying the MHD approach are (i) the validity of the continuum approximation and non-relativistic dynamics, and (ii) the quasi-neutrality of the plasma. It is further assumed that the collisional coupling between different ions constituting the plasma is sufficiently strong to treat the plasma as a single fluid (e.g. the temperature is the same for all ions).

The full set of MHD equations combines the induction equation (3.4) with a version of the Navier–Stokes equations that includes the Lorentz force $\mathbf{j} \times \mathbf{B}$. This results in equations for continuity, force balance, energy conservation, and magnetic field, respectively:

$$\frac{\partial \varrho}{\partial t} = -\nabla \cdot (\varrho \mathbf{v}), \tag{3.1}$$

$$\varrho \frac{\partial \mathbf{v}}{\partial t} = -\varrho(\mathbf{v} \cdot \nabla)\mathbf{v} - \nabla p + \varrho \mathbf{g} + \frac{1}{\mu_0}(\nabla \times \mathbf{B}) \times \mathbf{B} + \nabla \cdot \mathcal{T}, \tag{3.2}$$

$$\varrho \frac{\partial e}{\partial t} = -\varrho(\mathbf{v} \cdot \nabla)e - p\nabla \cdot \mathbf{v} + \nabla \cdot (\kappa \nabla T) + Q_\nu + Q_\eta, \tag{3.3}$$

$$\frac{\partial \mathbf{B}}{\partial t} = \nabla \times (\mathbf{v} \times \mathbf{B} - \eta \nabla \times \mathbf{B}). \tag{3.4}$$

Here **v** denotes the fluid velocity, **B** the magnetic field, e the specific internal energy, p the gas pressure, ϱ the mass density, and **g** gravity; Q_v is the viscous heating, and is given by

$$Q_v = \tau_{ik}\Lambda_{ik} \tag{3.5}$$

with deformation tensor

$$\Lambda_{ik} = \frac{1}{2}\left(\frac{\partial v_i}{\partial x_k} + \frac{\partial v_k}{\partial x_i}\right). \tag{3.6}$$

We have, for the viscous stress tensor $\mathcal{T}$

$$\mathcal{T}_{ik} = 2\varrho v \left(\Lambda_{ik} - \frac{1}{3}\delta_{ik}\nabla\cdot\mathbf{v}\right) \tag{3.7}$$

and the resistive (Ohmic) dissipation Q_η is given by

$$Q_\eta = \frac{\eta}{\mu_0}(\nabla\times\mathbf{B})^2. \tag{3.8}$$

The pressure p can be computed from the internal energy e through the equation of state

$$p = (\gamma - 1)\varrho e. \tag{3.9}$$

Here v, η, and κ are the values of the viscosity, magnetic diffusivity, and thermal conductivity; μ_0 denotes the permeability of the vacuum.

As an aside here, we point out one property of the magnetic field that guides our thinking about its effects in the force balance of Eq. (3.2): using a vector formula we see that

$$\frac{1}{\mu_0}(\nabla\times\mathbf{B})\times\mathbf{B} = -\nabla\frac{B^2}{2\mu_0} + \frac{1}{\mu_0}(\mathbf{B}\cdot\nabla)\mathbf{B}, \tag{3.10}$$

which shows that the Lorentz force is the sum of an isotropic pressure-like force and a tension force related to the curvature of the field. The ratio of the magnetic and gas pressure terms in Eq. (3.2) is commonly referred to as the plasma β-value:

$$\beta \equiv \frac{2\mu_0 p}{B^2}. \tag{3.11}$$

The natural approach to the dynamo problem would be to solve the MHD equations for a given setup that includes a magnetic field in order to see whether the magnetic field can be maintained by the velocity field. Unfortunately this approach is currently still strongly restricted by the available computing power (at least for the case of the solar dynamo). We briefly discuss in Section 3.5.3 the current status of 3D dynamo simulations; more discussion of this topic will be found in Vol. III of this series. For understanding the fundamental concepts we discuss the dynamo problem in the following sections using just the induction equation and basically asking whether a given velocity field $\mathbf{v}$ can sustain a magnetic field $\mathbf{B}$ against resistive decay. This is called the kinematic approach and is valid as long as the Lorentz force in the force equation can be neglected. The kinematic approach is very helpful for discussing the general properties that a velocity field has to have in order to qualify as a dynamo, which is the main focus of this introduction to dynamo theory.

3.2.2 Induction equation

The induction equation (3.4) arises from Ohm's law in combination with the non-relativistic approximation to the Maxwell equations. In its most general form Ohm's law is a relation between electric current, electric field, magnetic field, plasma motions, and electron pressure gradients. It is derived from an equation of motion for electrons in which the interaction with ions, defining the bulk motion of the plasma with velocity $\mathbf{v}$, is described through a collisional drag term related to the differential motion:

$$n_e m_e \frac{d\mathbf{v}_e}{dt} = n_e q_e(\mathbf{E} + \mathbf{v}_e \times \mathbf{B}) - \nabla p_e + n_e m_e \frac{\mathbf{v} - \mathbf{v}_e}{\tau_{ei}}. \tag{3.12}$$

Here $\mathbf{v}_e$ denotes the electron velocity, τ_{ei} the collision time between electrons and ions, q_e the electron charge, m_e the electron mass, n_e the electron density, and p_e the electron pressure. Using

$$\mathbf{j} = n_e q_e(\mathbf{v}_e - \mathbf{v}), \tag{3.13}$$

we can express $\mathbf{v}_e$ in terms of the electric current (the time derivative of $\mathbf{v}$ can be neglected since $m_e \ll m_i$, where m_i is the mass of an ion), which leads to Ohm's law:

$$\frac{\partial \mathbf{j}}{\partial t} + \nabla \cdot (\mathbf{v}\mathbf{j} + \mathbf{j}\mathbf{v}) + \frac{\mathbf{j}}{\tau_{ei}} = \frac{n_e q_e^2}{m_e}(\mathbf{E} + \mathbf{v} \times \mathbf{B}) + \frac{q_e}{m_e}\mathbf{j} \times \mathbf{B} - \frac{q_e}{m_e}\nabla p_e. \tag{3.14}$$

Here the first two terms on the left-hand side are related to the electron inertia; $\mathbf{v}\mathbf{j} + \mathbf{j}\mathbf{v}$ denotes a dyadic tensor with components $v_n j_m + v_m j_n$; a term quadratic in $\mathbf{j}$ has been absorbed into the electron pressure p_e. The second term on the right-hand side describes the Hall current, which becomes important if the collision time is longer than the electron gyration time, i.e. when $\tau_{ei} \omega_L > 1$, where $\omega_L = q_e B / m_e$ denotes the Larmor (or gyro) frequency. The Hall term leads to anisotropic plasma conductivity with respect to the magnetic field direction and is typically important in low-density plasmas, in which τ_{ei} can be very large. For stellar convection zones the Hall term, the electron pressure contribution, and the electron inertia are unimportant (unless high-frequency plasma oscillations are considered), leading to the simplified Ohm's law

$$\mathbf{j} = \sigma(\mathbf{E} + \mathbf{v} \times \mathbf{B}) \tag{3.15}$$

with plasma conductivity

$$\sigma = \frac{\tau_{ei} n_e q_e^2}{m_e}. \tag{3.16}$$

Using Ampère's law, $\nabla \times \mathbf{B} = \mu_0 \mathbf{j}$, yields for the electric field in the laboratory frame

$$\mathbf{E} = -\mathbf{v} \times \mathbf{B} + \frac{1}{\mu_0 \sigma} \nabla \times \mathbf{B}, \tag{3.17}$$

leading to the induction equation through one of Maxwell equations:

$$\frac{\partial \mathbf{B}}{\partial t} = -\nabla \times \mathbf{E} = \nabla \times (\mathbf{v} \times \mathbf{B} - \eta \nabla \times \mathbf{B}) \tag{3.18}$$

with magnetic diffusivity

$$\eta = \frac{1}{\mu_0 \sigma}. \tag{3.19}$$

In MHD the equations are typically expressed in terms of the magnetic field $\mathbf{B}$ and the flow $\mathbf{v}$, with electric fields and currents completely eliminated from the system. This is done primarily out of mathematical convenience, since formulating the problem in terms of currents leads to intractable equations involving integrals of the currents over the entire volume under study (see also e.g. Vasyliūnas, 2001, Parker, 2007, and the discussion in Section 10.4.1 for perspectives on this).

3.2.3 Advection, diffusion, and the magnetic Reynolds number

Let L be a typical length scale and U a characteristic velocity of the problem. Expressing the time in units of L/U and the spatial derivatives in the induction equation (3.4) in units of L leads to the dimensionless form of the induction equation,

$$\frac{\partial \mathbf{B}}{\partial t} = \nabla \times \left(\mathbf{v} \times \mathbf{B} - \frac{1}{R_m} \nabla \times \mathbf{B} \right) \tag{3.20}$$

with magnetic Reynolds number

$$R_m \equiv \frac{UL}{\eta}. \tag{3.21}$$

Table 3.1. *Characteristic values for the magnetic Reynolds number R_m and diffusion time scales for the core of the Earth, the Sun, and laboratory liquid sodium experiments*

Object	$\eta\,(\mathrm{m^2/s})$	$L\,(\mathrm{m})$	$U\,(\mathrm{m/s})$	R_m	P_m	τ_d
Earth (outer core)	2	10^6	10^{-3}	500	10^{-5}	10^4 years
Sun (molecular)	1	10^8	100	10^{10}	10^{-4}	3×10^8 years
Sun (turbulent)	10^8	10^8	100	100	~ 1	3 years
Lab experiment	0.1	1	10	100	10^{-5}	10 seconds

The limit $R_m \ll 1$ is referred to as the diffusion-dominated regime, in which the (dimensional) induction equation reduces to a diffusion equation of the form

$$\frac{\partial \mathbf{B}}{\partial t} = \eta \Delta \mathbf{B}. \tag{3.22}$$

Here we have made the additional simplifying assumption of a constant magnetic diffusivity η. Assuming that the magnetic field has a typical length scale L, we can estimate from the induction equation (3.4) a decay time scale given by

$$\tau_d \sim \frac{L^2}{\eta}. \tag{3.23}$$

The limit $R_m \gg 1$ is referred to as the advection-dominated regime; in it the induction equation reduces to the equation of ideal MHD (except for possible boundary layers where diffusivity could be still important),

$$\frac{\partial \mathbf{B}}{\partial t} = \nabla \times (\mathbf{v} \times \mathbf{B}). \tag{3.24}$$

Expanding the expression on the right-hand side of this ideal induction equation leads to

$$\frac{\partial \mathbf{B}}{\partial t} = -(\mathbf{v} \cdot \nabla)\mathbf{B} + (\mathbf{B} \cdot \nabla)\mathbf{v} - \mathbf{B}(\nabla \cdot \mathbf{v}). \tag{3.25}$$

While the first term on the right-hand side describes the advection of magnetic field, the last two terms describe the amplification of the field by shear (the second term) and compression (the third term).

Table 3.1 shows values for η, L, U R_m, P_m, and τ_d for different astrophysical objects and laboratory experiments. Here $P_m = \nu/\eta$ denotes the magnetic Prandtl number (the ratio of viscosity and magnetic diffusivity). For the Sun we have given two sets of values, one based on the molecular diffusivity and the other based on the turbulent diffusivity, which we introduce in more detail later. These values reflect the conditions close to the base of the convection zone; the Reynolds and Prandtl numbers for the photosphere are about 10^5 and 10^{-7}, respectively. The diffusion time scale based on the molecular diffusivity is of the same order as the age of the Sun. This number, however, only applies to the radiative zone of the Sun (the innermost 70% in radius), where the stratification is stable, with no convection. It is conceivable that a magnetic field (the primordial magnetic field) would have endured in the core of the Sun without additional support from induction effects if a magnetic field had been present in the interstellar cloud from which the Sun formed.

3.2.3.1 Alfvén's theorem: frozen-in flux

Let Φ be the magnetic flux through a surface F with the property that its boundary curve ∂F is moving with the fluid:

$$\Phi = \int_F \mathbf{B} \cdot \mathbf{df}, \tag{3.26}$$

where $\mathbf{df}$ is an element of the surface F. The total time derivative of Φ can be written as

$$\frac{d\Phi}{dt} = \lim_{\Delta t \to 0} \frac{1}{\Delta t} \left\{ \int_{F(t+\Delta t)} \mathbf{B}(t+\Delta t) \cdot \mathbf{df} - \int_{F(t)} \mathbf{B}(t) \cdot \mathbf{df} \right\}$$

$$= \lim_{\Delta t \to 0} \frac{1}{\Delta t} \left\{ \int_{F(t+\Delta t)} [\mathbf{B}(t+\Delta t) - \mathbf{B}(t)] \cdot \mathbf{df} \right.$$

$$\left. + \int_{F(t+\Delta t)} \mathbf{B}(t) \cdot \mathbf{df} - \int_{F(t)} \mathbf{B}(t) \cdot \mathbf{df} \right\}$$

$$= \int_F \frac{\partial \mathbf{B}}{\partial t} \cdot \mathbf{df} - \int_{\partial F} (\mathbf{v} \times \mathbf{B}) \cdot \mathbf{ds}$$

$$= \int_F \left[\frac{\partial \mathbf{B}}{\partial t} - \nabla \times (\mathbf{v} \times \mathbf{B}) \right] \cdot \mathbf{df} = 0. \tag{3.27}$$

Here $\mathbf{ds}$ is an element of ∂F and we have used the fact that the area elements $-\int_{F(t)} \mathbf{df}$, $\int_{\partial F} \mathbf{ds} \times \mathbf{v} \Delta t$ and $\int_{F(t+\Delta t)} \mathbf{df}$ form, in the limit $\Delta t \to 0$, a closed surface (see Fig. 3.1), so that $\nabla \cdot \mathbf{B} = 0$ leads to

$$\int_{F(t+\Delta t)} \mathbf{B} \cdot \mathbf{df} - \int_{F(t)} \mathbf{B} \cdot \mathbf{df} + \Delta t \int_{\partial F} \mathbf{B} \cdot (\mathbf{ds} \times \mathbf{v}) = 0. \tag{3.28}$$

Thus the magnetic flux through a surface F with a boundary ∂F that is co-moving with the fluid is conserved. The effect of a shear flow on a magnetic flux tube (a bundle of field lines, see Section 6.4) is illustrated in Fig. 3.2. In the limit of ideal MHD, magnetic field lines are "frozen into the fluid"; this is a useful way to illustrate induction effects. In general it does not make sense to talk about moving

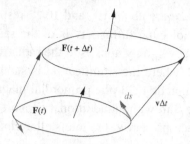

Fig. 3.1. The closed line ∂F is moving with the fluid (it is "frozen into the fluid"). The area elements $-\int_{F(t)} d\mathbf{f}$, $\int_{\partial F} d\mathbf{s} \times \mathbf{v}\Delta t$ and $\int_{F(t+\Delta t)} d\mathbf{f}$ form a closed surface.

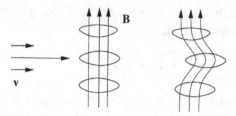

Fig. 3.2. Shearing of a flux tube by a velocity field in ideal MHD.

magnetic field lines since a field line is a mathematical construct with no physical identity. Since, however, in the case of ideal MHD a field line can be regarded as a material line moving with the plasma, the concept of a "moving" field line can be useful, but it has to be used with care.

3.3 The dynamo problem

3.3.1 Motivation

We discussed in Section 3.2.3 how in the absence of induction effects magnetic fields decay on a time scale $\tau_\mathrm{d} \sim L^2/\eta$ (Eq. 3.23). An indication of dynamo action is the presence of magnetic field on a time scale $t \gg \tau_\mathrm{d}$.

In the case of the Earth, studies of rock magnetism show that a magnetic field (comparable in strength with the current field) has been present over geological time scales of order 3.5×10^9 years, which is significantly longer than the diffusion time $\tau_\mathrm{d} \sim 10^4$ years. The magnetic field of the Earth must be maintained by dynamo action; any other type of permanent rock magnetism is ruled out because the temperature in the core is above the Curie point.

In the case of the Sun, magnetic activity can be traced back through ice core measurements of cosmogenic isotopes such as ^{10}Be. On the one hand, for the past 3×10^5 years these measurements show a persistent 11 year cycle disrupted by epochs of reduced solar activity spanning of order 100 to 200 years. Also, stellar observations indicate that solar-like magnetic activity is common among stars with outer convection zones. On the other hand, in Table 3.1 we see that the molecular diffusion time is about $\sim 10^9$ years (within an order of magnitude), which is comparable with the typical lifetime of a solar-like star. Unlike in the case of the geodynamo this is not enough evidence for dynamo action, and alternative explanations for the solar magnetism have been proposed that are based on the idea of a torsional oscillator periodically amplifying a primordial poloidal field frozen into the core of the Sun. However, this would require significant alterations in the internal rotation of the Sun, changing the sign of the rotational shear in a way that is not consistent with the almost steady differential rotation observed by helioseismology. As shown in Table 3.1, turbulent motions in the convection zone of the Sun (the outermost 30%) significantly reduce the effective diffusion time scale so that here also dynamo action is required to explain the observed activity.

3.3.2 A brief history of dynamo theory

Larmor (1919) was the first to ask how a body like the Sun can become a magnet. At that time he suggested already the possibility that the motion of an electrically conducting fluid across magnetic field lines in such a rotating body could create the currents required to maintain that field. In the following decades "advances" in dynamo theory were made mainly through the discovery of anti-dynamo theorems that rule out simple field and flow structures, which made the investigation of possible dynamos rather complicated and hindered theoretical progress for a few decades. The most famous anti-dynamo theorem on the impossibility of axisymmetric dynamo action was found by Cowling (1933), and many generalizations and additional theorems followed. The search for a general anti-dynamo proof was brought to an end with the examples of Herzenberg (1958) and Backus (1958), which demonstrated rigorously that dynamo action as defined in Section 3.3.3 below is possible. While these examples involved laminar flows (uniformly rotating spheres embedded into a conducting fluid), the first big breakthrough for astrophysical, turbulent, dynamos was made by Parker (1955), who realized that the dynamo problem can be simplified by averaging over the complicated non-axisymmetric component and deriving an equation for an (axisymmetric) mean field. This approach was later mathematically refined by Braginskii (1964) and Steenbeck *et al.* (1966). Mean-field theory allowed the fundamental study of dynamo theory and has been used since then as the main mathematical framework. For a summary of mean-field dynamo theory we refer to Rüdiger and Hollerbach (2004). Among the more recent developments is the addition of advection caused by the meridional circulation (see Section 3.3.7 below). With advances in computing power, 3D numerical simulations of the full MHD equations (with some approximations) have become a powerful tool for studying dynamos over the last two decades. In the following sections we focus primarily on mean-field theory in introducing the basic concepts of large-scale turbulent dynamo theory. At the end we discuss the limitations of mean-field theory and summarize some results from three-dimensional (3D) numerical simulations. For further reading we recommend also the textbooks on dynamo theory by Moffatt (1978) and Proctor and Gilbert (1995) and the review of nonlinear dynamo theory by Brandenburg and Subramanian (2005).

3.3.3 Definition of the dynamo problem

Let S be a volume filled with conducting fluid, bounded by a surface ∂S. We are looking for a smooth field $\mathbf{B}$ that is produced by currents contained within S, decays asymptotically as r^{-3} for $r \to \infty$, is solenoidal ($\nabla \cdot \mathbf{B} = 0$), and is a solution of the system

$$\frac{\partial \mathbf{B}}{\partial t} = \nabla \times (\mathbf{v} \times \mathbf{B} - \eta \nabla \times \mathbf{B}) \qquad \text{in } S,$$

$$\nabla \times \mathbf{B} = 0 \qquad\qquad\qquad\qquad \text{outside } S,$$

$$[\mathbf{B}] = 0 \quad (\mathbf{B} \text{ is continuous}) \qquad \text{across } \partial S; \qquad (3.29)$$

in the last line of the equation the square brackets denote a difference across a boundary. We assume further that the smooth velocity field $\mathbf{v}$ vanishes outside S and is tangential to the boundary ($\mathbf{n} \cdot \mathbf{v} = 0$ on ∂S) and that the kinetic energy within S has an upper bound:

$$E_{\text{kin}} = \int_S \frac{1}{2} \varrho \mathbf{v}^2 dV \leq E_{\text{max}} \qquad \forall\, t. \qquad (3.30)$$

The velocity field $\mathbf{v}$ acts as a dynamo if an initial condition $\mathbf{B} = \mathbf{B}_0$ exists for which the magnetic energy

$$E_{\text{mag}} = \int_{-\infty}^{\infty} \frac{1}{2\mu_0} \mathbf{B}^2 \, dV \qquad (3.31)$$

does not approach zero as $t \to \infty$.

The formal difference between the type of dynamo in which we are interested here and the self-exciting dynamos in power plants is the homogeneous distribution of conductivity, which does not put any constraints on the electric currents involved. For this reason these dynamos are also called homogeneous fluid dynamos. (In power plants homogeneous conductivity would lead to dramatic shortcircuiting! We could, however, regard electric wires as a special case of a highly inhomogeneous conductivity distribution.)

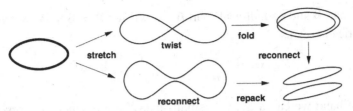

Fig. 3.3. Illustration of two possible flux-rope dynamos. In both cases the field amplification takes place during the stretch operation. The twist–fold (upper diagrams) and reconnect–repack (lower diagrams) steps are required to remap the amplified flux rope into the original volume element so that the process can be repeated. Magnetic diffusivity is essential to allow for the topology change required to close the cycle. Each cycle increases the field strength by a factor 2.

3.3.4 Basic ingredients of a dynamo

Figure 3.3 illustrates the basic ingredients required to amplify a closed magnetic field loop. After a full cycle, the magnetic field strength and the flux have doubled (two loops, each with the original magnetic flux magnitude) and the process is repeated. Even this very simple example illustrates some fundamental properties of a dynamo process. In order to be able to remap the magnetic field configuration into the original volume element, three-dimensional motions are required; amplification through stretching is possible in a strictly two-dimensional domain, but there is no way for the resulting field to return to the right-hand side of the image. The two examples in the figure also illustrate the crucial role of diffusivity in changing the topology of the field. The stretch–twist–fold mechanism (excluding diffusive steps) leads to loops of increased complexity, while the stretch–reconnect–repack process explicitly involves magnetic diffusivity and ends up with two flux ropes of similar topology. A reconnection step at the end of the stretch–twist–fold process leads a similar result. In the case of the stretch–twist–fold dynamo the sign of the twist does not matter. Note that neither mechanism requires a flow with a net kinetic helicity.

3.3.5 Large-scale and small-scale dynamos

The definition of the dynamo problem given in Section 3.3.3 focuses only on the asymptotic behavior of the magnetic energy and does not imply anything about the structure of the magnetic field. Most astrophysical objects have turbulent velocity fields with a correlation length scale l_c that is much smaller than the scale of the star or planet. Nevertheless these objects have a magnetic field that has a length scale much larger than l_c and also evolves over time scales much larger than the correlation time scale of the turbulence. If we decompose the magnetic field into a large-scale part $\overline{\mathbf{B}}$ and a small-scale part $\mathbf{B}'$, by writing $\mathbf{B} = \overline{\mathbf{B}} + \mathbf{B}'$, we can express the magnetic energy as

$$E_{\text{mag}} = \int \frac{1}{2\mu_0} \overline{\mathbf{B}}^2 \, dV + \int \frac{1}{2\mu_0} \overline{\mathbf{B}'^2} \, dV. \tag{3.32}$$

On the one hand we call a dynamo small-scale if the magnetic energy is mainly in small scales (the energy-carrying scales of the turbulence), meaning $\overline{\mathbf{B}}^2 \ll \overline{\mathbf{B}'^2}$. On the other hand, a large-scale dynamo produces a magnetic field with a scale comparable with the planet or star. In this case we have at least $\overline{\mathbf{B}}^2 \gtrsim \overline{\mathbf{B}'^2}$. While almost any chaotic (turbulent) velocity field is a small-scale dynamo for a sufficiently large magnetic Reynolds number R_{m}, a large-scale dynamo requires an inverse cascade that produces magnetic energy on a scale larger than the energy-carrying scale of the turbulence. This requires certain large-scale properties in the flow field such as a mean helicity and shear, as we discuss later. There is no sharp distinction between small- and large-scale dynamos and both can be present at the same time.

Three-dimensional numerical MHD simulations (Cattaneo, 1999; Vögler and Schüssler, 2007) have shown that almost all turbulent velocity fields lead to small-scale dynamo action, provided that the magnetic Reynolds number is large enough. For numerical reasons, most small-scale dynamo simulations have been performed in a setup in which the viscosity ν and the magnetic diffusivity η are similar, while in stellar convection zones typically $\nu \ll \eta$. The extent to which small-scale dynamos are affected by the magnetic Prandtl number $P_{\text{m}} = \nu/\eta$ is a current research topic. A small value of P_{m} implies that the dissipative scale of the velocity field is much smaller than the dissipative scale of the magnetic field. This means that on the smallest scale of the magnetic field the velocity field is still turbulent and therefore more dissipative; in the case of a large value of P_{m} the velocity field is smooth even on the smallest scales of the field and therefore more effective in amplifying field at those scales.

3.3.6 Slow and fast dynamos

We pointed out in Section 3.3.4 the important role of magnetic diffusivity in the dynamo process in allowing changes in field topology. A further distinction of dynamos is made with respect to the dependence of the growth rate of magnetic field on the magnetic Reynolds number R_{m} (Eq. 3.21). If the growth rate of a dynamo is given by a kinematic time scale L/v_{rms} that is asymptotically independent of R_{m} as $R_{\text{m}} \to \infty$ it is called a fast dynamo, while a dynamo is called a slow dynamo if the growth rate is explicitly influenced by the microscopic diffusivity and vanishes as $R_{\text{m}} \to \infty$. The distinction between fast and slow dynamos is only relevant in the initial kinematic growth phase of the magnetic field, in which the feedback of the Lorentz force on the flow field is not important (see Eq. 3.2). Fast dynamo action is only possible if the length scale l at which reconnection takes place decreases at least as $R_m^{-1/2}$ with increasing R_m, so that $\tau_d \sim l^2/\eta \sim L^2/(R_m\eta) \sim L/v_{rms}$. Fast dynamos are of particular interest for astrophysical systems since the magnetic Reynolds number is typically very large (e.g. 10^9 for the Sun).

For a comparative review of large- and small-scale dynamos in the solar convection zone as well as the fast dynamo problem we refer to Cattaneo and Hughes (2001).

3.3.7 Effect of differential rotation and meridional flow

Most spherical large-scale dynamos (i.e. those of the planets and Sun) show an almost dipolar axisymmetric field. We analyze here the simplest situation: the effect of an axisymmetric flow on an axisymmetric magnetic field. The two components of an axisymmetric flow are the differential rotation (the flow in the longitudinal or ϕ-direction) and the meridional circulation (the radial and latitudinal flow, i.e. the flow in the $r\theta$-plane). While these flows are not very important in the case of the outer core of the Earth, all stars with outer convection zones show strong differential rotation, and this is expected to have a significant role in stellar dynamos. A meridional circulation is coupled to differential rotation through the Coriolis force and was identified by Choudhuri *et al.* (1995) and Dikpati and Charbonneau (1999) as a major ingredient of the solar dynamo.

Any axisymmetric solenoidal ($\nabla \cdot \mathbf{B} = 0$) field can be written in spherical coordinates as

$$\mathbf{B} = B\mathbf{e}_\Phi + \nabla \times (A\mathbf{e}_\Phi), \tag{3.33}$$

where B is the toroidal field and the contour lines of $Ar \sin\theta$ are the field lines of the poloidal field. Together with the velocity field

$$\mathbf{v} = v_r \mathbf{e}_r + v_\theta \mathbf{e}_\theta + \Omega r \sin\theta\, \mathbf{e}_\Phi \tag{3.34}$$

this leads to induction equations in spherical coordinates for B and A:

$$\frac{\partial B}{\partial t} + \frac{1}{r}\left[\frac{\partial}{\partial r}(rv_r B) + \frac{\partial}{\partial \theta}(v_\theta B)\right] = r\sin\mathbf{B}_p \cdot \nabla\Omega + \eta\left[\Delta - \frac{1}{(r\sin\theta)^2}\right]B,$$

$$\tag{3.35}$$

$$\frac{\partial A}{\partial t} + \frac{1}{r\sin\theta}\mathbf{v}_p \cdot \nabla(r\sin\theta\, A) = \eta\left[\Delta - \frac{1}{(r\sin\theta)^2}\right]A, \tag{3.36}$$

where Δ denotes the Laplace operator, $\mathbf{B}_p = (B_r, B_\theta, 0)$ the poloidal component of the magnetic field, and $\mathbf{v}_p = (v_r, v_\theta, 0)$ the meridional flow (the poloidal component of the velocity). Here we have assumed further that η is uniform and constant.

The time evolution of the poloidal field is decoupled from the toroidal field and the poloidal field has only decaying solutions: $A \to 0$ for $t \to \infty$. The time evolution of the toroidal field is coupled to the poloidal field through the Ω-effect (or more exactly the $\nabla\Omega$-effect), which can act as a source term preventing decay of the toroidal field. Since $B_p \to 0$ for $t \to \infty$ this source term decays and we have also $B \to 0$ for $t \to \infty$.

This discussion shows that an axisymmetric magnetic field cannot be maintained by an axisymmetric flow. We show in Section 3.3.8 below that this statement remains valid for non-axisymmetric flows also. If we allow for a non-axisymmetric magnetic field, axisymmetric flows can act as a dynamo (see e.g. the dynamo proposed by Gailitis, 1970). However, purely toroidal motions in spherical geometry cannot form a dynamo if $\eta = \eta(r)$ (the toroidal theorem: Elsasser, 1946; Bullard and Gellman, 1954; Backus, 1958).

3.3.8 Cowling's anti-dynamo theorem

Given that in the solar convection zone the magnetic Reynolds number is of order 10^5–10^9 one would expect that it would be easy for fluid motions to overcome the resistive dissipation and that it would also be easy to identify a velocity field with a dynamo property. The difficulty arises from topological constraints on the magnetic field and the velocity field. Cowling's anti-dynamo theorem rules out most simple geometries for dynamo-generated magnetic fields:

No axisymmetric magnetic field with currents limited to a finite volume in space can be maintained by a flow with finite amplitude.

The original theorem of Cowling (1933) assumed a stationary magnetic field, and we present the basic idea of the proof for a poloidal field below. In the mean time the theorem has been proven in more general situations allowing also for time-dependent solutions.

In the case of a stationary magnetic field we have $\nabla \times \mathbf{E} = -\partial\mathbf{B}/\partial t = 0$ and we can express the electric field through a potential Ψ. In that case Ohm's law is given by $\mathbf{j} = \sigma(\mathbf{v} \times \mathbf{B} - \nabla\Psi)$. Having an axisymmetric field requires that the poloidal field $\mathbf{B}_p$ has at least one O-type neutral line (in the toroidal direction) on which the poloidal field strength vanishes. However, the toroidal electric current $\mu_0 \mathbf{j}_t = \nabla \times \mathbf{B}_p$ along this line does not vanish and has to be maintained by the toroidal component of $\mathbf{v} \times \mathbf{B} - \nabla\Psi$. The electric field does not contribute since axisymmetry requires $\partial\Psi/\partial\phi = 0$ and $(\mathbf{v} \times \mathbf{B})_t = \mathbf{v}_p \times \mathbf{B}_p$ vanishes since B_p vanishes at the neutral line. Therefore an axisymmetric poloidal field cannot be maintained by a flow. This proof can be generalized to the case of a neutral point of higher order, at which both B_p and its derivatives vanish. We refer here to textbooks such as Moffatt (1978) for an in-depth discussion.

3.4 Mean-field theory

In this section, we focus on the processes capable of maintaining a large-scale magnetic field. To this end we decompose the magnetic field into a large-scale "mean" field and small-scale components through an averaging procedure. We assume in the following that the averaging procedure obeys the Reynolds rules: for any functions f and g decomposed as $f = \overline{f} + f'$ and $g = \overline{g} + g'$, where the bar indicates the averaged and the prime the fluctuating quantity, we require that

$$\overline{\overline{f}} = \overline{f} \quad \Longrightarrow \quad \overline{f'} = 0, \tag{3.37}$$

$$\overline{f + g} = \overline{f} + \overline{g}, \tag{3.38}$$

$$\overline{\overline{f}g} = \overline{f}\,\overline{g} \quad \Longrightarrow \quad \overline{f'\overline{g}} = 0, \tag{3.39}$$

$$\overline{\partial f/\partial x_i} = \partial \overline{f}/\partial x_i, \tag{3.40}$$

$$\overline{\partial f/\partial t} = \partial \overline{f}/\partial t . \tag{3.41}$$

The averaging procedures that are of interest in the context of mean-field theory are the ensemble average (meaning that a chaotic system is averaged over several representations of the chaotic system) and the longitudinal average, in which $\overline{\mathbf{B}}$ reflects the average axisymmetric component of the large-scale magnetic field (a multipole series with $m = 0$).

3.4.1 Mean-field induction equation

We follow here a derivation of mean-field electrodynamics that follows closely Rädler (1980) and Rädler and Stepanov (2006). In order to derive an equation for the time evolution of the mean field we apply the averaging procedure to the induction equation (3.4), which leads to

$$\frac{\partial \overline{\mathbf{B}}}{\partial t} = \nabla \times \left(\overline{\mathbf{v}' \times \mathbf{B}'} + \overline{\mathbf{v}} \times \overline{\mathbf{B}} - \eta \nabla \times \overline{\mathbf{B}} \right). \tag{3.42}$$

The new term which enters this equation, in comparison with the original induction equation is the second-order correlation electromotive force (EMF),

$$\overline{\mathcal{E}} \equiv \overline{\mathbf{v}' \times \mathbf{B}'}. \tag{3.43}$$

While the fluctuating velocity component $\mathbf{v}'$ is assumed to be known (from a kinematic approach), $\mathbf{B}'$ has to be computed from the induction equation. An equation for $\mathbf{B}'$ can be derived by subtracting the mean-field induction equation (3.42) from the microscopic induction equation (3.4), which leads to

$$\frac{\partial \mathbf{B}'}{\partial t} = \nabla \times \left(\mathbf{v}' \times \overline{\mathbf{B}} + \overline{\mathbf{v}} \times \mathbf{B}' - \eta \nabla \times \mathbf{B}' + \mathbf{v}' \times \mathbf{B}' - \overline{\mathbf{v}' \times \mathbf{B}'} \right). \tag{3.44}$$

In general it is only possible to solve this equation by making strong assumptions, primarily because of the terms that are quadratic in the fluctuating quantities (the closure problem). We return to this in Section 3.4.3 below, first discussing the mean-field induction equation on the basis of more general symmetry properties which do not require an explicit solution for $\mathbf{B}'$.

3.4.2 Symmetry constraints on turbulent induction effects

For a given velocity field $\mathbf{v}$ (found using the kinematic approach) it follows from the induction equation (3.44) for a fluctuating field that $\mathbf{B}'$ is a linear functional of $\overline{\mathbf{B}}$, so that the electromotive force $\overline{\mathcal{E}}$ (Eq. 3.43) is also a linear functional of $\overline{\mathbf{B}}$. In general $\mathbf{B}'$ does not have to vanish if $\overline{\mathbf{B}}$ does, since the velocity field $\mathbf{v}'$ could also support a small-scale dynamo. For simplicity, however, we assume in the following that $\mathbf{B}'$ vanishes if $\overline{\mathbf{B}}$ does, assuming that any $\mathbf{B}'$ produced by a small-scale dynamo correlates only weakly with $\mathbf{v}'$ (if this is not the case, the following equations would have to be modified by adding an inhomogeneous source term not dependent on $\overline{\mathbf{B}}$). We can therefore express the components $\overline{\mathcal{E}}_i$ as

$$\overline{\mathcal{E}}_i(\mathbf{x}, t) = \int_{-\infty}^{\infty} d^3\mathbf{x}' \int_{-\infty}^{t} dt' \, \mathcal{K}_{ij}(\mathbf{x}, t, \mathbf{x}', t') \overline{B}_j(\mathbf{x}', t'). \tag{3.45}$$

The kernel $\mathcal{K}$ is dependent on the average quantities of the fluctuating velocity field $\mathbf{v}'$ as well as on the mean flow $\overline{\mathbf{v}}$. Assuming that the turbulence has finite correlation length and time scales l_c and τ_c, significant contributions in Eq. (3.45) only arise for $|\mathbf{x} - \mathbf{x}'| < l_c$ and $t - t' < \tau_c$. In order to derive simpler, interpretable, expressions we now introduce the concept of a scale separation by assuming that the length and time scales of the mean field are much larger than l_c and τ_c. Unfortunately, for most astrophysical applications this assumption is only marginally justified. Turbulence has a continuous spectrum ranging from a large energy-carrying scale down to the dissipative scale. The scale of the largest eddies is often not very different from the largest scales of the system, e.g. in the case of the solar convection zone the large scale of the turbulence spectrum is of the order of a pressure scale height at the base of the convection zone, which is about one-third of the convection zone depth. We refer to Section 3.5.1 below for a more detailed discussion of the limitations of the mean-field approach.

Under the assumption of a sufficient scale separation we can take care of the non-locality in space by using a Taylor expansion of $\overline{\mathbf{B}}$. For simplification we drop any non-locality in time here (see Section 3.5.1 for more discussion). This leads to the expansion

$$\overline{\mathcal{E}}_i = a_{ij}\overline{B}_j + b_{ijk}\frac{\partial \overline{B}_j}{\partial x_k} + \cdots. \tag{3.46}$$

From the induction equation (3.25) we see that the terms $\sim b_{ijk}$ result from the advection term, while the field-line stretching and compression terms lead to expressions $\sim a_{ij}$. We rewrite this expression by decomposing the tensors into symmetric and antisymmetric parts,

$$a_{ij} = \underbrace{\frac{1}{2}\left(a_{ij} + a_{ji}\right)}_{\alpha_{ij}} + \underbrace{\frac{1}{2}\left(a_{ij} - a_{ji}\right)}_{-\frac{1}{2}\varepsilon_{ijk}\gamma_k},$$

$$\frac{\partial \overline{B}_j}{\partial x_k} = \frac{1}{2}\left(\frac{\partial \overline{B}_j}{\partial x_k} + \frac{\partial \overline{B}_k}{\partial x_j}\right) + \underbrace{\frac{1}{2}\left(\frac{\partial \overline{B}_j}{\partial x_k} - \frac{\partial \overline{B}_k}{\partial x_j}\right)}_{-\frac{1}{2}\varepsilon_{jkl}(\nabla\times\overline{\mathbf{B}})_l}. \tag{3.47}$$

Here, we have indicated below the equations that the antisymmetric tensors can be expressed by a vector and the Levi–Civita tensor ε_{ijk}. This leads to a mathematically equivalent tensor equation in terms of the turbulent transport coefficients α, β, γ, and δ:

$$\overline{\mathcal{E}} = \alpha\overline{\mathbf{B}} + \gamma \times \overline{\mathbf{B}} - \beta\nabla \times \overline{\mathbf{B}} - \delta \times (\nabla \times \overline{\mathbf{B}}) + \cdots. \tag{3.48}$$

In Eq. (3.48) we have split the tensor in front of $\nabla \times \overline{\mathbf{B}}$ into a symmetric part "β" and an antisymmetric part "$\delta\times$", where δ is a vector. We have left out the term proportional to the symmetric part of $\partial \overline{B}_j/\partial x_k$ since it does not lead to physical effects that are distinct from what is already contained in α, β, γ, and δ. The relations to the original coefficients a_{ij} and b_{ijk} are given by

$$\alpha_{ij} = \tfrac{1}{2}\left(a_{ij} + a_{ji}\right), \qquad\qquad \gamma_i = -\tfrac{1}{2}\varepsilon_{ijk}a_{jk}$$
$$\beta_{ij} = \tfrac{1}{4}\left(\varepsilon_{ikl}b_{jkl} + \varepsilon_{jkl}b_{ikl}\right), \qquad \delta_i = \tfrac{1}{4}\left(b_{jji} - b_{jij}\right). \tag{3.49}$$

From the mathematical form of Eq. (3.48) we can see already that the effect of the γ-term is that of advection and that the β- and δ-terms can be interpreted as expressing an anisotropic magnetic diffusivity. The α-effect is completely new and the most important for large-scale dynamo action. We discuss the physical meaning of these terms in more detail in Section 3.4.3 below.

The vectors and tensors α, β, γ, and δ are large-scale correlations of the small-scale velocity field $\mathbf{v}'$. They have to reflect the same large-scale symmetries as the

MHD equations from which $\mathbf{v}'$ is derived. In the absence of any preferred direction they have to be homogeneous and isotropic, which means that the only possible construction elements are the isotropic homogeneous tensors δ_{ij} and ε_{ijk}. If the system has an additional preferred direction, such as gravity or rotation, then the directional quantities g_i and Ω_i also enter α, β, γ, and δ. In that case additional tensors can be formed from vector quantities through the dyadic product leading to expressions such as $a_{ij} = g_i \Omega_j$.

An additional property of the mean-field expansion (3.48) is that it relates the axial (pseudo) vector $\mathbf{B}$ to the polar (true) vector $\mathbf{E}$. Polar and axial vectors transform similarly under transformations that do not change the handedness of a coordinate system but transform differently under an improper rotation (an inversion plus a proper rotation). A coordinate inversion $\mathbf{x} \rightarrow -\mathbf{x}$ is an example of an improper rotation in three dimensions (but not in two dimensions). Let us consider two polar vectors $\mathbf{a}$ and $\mathbf{b}$ and an axial vector $\mathbf{c} = \mathbf{a} \times \mathbf{b}$. The above inversion leads to $\mathbf{a} \rightarrow -\mathbf{a}, \mathbf{b} \rightarrow -\mathbf{b}$ but $\mathbf{c} \rightarrow \mathbf{c}$. Note that the inversion of all the vector components with respect to the basis implies that a polar vector remains unchanged (since all the basis vectors are also pointing in the opposite direction), while an axial vector actually changes direction. Physical examples of axial (pseudo) vectors other than the magnetic field include angular momentum, torque, vorticity, and angular velocity. The concept of pseudo vectors can also be expanded to tensors of any rank. The most famous pseudo tensor of third rank is the Levi–Civita tensor ε_{ijk} and most of the examples of pseudo vectors mentioned above are related to polar vectors by ε_{ijk}. Additional pseudo tensors that are important here are those formed through a dyadic product, e.g. expressions such as $g_i \Omega_j$, which is a pseudo tensor of second rank since $\mathbf{g}$ is a polar and $\mathbf{\Omega}$ an axial vector. An expression such as $\Omega_i \Omega_j$ is again a true tensor, however.

All pseudo vectors have in common that their definition depends on the handedness of the coordinate system, which is pure convention (except for the weak force, where the chirality of the universe matters). The difference between vectors and pseudo vectors becomes important if one wants to exploit the symmetries of a physical system. Since electromagnetism does not have an intrinsic handedness, the mean-field expansion (3.48) has to be invariant under inversions (changing right-handed to left-handed coordinate systems). Therefore δ and α have to be pseudo tensors, while γ and β are true tensors. This sets constraints on the allowed combinations of the pseudo tensors Ω_i and ε_{ijk} and the true tensors g_i and the Kronecker delta δ_{ij} used in constructing α, β, γ, and δ.

Considering only terms that are not more than second order in g and Ω, we have the following expressions for turbulent induction effects:

$$\alpha_{ij} = \alpha_0 (\mathbf{g} \cdot \mathbf{\Omega}) \delta_{ij} + \alpha_1 \left(g_i \Omega_j + g_j \Omega_i \right), \quad \gamma_i = \gamma_0 g_i + \gamma_1 \varepsilon_{ijk} g_j \Omega_k,$$

$$\beta_{ij} = \beta_0 \, \delta_{ij} + \beta_1 \, g_i g_j + \beta_2 \, \Omega_i \Omega_j, \qquad \delta_i = \delta_0 \Omega_i. \qquad (3.50)$$

The scalar coefficients $\alpha_0, \ldots, \delta_0$ are typically functions of the rms velocity and correlation time scale of the turbulence field. Since they can be also zero, the following discussion can give only necessary conditions for the existence of the various turbulent induction effects.

In the case of isotropic homogeneous turbulence the only term that can exist is $\beta_{ij} = \beta_0 \delta_{ij}$, which is interpreted as the isotropic turbulent diffusivity. For most astrophysical systems we have $\beta_0 \gg \eta$ as consequence of the large magnetic Reynolds number. In the presence of stratification due to a gravitational field $\mathbf{g}$, the horizontal and vertical diffusivities can differ, while a rotation $\mathbf{\Omega}$ can introduce anisotropy with respect to the axis of rotation. The δ-effect, which contains the antisymmetric part of the diffusivity tensor, can exist in the case of rotation.

The advection-like γ-effect can be present if there is stratification owing to radial transport or, at higher orders, as a consequence of both stratification and rotation in the form of longitudinal transport. The latter is of interest since it can mimic the effects of differential rotation. The most important contribution to the radial transport in the case of stratification results from anisotropies between upflows and downflows. Since upflows have to expand and downflows have to compress as consequence of stratification, downflows have typically a smaller filling factor and larger velocity amplitudes. The net effect of this anisotropy on the magnetic field is typically a downward-directed transport, also called turbulent pumping.

Unlike the other effects, the α-effect can only exist if stratification and rotation are present at the same time. The reason for this constraint and the physics behind the α-effect will become more evident in the next subsection, where we relate the turbulent induction effects directly to properties of the turbulent velocity field.

3.4.3 Dynamo coefficients from second-order correlation approximation (SOCA)

After our more general discussion of turbulent induction effects from the point of view of symmetry in Section 3.4.2 we return now to the path we left at the end of Section 3.4.1 and explicitly solve Eq. (3.44) for the fluctuating field, which requires in general significant simplifications. To illustrate this approach we will neglect the mean flow $\overline{\mathbf{v}}$. The higher-order moments are obviously the terms most difficult to treat and can be neglected if $B' \ll \overline{B}$. Then the induction equation for the fluctuating field reduces to

$$\left(\frac{\partial}{\partial t} - \eta \Delta\right) \mathbf{B}' = \nabla \times \left(\mathbf{v}' \times \overline{\mathbf{B}}\right). \tag{3.51}$$

The condition $B' \ll \overline{B}$ is fulfilled if either $R_{\mathrm{m}} \ll 1$ or the Strouhal number $S = \tau_{\mathrm{c}} v / l_{\mathrm{c}} \ll 1$. Unfortunately, neither condition is fulfilled in stellar convection zones. We have $R_{\mathrm{m}} \gg 1$ and the second condition is hardly fulfilled since $S \sim 1$;

surprisingly, this approach yields reasonable results nevertheless (which might indicate that contributions from the higher-order terms correlate only weakly with $\mathbf{v}'$ and drop out when $\overline{\mathcal{E}}$ is calculated). Integrating Eq. (3.51) leads in the limit of large R_m to

$$\mathbf{B}'(t) = \int_{-\infty}^{t} \nabla \times [\mathbf{v}'(\tau) \times \overline{\mathbf{B}}(\tau)]d\tau. \tag{3.52}$$

The turbulent electromotive force is given by

$$\overline{\mathcal{E}} = \int_{-\infty}^{t} \overline{\mathbf{v}'(t) \times \{\nabla \times [\mathbf{v}'(\tau) \times \overline{\mathbf{B}}(\tau)]\}}d\tau. \tag{3.53}$$

This expression for $\overline{\mathcal{E}}$ accounts for non-locality in time and requires the computation of two-time correlations of the velocity field. For further simplification we reduce the correlations to single-time correlations by making the following approximation:

$$\int_{-\infty}^{t} \overline{v_i'(t)v_k'(\tau)}d\tau = \tau_c\overline{v_i'(t)v_k'(t)}. \tag{3.54}$$

Formally, expressions containing the correlation between the velocity and a velocity gradient could have a different τ_c. For reasons of simplification, in the following we use a single τ_c for all expressions.

Expansion of these terms and rearrangement according to the mean-field expansion (3.48) leads to the following second-order correlation approximation (SOCA) expressions:

$$\alpha_{ij} = \frac{1}{2}\tau_c\left(\overline{\varepsilon_{ikl}v_k'\frac{\partial v_l'}{\partial x_j}} + \overline{\varepsilon_{jkl}v_k'\frac{\partial v_l'}{\partial x_i}}\right), \qquad \gamma_i = -\frac{1}{2}\tau_c\frac{\partial}{\partial x_k}\overline{v_i'v_k'},$$

$$\beta_{ij} = \frac{1}{2}\tau_c\left(\overline{\mathbf{v}'^2}\delta_{ij} - \overline{v_i'v_j'}\right), \qquad \delta_i = 0. \tag{3.55}$$

The fact that we do not get a δ-effect here is a consequence of the (optional) simplification introduced through Eq. (3.54). Computing the traces of α and β gives

$$\alpha_{ii} = \tau_c\,\overline{\varepsilon_{ikl}v_k'\frac{\partial v_l'}{\partial x_i}} = -\tau_c\,\overline{v_k'\varepsilon_{kil}\frac{\partial v_l'}{\partial x_i}}$$

$$= -\tau_c\,\overline{\mathbf{v}' \cdot (\nabla \times \mathbf{v}')} \tag{3.56}$$

$$\beta_{ii} = \tau_c\,\overline{\mathbf{v}'^2}.$$

This shows that the magnitude of the turbulent magnetic diffusivity is directly related to the intensity of the turbulence. The α-effect, in contrast, is related to the

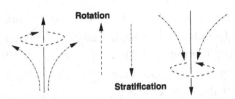

Fig. 3.4. Mean helicity as a consequence of stratification and rotation. The stratification leads to a correlation between the horizontal divergence and the vertical velocity. The influence of the Coriolis force on the horizontal flows causes a non-zero mean kinetic helicity.

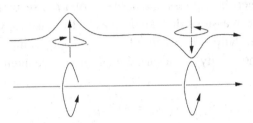

Fig. 3.5. Relation between kinetic helicity and the α-effect. The upper diagram depicts an originally toroidal field line being deformed by an upflow (left) and downflow (right), both with negative helicity, as indicated by the curved arrows. The lower panel shows the toroidal and newly formed poloidal components of the resulting deformed field. The assumed negative kinetic helicity (left-hand screw) induces a poloidal field with a current parallel to the toroidal field, leading to a positive α-effect. A net helicity is required to have up- and downward motions contributing the same way.

kinetic helicity of the flow, which is given by the expression

$$H_{\mathrm{k}} = \overline{\mathbf{v}' \cdot (\nabla \times \mathbf{v}')}. \tag{3.57}$$

In a helical flow the velocity and vorticity are aligned with each other. Positive helicity corresponds to a right-handed screw: looking in the direction of the flow, the fluid is moving clockwise. While in a turbulent flow H_{k} is in general locally different from zero, the α-effect requires that, moreover, the average of the kinetic helicity does not vanish. A non-zero mean kinetic helicity is in general the consequence of stratification and rotation together, as we show schematically in Fig. 3.4. Rising convective motions lead to divergent horizontal flows (owing to density stratification); these are turned into a left-handed screw through the Coriolis force, leading to a negative flow helicity. Since downflows are accompanied by horizontal convergent flows, they also have a negative helicity, leading to a negative mean helicity of the overall flow. In Fig. 3.5 we illustrate the relation between kinetic helicity and the α-effect for the case of negative kinetic helicity. A magnetic field originally contained within the plane shown (e.g. the toroidal field in the case of the Sun) is deformed and twisted, leading to a component of field outside the plane (poloidal). The net helicity ensures that the contributions of the up- and downwelling motions do not compensate each other and so give a net contribution. Evaluating the expression for the α-effect, Eq. (3.59) below, for the northern hemisphere of the solar convection zone leads to a positive α-effect. A net correlation between vertical velocity and horizontal flow divergence exists only if the medium is stratified or close to boundaries, which is in agreement with the finding of Section 3.4.2 that the α-effect requires an additional preferred direction beside rotation.

To simplify further the expressions derived from the second-order correlation effect (SOCA), we can assume that the deviations from isotropy are small enough to neglect cross-correlations between different velocity components, but we will still allow for inhomogeneity (i.e. a variation of turbulence intensity with location), so that

$$\overline{v_i' v_k'} = \tfrac{1}{3}\overline{\mathbf{v}'^2}\delta_{ik} . \tag{3.58}$$

Under these assumptions the tensors $\boldsymbol{\alpha}$ and $\boldsymbol{\beta}$ become diagonal and can be expressed by scalar quantities

$$\alpha = \tfrac{1}{3}\alpha_{ii} = -\tfrac{1}{3}\tau_c \overline{\mathbf{v}' \cdot (\nabla \times \mathbf{v}')} = \tfrac{1}{3}H_k , \qquad \eta_t = \tfrac{1}{3}\beta_{ii} = \tfrac{1}{3}\tau_c \overline{\mathbf{v}'^2} . \tag{3.59}$$

The γ-effect simplifies to

$$\gamma = -\tfrac{1}{2}\nabla \eta_t . \tag{3.60}$$

The assumptions required for these expressions are (a) the validity of the kinematic approach (i.e. a sufficiently weak field), (b) $B' \ll \overline{B}$ (achieved by either $R_m \ll 1$ or $S = \tau_c v/l_c \ll 1$), (c) $\overline{\mathbf{v}} = 0$, (d) the introduction of τ_c, Eq. (3.54), and (e) isotropic, non-mirror-symmetric, turbulence (with weak inhomogeneity). While the assumptions (d) and (e) are not essential and are used primarily to simplify the expressions, assumptions (a) and (b) are crucial and yet at best marginally justified in most astrophysical applications. If (c) is not fulfilled then the mean flow can lead to additional contributions if the advection time scale is comparable with the turbulent correlation time scale. The general form of Eq. (3.48) remains valid even when these conditions are not fulfilled; however, higher-order terms might be relevant if scale separation is not well justified.

The expression in Eq. (3.60) for the γ-effect is often called turbulent diamagnetism, since it expresses a tendency to expel magnetic field from regions with increased turbulence intensity. In a stellar convection zone the turbulence intensity varies mainly in the radial direction, which leads to radial transport. Toward the base of the convection zone the transport is downward directed, since there η_t drops significantly (the amplitude of convective motions decreases since the fraction of energy flux transported by convection decreases smoothly to zero as the base of the convection zone is approached).

As discussed above, the mean kinematic helicity required for the α-effect follows from the correlation between horizontal divergence and vertical motions resulting from stratification. A more detailed calculation for the turbulent α-effect leads to the result

$$\alpha \approx \alpha_0 (\mathbf{g} \cdot \mathbf{\Omega}) = \tau_c^2 v_{rms}^2 \mathbf{\Omega} \cdot \nabla \ln(\varrho v_{rms}) \tag{3.61}$$

for the diagonal elements of the α-tensor.

3.4.4 Energy equation for the mean field

For later discussion it will be helpful now to derive an energy equation for the volume-integrated energy of the mean field. Scalar multiplication of the mean-field induction equation (3.42) with $\overline{\mathbf{B}}$ and integration over the entire volume leads to

$$\frac{d}{dt} \int \frac{\overline{\mathbf{B}}^2}{2\mu_0} \, dV = -\mu_0 \int \eta \overline{\mathbf{j}}^2 \, dV - \int \overline{\mathbf{v}} \cdot (\overline{\mathbf{j}} \times \overline{\mathbf{B}}) \, dV + \int \overline{\mathbf{j}} \cdot \overline{\mathcal{E}} \, dV. \tag{3.62}$$

The first term on the right-hand side reflects resistive dissipation, the second term measures the work done by mean flows against the mean Lorentz force (e.g. differential rotation), and the last term describes the energy converted into the mean field by turbulent induction effects. This sustains the mean field if $\overline{\mathcal{E}}$ on average points in the direction of the mean current.

3.4.5 Destruction of magnetic field: turbulent diffusivity

The derivation of the mean-field induction equation leads to an additional dissipative term that significantly enhances the effective magnetic diffusivity in the presence of turbulence. In the case of isotropic homogeneous turbulence, the turbulent diffusivity is given by the scalar quantity (Eq. 3.59)

$$\eta_t = \tfrac{1}{3} \tau_c \overline{\mathbf{v}'^2} \sim L v_{rms} \sim R_m \eta \gg \eta. \tag{3.63}$$

Formally this arises from the advection term in the induction equation (3.25), which is a non-dissipative transport term. The only dissipative term in the microscopic induction equation (3.4) relates to the microscopic diffusivity η, where the microscopic dissipation rate is $-\mu_0 \eta \mathbf{j}^2$. The effective dissipation rate in the mean-field energy equation (3.62) is given by $-\mu_0 (\eta + \eta_t) \overline{\mathbf{j}}^2$, however. Turbulent diffusivity parameterizes a turbulent transport process that transports magnetic energy through advection and reconnection (see Chapter 5 for details regarding reconnection), from the large scale L to the small scale l at which the energy is dissipated. The energy flux through the turbulent cascade is determined by the large scale, and the small scale adjusts to allow for the required dissipation:

$$\eta j_l^2 \sim \eta_t \bar{j}^2 \longrightarrow \frac{B_l}{l} \sim \sqrt{R_m} \frac{\bar{B}}{L}. \tag{3.64}$$

While, in the case of purely hydrodynamic turbulence, universal scaling relations such as $v_l \sim l^{1/3}$ exist, this is in general not the case for MHD turbulence. We refer to Chapter 7 for a more detailed discussion of MHD turbulence.

Owing to the large magnetic Reynolds numbers in astrophysical systems, the high turbulent diffusivity significantly raises the bar for the creation of large-scale magnetic field.

3.4.6 Creation of magnetic field: turbulent dynamos

Our main interest here is to understand how fluid motions can support an axisymmetric mean field. Since the total magnetic field $\mathbf{B} = \bar{\mathbf{B}} + \mathbf{B}'$ is in general non-axisymmetric, the maintenance of an axisymmetric mean field is not excluded by Cowling's theorem. However, if the mean-field induction equation is mathematically identical to the microscopic induction equation (3.4), Cowling's theorem also rules out the support of an axisymmetric mean field. The effects that change the mathematical form of the mean-field induction equation and therefore allow one to circumvent Cowling's anti-dynamo theorem for $\bar{\mathbf{B}}$ are the α-effect, the δ-effect, and possibly also the β-effect for special cases of anisotropy. We focus the discussion here mainly on the former two.

3.4.6.1 Generalized Ohm's law

A different way to interpret the turbulent induction effects is by looking at Ohm's law for the mean current. Averaging Ohm's law (3.15) leads to

$$\bar{\mathbf{j}} = \sigma(\bar{\mathbf{E}} + \bar{\mathbf{v}} \times \bar{\mathbf{B}} + \bar{\mathcal{E}}). \tag{3.65}$$

Inserting the expression for the EMF (Eq. 3.43) and bringing all terms proportional to $\bar{\mathbf{j}}$ (the β- and δ- effects) to the left-hand side allows to write the mean-field Ohm's law as

$$\bar{\mathbf{j}} = \tilde{\sigma}\left(\bar{\mathbf{E}} + \bar{\mathbf{v}} \times \bar{\mathbf{B}} + \gamma \times \bar{\mathbf{B}} + \alpha\bar{\mathbf{B}}\right) \tag{3.66}$$

where the anisotropic mean-field conductivity tensor $\tilde{\sigma}$ contains contributions from the molecular conductivity and the β- and δ- effects.

The crucial point in the proof of Cowling's anti-dynamo theorem is the impossibility of maintaining a current along the neutral line of the poloidal field even in the presence of a toroidal field. Ohm's law for the mean field, Eq. (3.66), shows clearly that the α-effect can maintain such a current if a toroidal field is present and therefore allows for the maintenance of an axisymmetric mean field by fluid motions. It is not the only α-effect term which potentially has such a property. If we set $\alpha = 0$ on the right-hand side of Eq. (3.66) then the remaining terms would drive currents perpendicular to the mean field. If the conductivity tensor $\tilde{\sigma}$ is sufficiently anisotropic, these currents can have again components aligned with the mean field and can therefore also circumvent Cowling's anti-dynamo theorem. The latter possibility is a possible dynamo scenario involving the δ-effect.

3.4.6.2 The α^2-dynamo

The additional induction effect introduced by the α-effect can be most easily understood for an isotropic homogeneous α, which leads in the mean-field induction equation (Eq. 3.42 with Eq. 3.48) to the additional term

$$\frac{\partial \overline{\mathbf{B}}}{\partial t} = \cdots + \nabla \times \left(\alpha \overline{\mathbf{B}} \right) = \cdots + \alpha \mu_0 \overline{\mathbf{j}} . \tag{3.67}$$

The induced magnetic field is proportional to the mean current. Since the current maintaining a poloidal field is toroidal and the current maintaining a toroidal field is poloidal, this process converts poloidal magnetic field into toroidal field and vice versa. We have therefore the following dynamo scenario:

$$\mathbf{B}_t \xrightarrow{\alpha} \mathbf{B}_p \xrightarrow{\alpha} \mathbf{B}_t . \tag{3.68}$$

Since the α-effect is responsible for the conversion between poloidal and toroidal components in both directions, this type of dynamo is called an α^2-dynamo. Inspecting the direction of the currents maintaining the poloidal and toroidal field shows that after completion of a full cycle, as indicated in Eq. (3.68), the magnetic field does not change sign. The α^2-dynamo generally produces stationary solutions with poloidal and toroidal fields of comparable amplitude.

If the α-tensor is anisotropic, however, the process regenerating $\mathbf{B}_t$ from $\mathbf{B}_p$ can have a different amplitude from the process regenerating $\mathbf{B}_p$ from $\mathbf{B}_t$. In this case either $\mathbf{B}_t$ or $\mathbf{B}_p$ could dominate. The off-diagonal elements of the α-tensor can lead to effects similar to advection effects, i.e. they modify the advection in such a way that the effective advection velocity is different for the three field components.

Inserting the expression $\overline{\mathcal{E}} = \alpha \overline{\mathbf{B}}$ into the mean-field energy equation (3.62) shows that the α-effect sustains the mean field if the mean-field current helicity $\int \overline{\mathbf{j}} \cdot \overline{\mathbf{B}} \, dV$ has the same sign as α, i.e. the dynamo-generated magnetic field is

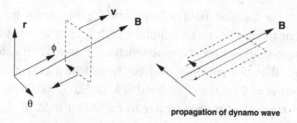

Fig. 3.6. Dynamo waves in an $\alpha\Omega$-dynamo. The poloidal field (broken line) produced by the α-effect is sheared by the velocity field. The resulting "new" toroidal field amplifies the original toroidal field on the poleward side and weakens it on the equatorward side. As a result the toroidal field pattern moves poleward in the $-\theta$ direction.

helical. This condition is satisfied because the α-effect induces a magnetic field parallel to the mean current.

3.4.6.3 The $\alpha\Omega$- and $\alpha^2\Omega$-dynamos

We discussed in Section 3.3.7 the effect of differential rotation, which leads to the production of toroidal field from poloidal field. The system of equations (3.35) and (3.36) is not able to maintain an axisymmetric field since the poloidal field is always decaying. This decay can be prevented in the presence of an α-effec. A dynamo in which differential rotation is regenerating toroidal field from poloidal field and the α-effect is regenerating poloidal field from toroidal field is called an $\alpha\Omega$-dynamo and is described by

$$\mathbf{B}_t \xrightarrow{\alpha} \mathbf{B}_p \xrightarrow{\Omega} \mathbf{B}_t . \tag{3.69}$$

The $\alpha\Omega$-dynamo is an approximation that is valid if the Ω-effect is much stronger than the α-effect in producing poloidal field. The mathematical equations describing this type of dynamo are very similar to Eqs. (3.35) and (3.36) but with η replaced by $\eta + \eta_t$ and an additional term αB in Eq. (3.36). This type of dynamo leads to magnetic field configurations that are dominated by the toroidal field component. If both the Ω- and α- effects contribute significantly to poloidal field production, the dynamo is called an $\alpha^2\Omega$-dynamo.

Unlike the α^2-dynamo, in general $\alpha\Omega$-dynamos have oscillating solutions with magnetic field patterns propagating along lines of constant Ω (the Parker–Yoshimura rule: Yoshimura, 1975). The reason for this is illustrated in Fig. 3.6. A positive radial gradient $\partial\Omega/\partial r$ in the rotation profile and a positive α leads to a poleward propagation of the toroidal field pattern. The propagating field pattern of an $\alpha\Omega$-dynamo is often referred to as a dynamo wave.

In the case of an $\alpha\Omega$-dynamo the primary energy source for the mean field is the work done against the Lorentz force (the second term on the right-hand side of Eq. (3.62)).

3.4.6.4 The $\Omega \times J$-dynamo

The δ-effect in the mean-field expansion Eq. (3.48) is also referred to as the $\Omega \times \mathbf{J}$-effect since it leads to a field regeneration of the form

$$\frac{\partial \overline{\mathbf{B}}}{\partial t} = \nabla \times [\delta \times (\nabla \times \overline{\mathbf{B}})] \sim \nabla \times (\mathbf{\Omega} \times \overline{\mathbf{j}}) \sim \frac{\partial \overline{\mathbf{j}}}{\partial z}. \tag{3.70}$$

The currents of a poloidal field are toroidal and therefore $\partial \overline{\mathbf{j}}/\partial z$ is also toroidal; the currents of a toroidal field are poloidal and therefore $\partial \overline{\mathbf{j}}/\partial z$ is also poloidal. As in the case of the α-effect, the δ-effect can convert toroidal into poloidal field and vice versa. However, unlike the α-effect, a δ^2-dynamo is not possible. This becomes evident from the mean-field energy equation (3.62) since, in the case of the δ-effect,

$$\overline{\mathbf{j}} \cdot \overline{\mathcal{E}} = \overline{\mathbf{j}} \cdot (\delta \times \overline{\mathbf{j}}) = 0. \tag{3.71}$$

This restriction, however, does not rule out dynamos such as the $\delta\Omega$-dynamo, in which the Ω-effect pumps energy into the mean field (the second term on the right-hand side of Eq. (3.62)) while the δ-effect is required to produce poloidal from toroidal field (a change in topology).

We should mention that the δ-effect is controversial since not all the approaches used to calculate the mean-field transport coefficients lead to a δ-effect. Furthermore, in most astrophysical systems the α-effect is normally the dominant mechanism.

3.4.7 Dynamos and magnetic helicity

The magnetic helicity is defined as the volume integral of the scalar product of the magnetic field $\mathbf{B}$ and its vector potential $\mathbf{A}$:

$$H_m = \int \mathbf{A} \cdot \mathbf{B} \, dV. \tag{3.72}$$

The magnetic helicity is an integral measure of the magnetic field topology. Since the vector potential $\mathbf{A}$ allows for some gauge freedom, magnetic helicity reflects that ambiguity. It is nevertheless a very helpful quantity since it is conserved under ideal MHD. We refer to Chapter 4 for a more detailed discussion of magnetic helicity and its topological meaning.

Assuming that there are no helicity fluxes across the domain boundaries, the induction equation provides a conservation law for magnetic helicity:

$$\frac{d}{dt} \int \mathbf{A} \cdot \mathbf{B} \, dV = -2\mu_0 \eta \int \mathbf{j} \cdot \mathbf{B} \, dV, \tag{3.73}$$

where the quantity $\int \mathbf{j} \cdot \mathbf{B} \, dV$ is the current helicity of the magnetic field. Magnetic helicity is strictly conserved for ideal MHD. For astrophysical systems with very large magnetic Reynolds numbers it can evolve only on a very slow resistive time scale. In this case helicity conservation can be still a significant constraint as we will see in Section 3.5.2.

Deriving a similar equation for the mean field and subtracting that equation from Eq. (3.73) allows us to split the helicity evolution into large- and small-scale parts, leading to

$$\frac{d}{dt} \int \overline{\mathbf{A}} \cdot \overline{\mathbf{B}} \, dV = +2 \int \overline{\mathcal{E}} \cdot \overline{\mathbf{B}} \, dV - 2\mu_0 \eta \int \overline{\mathbf{j}} \cdot \overline{\mathbf{B}} \, dV, \qquad (3.74)$$

$$\frac{d}{dt} \int \overline{\mathbf{A}' \cdot \mathbf{B}'} \, dV = -2 \int \overline{\mathcal{E}} \cdot \overline{\mathbf{B}} \, dV - 2\mu_0 \eta \int \overline{\mathbf{j}' \cdot \mathbf{B}'} \, dV. \qquad (3.75)$$

Inserting the expression $\overline{\mathcal{E}} = \alpha \overline{\mathbf{B}}$ shows that at large scales the α-effect induces magnetic helicity with the same sign as the α-effect, while the small-scale helicity has the opposite sign. In the case of stationarity, which is only achieved on a slow resistive time scale, the same also applies to the current helicity. We show later that the accumulation of current helicity on the small scale can lead to a significant reduction in the α-effect.

3.4.8 General comments on large-scale turbulent dynamos

All large-scale turbulent dynamo scenarios discussed here contain at least one process (either α or δ) related to a pseudo vector or tensor. According to Eq. (3.50) this requires a preferred direction expressed by a pseudo vector, which is a rotation Ω in most astrophysical systems (shear flows could also define such a pseudo vector through $\nabla \times \overline{\mathbf{v}}$). For a dynamo of type α^2 or $\alpha^2 \Omega$, rotation alone is not sufficient since the α-effect requires the existence of an additional symmetry direction. In most cases this additional preferred direction is due to stratification, but the influence of boundary conditions can be sufficient: turbulence resulting from isotropic forcing and rotation has an α-effect near boundaries, since the normal vector at the boundary introduces a preferred direction in the velocity field. In the case of the geodynamo, such boundary effects at the core–mantle boundary are crucial since the system is only weakly stratified (incompressible).

The turbulent induction effects that enable large-scale dynamo action do not necessarily have to convert energy to maintain the mean field. In the presence of strong differential rotation most energy conversion is done through the large-scale shear flow, and the role of the α- or δ- effect is reduced to that of changing the field topology (the production of poloidal field from toroidal field).

The mean-field dynamos that we have discussed here are all fast dynamos. Even though the α- and β- effects require reconnection and are therefore indirectly dependent on a microscopic diffusivity η, the expressions that we derived for α, Eq. (3.59), do not contain η explicitly and are independent of R_m (within the limits of the approximations we made). We will show in Section 3.5.2 that under certain conditions this conclusion does not hold: α can become asymptotically limited by the microscopic value of η.

Another more fundamental question in dynamo theory is related to the linear homogeneous structure of the induction equation, which means that $\mathbf{B} = 0$ is always a valid solution. In a dynamo, however, the solution $\mathbf{B} = 0$ is unstable and any small perturbation will lead to exponential growth (the linear kinematic phase) until nonlinear feedback through the Lorentz force becomes important. We want to mention that there are also other processes that can produce large-scale fields if the more general form Eq. (3.14) of Ohm's law is used. Keeping the electron pressure term and re-deriving the induction equation leads to an additional term $\sim \nabla n_e \times \nabla p_e$. Unlike the other terms, which contribute only if there is already a magnetic field present, this term is a real source provided that the gradients of the electron density and the temperature are not parallel (e.g. the ionization fronts in an inhomogeneous medium). However, these fields are very weak and additional amplification through dynamo processes is required nevertheless.

3.5 Limitations of mean-field approximation, 3D simulations

3.5.1 Comments on general applicability

In Section 3.4.3 we computed expressions for the dynamo coefficients using the second-order correlation approximation (SOCA). In the case of large magnetic Reynolds numbers, which is typical for stellar convection zones, this approximation is only justified when the Strouhal number $S = v\tau_c/l_c$ is much smaller than unity. Typical values for convection zones are close to unity since the basic mixing-length relation $l_c \sim H_p \sim v\,\tau_c$ holds. Therefore the justification of this approach is at best marginal. Nevertheless, the expressions (3.56) derived from SOCA give a reasonable agreement when tested in numerical simulations (Käpylä *et al.*, 2006), which indicates that the applicability of the expressions derived from SOCA might go beyond the regime with $S = v\tau_c/l_c \ll 1$. We emphasize that the expressions (3.56) can remain valid even if $B' \gg \overline{B}$ provided that $\mathbf{v}'$ correlates only weakly with $\nabla \times (\mathbf{v}' \times \mathbf{B}')$.

The constraints on the general applicability of the mathematical form of the mean-field expansion (3.48) are less severe, since the latter follows from very general symmetry principles. A requirement for the fast convergence of the spatial part of this expansion is sufficient scale separation, which has to be tested in a given application. If locality in time is not justified, terms such as $\overline{\mathcal{E}} \sim \alpha \overline{\mathbf{B}}$ have to be replaced by terms of the form

$$\overline{\mathcal{E}} \sim \int_{-\infty}^{t} \alpha(t, t') \overline{\mathbf{B}}(t') dt'. \tag{3.76}$$

An alternative but equivalent approach would be to solve an evolution equation for $\overline{\mathcal{E}}$.

In the case of the solar dynamo we have $\tau_c \sim 10$ days, deep in the convective envelope, while the mean field is evolving on a time scale of several years. The scale separation in time is about two orders of magnitude and well justified. Convective cells at the base of the convection have a length scale comparable with the density scale height $l_c \sim 10^8$ m $\sim 0.15 R_\odot$. Scale separation in the length scales is only marginally justified, however. Therefore higher-order derivatives of the mean field could be required (leading to hyperdiffusive behavior). In the case of the geo-dynamo the scale separation in length scales is comparable with that in the solar dynamo, but the mean field shows variability on the time scale τ_c itself; mean-field theory in the form of Eq. (3.48) is here only helpful for understanding the structure of the temporally averaged field (averaging the dipole moment for 10^3 to 10^4 years removes most of the variability except reversals).

One of the biggest limitations, in general, is that it is not easy to compute the mean-field coefficients $\alpha, \ldots, \delta$ from first principles. The first problem is that one needs a velocity field $\mathbf{v}'$ and the second problem is that one has to solve Eq. (3.44) in order to be able to compute the electromotive force $\overline{\mathcal{E}}$ (Eq. 3.43). We presented a very crude approach for the second step in Section 3.4.3. Approaches that combine both steps use, in general, a spectral representation and compute the anisotropies and inhomogeneities resulting from rotation and stratification from the Navier–Stokes equations using a quasi-linear approach. Expressions for the dynamo coefficients obtained this way have been compared with numerical simulations (see below) and tend to show qualitative agreement. Differences are caused by both the limitation of the mean-field approach and the limitations of the numerical simulations. More details on different ways to compute dynamo coefficients can be found in Rädler and Stepanov (2006), Rädler and Rheinhardt (2006), and Rüdiger and Hollerbach (2004).

3.5.2 Non-kinematic effects

A much more detailed summary of nonlinear dynamo theory and advances in this field can be found in Brandenburg and Subramanian (2005). In Section 3.4 we followed the kinematic approach. We assumed that a turbulent velocity field $\mathbf{v}'$ was given and focused only on solving the induction equation. Since this equation is linear, solutions show either exponential decay or growth. The kinematic approach is helpful in understanding the onset of dynamo action but leaves out the interesting question of the saturation of a dynamo. This can be addressed only if the Lorentz force is included in the Navier–Stokes equation, which requires in general solving the full system of MHD equations. The Lorentz force modifies the small-scale flows, leading to turbulent induction effects as well as large-scale mean flows (due to e.g. differential rotation). The influence of the Lorentz force on the mean flows is referred to very often as macroscopic feedback (or the Malkus–Proctor effect, Malkus and Proctor, 1975). The components of the Lorentz force that change large-scale flows are given by

$$\overline{\mathbf{j} \times \mathbf{B}} = \overline{\mathbf{j}} \times \overline{\mathbf{B}} + \overline{\mathbf{j}' \times \mathbf{B}'} . \tag{3.77}$$

While the first term on the right-hand side is directly accessible in a mean-field model, the second term requires the calculation of the fluctuating fields, which are in general not known. Most nonlinear models with macroscopic feedback, such as the models of Covas *et al.* (2000), Bushby (2005), and Rempel (2006), only consider the contribution from the mean field (the first term in (3.77)).

The microscopic feedback on the small-scale motions has been included through ad hoc quenching parameterizations such as

$$\alpha = \frac{\alpha_k}{1 + \left(B/B_{eq}\right)^2} . \tag{3.78}$$

Here α_k denotes the kinematic expression from Eq. (3.59) and B_{eq} the equipartition field strength, $\varrho v_{rms}^2/2 = B_{eq}^2/(2\mu_0)$. The rationale behind this expression is that the small-scale motions are reduced in amplitude once the magnetic energy density is similar to the kinetic energy density of turbulent motions.

Vainshtein and Cattaneo (1992) and Cattaneo and Hughes (1996) suggested that this expression should be modified to

$$\alpha = \frac{\alpha_k}{1 + R_m\left(B/B_{eq}\right)^2} ; \tag{3.79}$$

this is referred to also as catastrophic α-quenching, owing to the large value of R_m in astrophysical systems and would basically rule out an α-effect for any field strength of interest.

The saturation of the α-effect is closely related to the evolution of magnetic helicity. Pouquet *et al.* (1976) showed that the small-scale current helicity leads to a magnetic correction of the α:

$$\alpha = -\frac{1}{3}\tau_c \left(\overline{\boldsymbol{\omega}' \cdot \mathbf{v}'} - \frac{1}{\varrho}\overline{\mathbf{j}' \cdot \mathbf{B}'}\right) . \tag{3.80}$$

Thus build-up of small-scale helicity according to Eq. (3.75) creates a magnetic α-effect (the second term on the right-hand side of Eq. (3.80)) that offsets the kinematic α-effect (the first term). Since we do not have observational evidence for catastrophic quenching in astrophysical dynamos it was suggested by Brandenburg and Sandin (2004) that small-scale helicity losses might alleviate the catastrophic quenching. In the case of the Sun there is observational evidence for losses of negative helicity in the northern hemisphere and of positive helicity in the southern hemisphere.

3.5.3 *3D dynamo simulations*

With the significant advances in computing power, 3D dynamo simulations solving the full set of MHD equations have become a very powerful tool in dynamo theory. The latter remains, however, a significant challenge with varying success for different objects. The main difficulty in simulating large-scale dynamos results from the fact that we require to capture the large scales and at the same time the significantly smaller scales of the turbulence. In a local-turbulence simulation the largest scales are typically not far above the energy-carrying scales of the turbulence, which for the solar dynamo are at least one order of magnitude smaller than the global scale of the system (the solar radius). Furthermore, large-scale dynamos evolve typically on time scales much longer than the overturning time scale of turbulent motions (e.g. for the solar dynamo we have $P_{\text{cyc}} \approx 22 \, \text{years} \approx 8000 \, \text{days}$ and $\tau_{\text{c}} \approx 10 \, \text{days}$). This makes studying a large-scale dynamo much more challenging than a local simulation of turbulence, which is typically run only over a few overturning time scales. A further complication comes from the fact that most astrophysical systems (except galaxies) have a very low magnetic Prandtl number R_{m}, meaning that $R_{\text{m}} \ll R_{\text{e}}$. Since dynamo action requires a minimum R_{m} to be exceeded, relatively large values of R_{e} are therefore required in general (this problem is often avoided by forcing R_{m} to be close to unity). Owing to these difficulties, large-scale dynamos (or at least some of the ingredients) have also been studied using more local approaches, as we now briefly outline in Section 3.5.3.1.

3.5.3.1 *Local simulations*

Local simulations place a computational domain significantly smaller than the full system (e.g. 10% of the size of the star) at a given latitude. The latitude is defined through the orientation of the axis of rotation with respect to the vertical, i.e. the direction of gravity at that latitude. Since simulations cannot include the density contrast, which is about 10^6 in the solar convection zone, only a fraction of the radial extent of the convection zone is covered. The depth location of the box is indirectly imposed through the Coriolis number (defined as $Co = 2\Omega\tau_{\text{c}}$) measuring the influence of rotation (in the solar convection zone the Coriolis number increases from less than 1 at the surface to about 10 at the base of the convection zone). This approach is also called the f-plane approximation.

A setup like this allows the study of the ingredients of a large-scale dynamo, such as the α- and γ- effects (turbulent pumping), without actually modeling the large-scale field. As a consequence local simulations are not necessarily of a dynamo (except for small-scale dynamo action) but, rather, of magneto-convection, in which a large-scale mean field is imposed to "measure" the dynamo coefficients. Ossendrijver *et al.* (2001, 2002) and Käpylä *et al.* (2006) computed dynamo coefficients as function of latitude and Coriolis number (expressing the influence of rotation, mainly in the depth dependence) and found a rather good correspondence between the numerically computed dynamo coefficients and those computed from the second-order correlation approximation (3.56), in the case of moderate rotation, and qualitative agreement in the case of fast rotation.

3.5.3.2 *Global simulations*

Despite the computing power available today, global simulations of dynamos remain very challenging. While self-consistent dynamo models for the geodynamo were presented first about a decade ago by Glatzmaier and Roberts (1995), building upon earlier work of Glatzmaier and Gilman (1982), Gilman (1983), and Glatzmaier (1984), a self-consistent dynamo model for the solar dynamo has not been achieved so far. In the case of the geodynamo, the relatively low magnetic Reynolds number, of order a few 100, works in favor of numerical simulations since it is possible to resolve all relevant magnetic scales of the dynamo. The flow Reynolds number is, however, still significantly larger and so subgrid-scale modeling is required for the velocity field. As long as the unresolved scales in the velocity field exhibit only a diffusive behavior for the magnetic field it is possible to capture all relevant induction effects in a 3D simulation. For a summary of the main dynamo mechanism of the geodynamo see Olson *et al.* (1999).

Unlike the case of the geodynamo, numerical models for the solar dynamo are far from resolving all the relevant magnetic scales. Current simulations of global solar convection are of order 1000^3 in grid size and can be run in a few years of real time. Expanding the time covered by at least a factor 10 (to cover a cycle) and increasing the resolution by a factor 3 would increase the demand on computing power by roughly a factor 10^3. An additional major problem in 3D simulations of the solar dynamo is modeling the transition layer at the base of the solar convection zone (the tachocline), which shows strong radial shear in differential rotation and most likely plays a key role in the dynamo process. The transition toward the more stable radiative interior of the Sun introduces time scales far beyond the convective time scales governing the convection zone of the Sun.

The early work of Gilman (1983) showed cyclic dynamos in spherical 3D simulations; however, more turbulent simulations at higher resolution do not yield periodic solutions for the solar parameters (Brun *et al.*, 2004; Browning *et al.*, 2006). Nevertheless there is significant progress in understanding important ingredients of the dynamo process, with increasing detail. For a review of the dynamics of the solar convection zone from 3D simulations we refer to Miesch (2005). Unfortunately, knowledge of the basic dynamo ingredients is not sufficient to determine the primary dynamo process, since all these ingredients (except differential rotation) have an uncertain amplitude and spatial distribution within the solar convection zone. This leads to quite a variety of different dynamo scenarios, as summarized by Ossendrijver (2003) and Charbonneau (2005).

4

Magnetic field topology

DANA W. LONGCOPE

One of the most idiosyncratic aspects of space physics is the central role assigned to magnetic field lines. Particularly in studies of the Sun, the heliosphere, and the magnetosphere, magnetic field lines are treated as fully fledged physical objects with their own dynamics. The electrical current, when needed, is derived *from* the magnetic field lines. These practices appear to be at odds with the basic approach, followed in elementary electrodynamics, of deriving the magnetic field *from* a current distribution and treating magnetic field lines as at best fictitious curiosities. However, physical laws such Ampère's law (excluding the displacement current),

$$\nabla \times \mathbf{B} = \mu_0 \mathbf{j}, \tag{4.1}$$

do not attribute a causative nature to either side of the equality; they simply state the equality of two quantities. So either approach to satisfying Eq. (4.1), beginning with either $\mathbf{j}$ or $\mathbf{B}$, must be valid.

This curious approach is adopted in space physics because magnetic fields are found in highly conductive plasmas. Denoting by $\mathbf{E}$ and $\mathbf{B}$ the electric and magnetic fields in the stationary (laboratory) frame, a fluid element traveling at velocity $\mathbf{v}$ will experience a local electric field $\mathbf{E}' = \mathbf{E} + \mathbf{v} \times \mathbf{B}$. If the fluid is a sufficiently good conductor, evolving sufficiently slowly, then $\mathbf{E}'$ will be extremely small. For example, Ohm's law in a conductor of conductivity σ is $\mathbf{E}' = \mathbf{j}/\sigma$, so $\mathbf{E}'$ will be small when σ is large compared with something else with the same units. The so-called magnetic Reynold's number, $R_{\mathrm{m}} = vL\sigma\mu_0$, serves as a dimensionless version of the conductivity. When it is large compared with unity, the conductivity is genuinely large. In many space plasmas $R_{\mathrm{m}} \gg 10^6$ (cf. Table 3.1). In this chapter we will assume $\mathbf{E}'$ to be so small that it can be neglected entirely and so we have $\mathbf{E} = -\mathbf{v} \times \mathbf{B}$. Using this equation in Faraday's law leads directly to the ideal induction equation (3.24). The term "ideal" refers here to the assumed absence of effects which might support a local electric field $\mathbf{E}'$, such as the presence of electrical resistivity, $1/\sigma$.

Equation (3.24) may be used as a recipe for time-advancing the magnetic field. This time advance can be achieved without explicit reference to the electric field or current. There is also an equation for time-advancing the velocity, Eq. (3.2), which can be written so that it only involves **B**. This ability to formulate the equations entirely in terms of **B** leads to the centrality of **B** in much of space physics.

Following this approach has led space physicists to develop a body of literature, of analysis techniques, and of mathematical theorems concerning only magnetic field lines. Magnetic field lines can be studied using the ideal induction equation (3.24) alone. For example, field lines can be grouped topologically, according to whether they are open or closed and to which regions closed field lines are anchored. Between distinct groups of field lines lie surfaces called separatrices, whose mutual intersections are curves called separators. Points where the magnetic field vanishes, called null points, play a crucial role in defining these boundary elements. The field line topology throughout a volume can be characterized by a single quantity, its magnetic helicity.

These concepts concerning magnetic field line topology pervade space physics, from the solar interior to the magnetosphere and to the outer heliosphere. They are equally useful under static or dynamic conditions, in plasmas of high or low β, with or without non-thermal populations, provided only that non-ideal electric fields remain negligible. Even cases where this condition does not hold, such as magnetic reconnection (see Chapter 5), can be studied using field line topology restricted to the regions outside the reconnection region. Indeed, the simple characterization of reconnection as the "breaking" and "rejoining" of magnetic field lines is a construction using field line topology. This chapter presents an introduction to the analysis of magnetic field line topology in which it is assumed that in low-R_m regions the field can be cut and spliced. While the topic is general and applicable in most areas of space physics, the concepts are illustrated with specific examples.

4.1 Magnetic field lines

A magnetic field line, sometimes called a line of force, is a space curve $\mathbf{r}(\ell)$ which is everywhere tangent to the local magnetic field vector $\mathbf{B}(\mathbf{x})$. This description can be cast as the differential equation

$$\frac{d\mathbf{r}}{d\ell} = \frac{\mathbf{B}(\mathbf{r}(\ell))}{|\mathbf{B}(\mathbf{r}(\ell))|}, \tag{4.2}$$

whose solution, starting from some initial point $\mathbf{r}(0)$, is a magnetic field line. Choosing different initial points $\mathbf{r}(0)$ permits the creation of a potentially infinite number of different field lines; some field line passes through almost every point in space. Note that the definition in Eq. (4.2) yields $|d\mathbf{r}/d\ell| = 1$, demonstrating that the parameter ℓ is the arc length measured forward along the field line from $\mathbf{r}(0)$.

A field line is a curve and therefore has zero volume. A *flux tube* may be constructed by bundling together a group of field lines. The net flux, Φ, of the tube is the integral $\int \mathbf{B} \cdot d\mathbf{a}$ over any surface pierced by the entire tube. Since $\nabla \cdot \mathbf{B} = 0$ the tube must have the same flux at every cross section.

The only way, in general, to find a field line is to integrate the differential Eq. (4.2). A useful shortcut is available in cases with a symmetry direction (i.e. in two dimensions). In cases of azimuthal symmetry, for example, an arbitrary magnetic field can be expressed in cylindrical coordinates r, ϕ, z as

$$\mathbf{B}(r, z) = \nabla f \times \nabla \phi + B_\phi(r, z)\hat{\boldsymbol{\phi}} = r^{-1}\nabla f \times \hat{\boldsymbol{\phi}} + B_\phi(r, z)\hat{\boldsymbol{\phi}}, \qquad (4.3)$$

where the flux function f is related to the magnetic vector potential by $A_\phi(r, z) = r^{-1}f(r, z)$ and $\hat{\boldsymbol{\phi}}$ is a unit coordinate vector.[†] A useful property of the flux function f is that it is constant along field lines,

$$\frac{df}{d\ell} = \frac{\mathbf{B}}{|\mathbf{B}|} \cdot \nabla f \propto (\nabla f \times \nabla \phi) \cdot \nabla f = 0. \qquad (4.4)$$

Plotting contours (i.e. level sets) of $f(r, z)$ is thus a quick way of plotting representative field lines within the meridional plane.

There is, alas, no equally simple and useful method of quickly identifying and rendering field lines in three-dimensional magnetic fields lacking symmetry. This limitation may go a long way toward explaining the popularity of two-dimensional models even in our manifestly three-dimensional universe. Lacking a short cut, one must resort to solving Eq. (4.2) for a selection of initial points $\mathbf{r}(0)$. Experience has shown that the appearance of the field as outlined by selected field lines depends on how these initial points are chosen.

4.1.1 Physical significance of magnetic field lines

A solution to the field line equation (4.2) can in principle be found for a magnetic field at any instant. What is not immediately evident is why such a curve should be physically significant, even if one concedes that the magnetic field itself is significant. There is, in fact, no single reason that field lines will be significant under general circumstances – this is why students are often warned not to attribute undue importance to them. There are, however, numerous circumstances arising in space physics whereby a magnetic field line can achieve a degree of reality. The

---

[†] The analogous expression for planar symmetry along z is $\mathbf{B}(x, y) = \nabla A \times \hat{\mathbf{z}} + B_z(x, y)\hat{\mathbf{z}}$, where $A(x, y)$ is the flux function.

following is a brief list of the most common, applicable to a wide variety of plasma regimes from (i) general, to (ii) the fluid regime, to (iii) MHD, to (iv) ideal MHD.

(i) *General: single particle motion.* If subject to no other forces than that due to a relatively stationary magnetic field, a charged particle will remain close to a single field line as it gyrates about it according to its mass and charge. Drifts will displace the particle's guiding center by several gyroradii after it has traversed a length comparable with the field's curvature radius or gradient scale. The global scales of space plasmas are typically very much greater than the gyroradii of their electrons and, to a lesser extent, of their heavier ions (the Earth's geomagnetic ring current is a counterexample to this, however). Waves in the field may scatter particles (important in e.g. the Earth's radiation belts), but this too is generally unimportant. Field lines therefore serve as excellent approximations of the electron orbits.

    The heliosphere, for example, includes a population of high-energy electrons (halo electrons) for which collisions are so rare that each one remains effectively confined to a single field line from the Sun to beyond the Earth's orbit (Feldman *et al.*, 1975). Another example is that of solar flares, which produce an accelerated population of electrons that follow the field line on which they are produced either downward until they impact the much denser chromosphere or upward out into the heliosphere.

(ii) *Fluid regime: thermal conductivity and solar coronal loops.* In a diffuse high-temperature plasma, thermal energy is conducted principally by electrons. When electrons are strongly magnetized (i.e. the cyclotron frequency is much greater than the collision frequency), their orbits will follow field lines over long distances between collisions; at these they scatter a perpendicular distance no greater than a single gyro-radius. The huge disparity between the parallel and perpendicular scattering distances makes the thermal conductivity highly anisotropic (Braginskii, 1965). Consequently, heat is conducted parallel to the magnetic field far more readily than perpendicularly to the field.

    Owing to this anisotropic conductivity, heat deposited somewhere in a plasma is rapidly and efficiently conducted to all points on the same field line, at least while collision frequencies remain relatively low. In the solar corona, the plasma β is also generally low, so plasma flows are mechanically confined by the field. This means that a bundle of field lines will behave as a one-dimensional autonomous atmosphere (Rosner *et al.*, 1978), at least as long as reconnection is relatively unimportant. High-resolution images of the corona made in the soft X-ray (SXR) or extreme ultraviolet (EUV) regions show numerous thin *coronal loops*, each believed, for the reasons just described, to be a single bundle of field lines.

    Coronal loop images in soft X-rays or in the extreme ultraviolet (e.g. Fig. 8.9) provide one of the best observational indicators of magnetic field line topology in the solar corona. In the end, though, a coronal loop is not a magnetic field line or even a flux tube but a column of plasma characterized by its excess emission. The neighboring corona is filled with other field lines, which, as far as we know, are magnetically identical but which appear at different intensities in these images. One interpretation of the various coronal imaging observations is that the corona has a tendency to form density enhancements along selected bundles of field lines, which then appear in imaging instruments as loops. No complete explanation has yet emerged as to why some bundles are selected while the majority are not (see Chapter 8).

(iii) *MHD: Alfvén wave propagation.* Low-frequency waves in a magnetized plasma comprise three branches: slow magnetosonic, fast magnetosonic, and shear Alfvén waves. The group velocity of the shear Alfvén waves is exactly parallel to the local magnetic field. In the limit of very short wavelengths, any small localized disturbance will therefore propagate along a path following a magnetic field line. This means that a given field line will "learn" of perturbations anywhere along itself at the Alfvén speed. In this sense the magnetic field line has a dynamical integrity similar to that of a piece of string. Indeed, it is common to derive the Alfvén speed intuitively using the analogy to a string under tension.

When equilibrium is established in a magnetic field, the distribution of current and pressure is dictated by equations whose characteristics are the field lines. For example, in a plasma in equilibrium with weak Lorentz forces, gas pressure, and gravity (a force-free equilibrium) the electrical current must be proportional to the field: $\nabla \times \mathbf{B} = \alpha \mathbf{B}$, where $\mathbf{B} \cdot \nabla \alpha = 0$. This means that the field line twist parameter, $\alpha(\mathbf{x})$, must be constant along each field line. In this way the field lines are the mathematical characteristics for the equilibrium equation (see Parker (1979) for a discussion of characteristics in magnetostatic equations).

(iv) *Ideal MHD: frozen-in field lines.* The concept of frozen-in field lines (cf. Section 3.2.3) was first put forward by Alfvén (1943), developed further in following years (Sweet, 1950; Dungey, 1953b) and to a high level of sophistication by Newcomb (1958). It is presented in most plasma physics textbooks (Moffatt, 1978; Parker, 1979; Priest, 1982; Sturrock, 1994), in many review papers (Stern, 1966; Axford, 1984; Greene, 1993; Longcope, 2005), and among the preliminaries of topologically oriented investigations (Vasyliūnas, 1975a, b; Hesse and Schindler, 1988; Hornig and Schindler, 1996).

At its simplest, the frozen-in-field-line theorem states that if two fluid elements lie on a common field line at one time then they lie on a common field line at all times past and future. This follows directly from the ideal induction equation (3.24) and from the fact that all fluid elements move at the velocity $\mathbf{v}$ that appears in it.

Tracing field lines and following fluid elements are processes akin to integration, each of which is countered by a kind of differentiation. Fluid-element-following is countered by the differential operator known as the *advective derivative*,

$$\frac{d}{dt} = \frac{\partial}{\partial t} + \mathbf{v} \cdot \nabla, \tag{4.5}$$

where the partial derivative acts on fixed Eulerian coordinates. Applying this to a position vector, expressed in Eulerian coordinates, quickly returns the elementary fact that fluid elements move at velocity $d\mathbf{r}/dt = \mathbf{v}$. The operator inverse to field line tracing can be formed similarly, but it turns out to be most convenient to parameterize field lines using the column mass per unit flux, μ, rather than using the integrated length ℓ. In this parameterization the inverse of field line tracing is the differential operator

$$\frac{d}{d\mu} = \frac{1}{\rho}\mathbf{B}\cdot\nabla, \tag{4.6}$$

where ρ is the mass density of the plasma. Applying this derivative to a position vector yields a form of the field line equation which becomes Eq. (4.2) after using the relation $d\mu/d\ell = \rho/B$.

The operations of tracing and following will be interchangeable if their inverses are interchangeable. Differential operations are interchangeable, i.e. they commute, if their commutator, $[\hat{A}, \hat{B}] \equiv \hat{A}\hat{B} - \hat{B}\hat{A}$, is exactly zero.[†] A series of algebraic steps allows the commutator of the operators above to be rewritten in the form

$$\left[\frac{d}{dt}, \frac{d}{d\mu}\right] = \left\{\frac{d}{dt}\left(\frac{1}{\rho}\mathbf{B}\right) - \frac{d}{d\mu}\mathbf{v}\right\}\cdot\nabla$$

$$= \frac{1}{\rho}\left\{\frac{\partial\mathbf{B}}{\partial t} + (\mathbf{v}\cdot\nabla)\mathbf{B} + (\nabla\cdot\mathbf{v})\mathbf{B} - (\mathbf{B}\cdot\nabla)\mathbf{v}\right\}\cdot\nabla \tag{4.7}$$

after using mass conservation, $d\rho/dt = -\rho(\nabla\cdot\mathbf{v})$. The last term in the braces can be seen, using vector identities, to vanish identically as long as $\mathbf{B}$ and $\mathbf{v}$ satisfy the ideal induction equation (3.24).

The steps above demonstrate that differentiation along a field line is interchangeable with differentiation along a flow trajectory. From this it follows that a field line linking two fluid elements can be traced either before or after following the flow of those elements. This constitutes a restatement of the theorem introduced above. One can thereafter imagine "labeling" all the fluid elements along a given field line and then following those fluid elements as they move at their own velocities $\mathbf{v}$. These material elements, which are manifestly real, will trace out a single field line at all times, so in some sense that field line must be real as well.

The brief derivation above holds regardless of how the flow velocity is defined – it never uses the momentum equation or any equation of state. It does depend critically on the ideal induction equation (3.24), which is lacking in non-ideal effects such as resistivity. Were $\mathbf{E}' \neq 0$ at any point the term in braces in Eq. (4.7) would not vanish, nor would the commutator, and the field lines would lose their reality. Theories of magnetic reconnection (see Chapter 5) concern situations where $\mathbf{E}' \neq 0$, but only within a small *diffusion region*. In such cases all the concepts we develop in this chapter will apply throughout space with the sole exception of diffusion regions.

A related idea, called the frozen-in-flux or Alfvén's theorem, states that the net flux crossing a material surface element cannot change (see Section 3.2.3 for a derivation). This follows from Faraday's law, $\partial\mathbf{B}/\partial t = -\nabla\times\mathbf{E}$, applied to the perimeter of the surface element, given that $\mathbf{E}' = \mathbf{0}$. Alfvén's theorem serves a similar purpose for flux tubes that the frozen-in-field-lines theorem does for field lines alone.

[†] Commutators are commonly associated with quantum mechanics, but they are broadly useful for exploring the interchangeability of operators. Any elementary text in quantum mechanics will introduce the basic mechanics of their computation.

4.1.2 Null points, separatrices, and separators

For a continuous magnetic field $\mathbf{B}(\mathbf{x})$, the field line equation (4.2) is singular only where the magnetic field vector vanishes (Arnold, 1973). In a general field, $\mathbf{B}(\mathbf{x})$ will vanish only at isolated points called *null points*, which constitute key topological features of a magnetic field. In many cases the structure of the field can be characterized entirely in terms of the so-called *magnetic skeleton* formed by the null points and field lines attached to them (Priest *et al.*, 1997; Longcope and Klapper, 2002; Beveridge and Longcope, 2005).

In the vicinity of a null point $\mathbf{x}_0$ the magnetic field has the generic[†] form of a Taylor expansion:

$$B_i(\mathbf{x}) = \sum_{j=1}^{3} (x_j - x_{0,j}) M_{ij} + \cdots, \tag{4.8}$$

where $M_{ij} \equiv \partial B_i / \partial x_j |_{\mathbf{x}_0}$ is the field's Jacobian matrix. A null point for which the matrix $\mathbf{M}$ vanishes entirely is termed *higher order* and occurs only in special circumstances. Excepting these cases, the field lines have a simple behavior in the neighborhood of null points which may be characterized entirely by the eigenvectors and eigenvalues of the Jacobian matrix. In particular, it is possible to assign any null (excepting non-generic cases) as positive or negative, according to the number of its eigenvalues with positive real parts. A thorough analysis of this categorization is given in Parnell *et al.* (1996), from which a few important results are summarized here.

The matrix $\mathbf{M}$ will have three eigenvalues, which may be either all real or both real and complex. In the first case the eigenvalues can be ordered so that $\lambda_1 \leq \lambda_2 \leq \lambda_3$; in the second there must be one real eigenvalue λ_r and a complex conjugate pair λ_c and λ_c^*. Since $\nabla \cdot \mathbf{B} = \sum_i M_{ii} = 0$ the three eigenvalues must sum to zero. A null point for which one eigenvalue is negative and the other two are positive (or have positive real parts) is called a *positive* null point.[‡] (The null is therefore called positive if $\det \mathbf{M} < 0$, which is confusing until one considers the sense of field lines in the fan surface; see Fig. 4.1.) The real and imaginary parts of the eigenvectors from the two positive eigenvalues span a plane within which all field lines originate at the null point. Following these field lines beyond the immediate neighborhood, one sees that they form a surface called the *fan*

[†] The term *generic*, and the related term *structurally stable*, have a very precise mathematical sense. A full definition, as given in e.g. Guckenheimer & Holmes (1983), is not possible here. The terms basically refer to situations whose qualitative form (i.e. topology) is not destroyed by small changes. One or other is invoked to rule out specially constructed cases, such as perfect symmetry, to which a given statement will not apply.

[‡] An alternative term, *B-type null*, first appeared in the magnetospheric literature (Yeh, 1976) and has also been used in the solar physics literature (Greene, 1988; Lau and Finn, 1990; Priest and Titov, 1996). The term "positive null" is used here since it is more descriptive than "B-type".

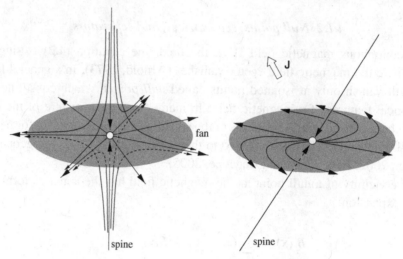

Fig. 4.1. Schematic depictions of positive null points. Two *spine field lines* directed toward the *null point* (the white circle) appear as dark lines with arrow heads next to the null. The central portion of the horizontal *fan surface* is colored grey and contains fan field lines directed outward like spokes on a wheel. The left-hand case is the simpler: a potential null point which is cylindrically symmetric (with two identical real eigenvalues, $\lambda_2 = \lambda_3$). The thinner solid and broken lines show a few of the field lines on either side of the fan surface. The right-hand case is a non-potential null whose spines are not orthogonal to the fan surface. This is based on a non-force-free case from Parnell *et al.* (1996) with the current (white arrow) at an angle to the spine. (Reproduced from Longcope, 2005.)

surface of the null. There are also two *spine field lines* which terminate at the null in directions both parallel and antiparallel to the eigenvector of the negative eigenvalue (see Fig. 4.1). A *negative null* has the opposite structure: one positive eigenvalue and two with negative real parts, a fan surface of field lines ending at the null, and two spine field lines originating at the null.

Cases where one or more eigenvalue real parts vanish cannot be classified as either positive or negative. Such cases are not generic (they will not survive an arbitrary small perturbation to the field) but do occur in cases of symmetry, such as two-dimensional models. An X-type null, or X-point, is one where one eigenvalue vanishes, namely λ_2, and the other two have equal magnitude and opposite sign: $\lambda_1 = -\lambda_3$. This is the standard heteroclinic point in two-dimensional fields; however, as alluded to previously, they do not generally occur as such in three-dimensional fields. If the real parts of two eigenvalues vanish then so must the third, since they must sum to zero. Barring a higher-order null (where all three eigenvalues are identically zero), this must be an O-type null with two purely complex eigenvalues, $\lambda_c = i\gamma$, and a symmetry direction for which $\lambda_r = 0$.

Fan field lines and spine field lines are notable exceptions to the general rule that field lines have no beginning or ending – it seems that certain field lines terminate at null points. Since these are only field lines, and not flux tubes, their termination does not create a divergence in **B**: integrating $\nabla \cdot \mathbf{B}$ over a sphere enclosing the null shows that the total inward flux must equal the total outward flux. The spine field lines intersect this sphere at two points, and the fan surface intersects it along a curve. The net flux within either a point or a curve is zero so the seeming imbalance between two spines and an infinite number of fan field lines does not create a measurable effect: both account for exactly zero flux.

A fan surface divides the volume of field lines into two regions (or domains), thereby serving as one type of *separatrix* in three-dimensional magnetic fields. The spines, however, are one-dimensional curves and therefore do not form separatrices. This makes three-dimensional null points topologically different from two-dimensional X-points since all four field lines connecting to an X-point are dubbed separatrices, regardless of their orientation. If a three-dimensional null is transformed continuously into an X-point by taking its intermediate eigenvalue to zero then the spine and fan both become separatrices in the limit $\lambda_2 \rightarrow 0$. Which of the X-point's separatrices was formerly a spine depends on the former null type and from which side λ_2 approached zero.

An even more exotic kind of field line occurs when the fan surface of a positive null intersects the fan surface of a negative null, as shown in Fig. 4.2. When the intersection is transversal, as it must be in a non-singular magnetic field, it forms a *null–null line* (Lau and Finn, 1990). Intersections of separatrices are called *separators*, in general, and, since a fan surface is one type of separatrix, null–null lines are one type of separator. There are, however, other forms of separators including finite-width current sheets, called current ribbons, which occur when the separatrix intersection is not transversal (Longcope and Cowley, 1996).

A null–null line has two termini since it must begin at a positive null and end at a negative null. Since they lie at the intersection of separatrices, null–null lines are more natural analogs of two-dimensional X-points than are the three-dimensional null points themselves. In this analogy, however, it must be borne in mind that while an X-point may be identified locally, a null–null line is locally indistinguishable from the nearby field lines; its uniqueness derives only from the global topology. It is often, although not always, the case that the field vectors in the vicinity of a null–null line have an X-type shape (see the left-hand inset in Fig. 4.2). The exact location of the X in such a slice depends critically on the orientation of the plane. A plane with a slightly different orientation will be crossed perpendicularly by a different field line, and that field line will be the X-point for that choice of plane. This local criterion cannot, therefore, be used to identify a null–null line; the only

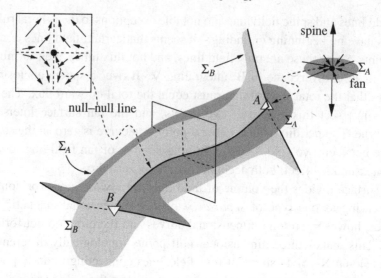

Fig. 4.2. The structure of a null–null line. Positive and negative null points B and A have fan surfaces Σ_B and Σ_A, shown in light and dark shades of grey, respectively. These intersect transversally along a null–null line (bold). Surface Σ_B is then bounded by the two spines from null point A (bold); similarly for Σ_A and the spines from B. The field in neighborhood of the null–null line is illustrated in the inset. This shows the direction of **B** within a plane pierced normally by the null–null line at the dark circle. The fan surfaces cross the plane along the broken lines. (Reproduced from Longcope, 2005.)

way to locate such a line is by following field lines in both directions. (Longcope (1996) presents an algorithm for this.)

For the purposes of illustration, consider a dipole representing a planetary field, say that of the Earth, pointing exactly southward and embedded in a perfectly southward interplanetary magnetic field (IMF), in the absence of a solar wind; see the upper panel of Fig. 4.3. This configuration is purely axisymmetric and can therefore be expressed through a flux function, as in Eq. (4.3), with

$$f(r, \theta) = B_E \frac{R_E^3 \cos^2 \theta}{r} + \tfrac{1}{2} B_{IMF} r^2 \cos^2 \theta, \qquad (4.9)$$

where r is the radial distance from Earth's center and θ is the polar angle (co-latitude). The magnetic field generated by this flux function will vanish at $r = (B_E/B_{IMF})^{1/3}$, $\theta = \pi/2$, where $\nabla f = 0$. Owing to the axisymmetry this constitutes a ring-shaped continuum of nulls rather than distinct points; also owing to the symmetry, they are X-type nulls, i.e. X-points.

Any departure from axisymmetry, for example an IMF not parallel to the Earth's dipole (as in the lower panel in Fig. 4.3) will destroy the ring of X-points. In its place

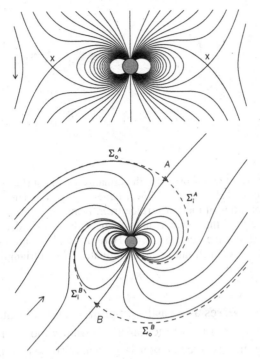

Fig. 4.3. The upper panel shows field lines from a southward dipole (similar to that of Earth) immersed in a uniform southward surrounding field (such as the interplanetary magnetic field in the absence of a solar wind; see the arrow). This special axisymmetric case is plotted here within the noon–midnight meridian. The field lines are rendered by contouring the flux function $f(r, \theta)$ from Eq. (4.9). The ring of null points intersects the page at the two locations labeled X. The lower panel shows the case of a northward, tilted, external field (arrow). A selection of field lines within the plane of the paper (again equivalent to the noon–midnight meridian) is plotted. Figure 4.4 shows the 3D structure of the fan surfaces. The symmetry is broken, of course, if there is a plasma flow in the surrounding medium like the solar wind around the Earth; compare Fig. 10.3.

there will be two null points connected by a pair of null–null lines in a closed loop (Yeh, 1976). The null point labeled B in Fig. 4.3 is positive. The solid field lines leaving it are its spines, one of which connects to the Earth's southern hemisphere (but not to the magnetic south pole). The fan surface includes the broken field lines labeled Σ_i^B and Σ_o^B, which leave the null point perpendicularly to the spines. The full surfaces are plotted in the two panels of Fig. 4.4. Corresponding portions of the fan surfaces from the negative null (A) are also shown. These portions intersect along the null–null lines (not shown). The outer portions of the separatrices extend into the IMF as approximately cylindrical surfaces, whose cross sections are shown in the right-hand panel of Fig. 4.4.

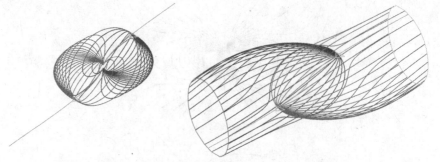

Fig. 4.4. Perspective views, from near the equator, of fan surfaces from the two null points for the field shown in the lower panel of Fig. 4.3. The left-hand panel shows the inner portions of the fan surface from A and B. The broken field lines labeled Σ_i^A and Σ_i^B in the lower panel of Fig. 4.3 are included in these. The right-hand panel shows the outer portions. Here the surfaces are truncated at the faces of a viewing box. Each fan surface originates from a triangle. See also the color-plate section.

The example above serves as a vastly simplified version of the Earth's magnetosphere, neglecting the rotation of the Earth, the solar wind, currents, and plasma pressure among other things (see Chapter 10). Each of these is important in the real magnetosphere, so what can be learned from such a gross simplification? The fact is that the topology of magnetic field lines is fairly robust and will generally persist under continuous changes to the magnetic field. Picture some continuous method of introducing a neglected effect, such as increasing the model-Earth's rotation rate from zero to that of the real Earth. This process will change the magnetic field of the model continuously, and quantities such as the number and location of the null points will also change in a continuous manner. The number of null points is, however, an integer (2 in this case) and cannot change continuously. Instead, it must remain fixed even though the magnetic field changes, except for the special events considered below. This thought experiment illustrates the power of magnetic topology. Continuing along the same line of thought leads to the expectation that a real magnetosphere of any magnetized planet includes two magnetic null points, one positive and one negative, just as the simplified model does; their locations will almost certainly be very different from the simplified models of Figs. 4.3 and 4.4 (see Chapter 13).

An exception to the above discussion occurs at discrete (as opposed to continuous) events known as *bifurcations*. These discrete events punctuate an otherwise continuous evolution in which a topological characteristic, such as the number or type of null points, changes discontinuously. There is a rich literature on the identification and classification of bifurcations in general vector fields (Guckenheimer and Holmes (1983) offer one account of this general theory).

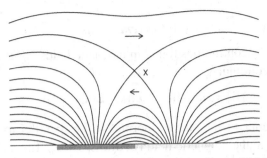

Fig. 4.5. A simple model of a solar coronal magnetic arcade. Field lines connect a positive region (light shading) to a negative region (dark shading). The overlying large-scale field is oppositely directed. It is separated from this by separatrices from a single null point X.

The theory of bifurcations makes extensive use of a quantity called the *topological degree* (Greene, 1992), a generalization of the Poincaré index for two-dimensional vector fields. The invariance of the topological index places severe constraints on bifurcations. For example, a single, isolated, positive null point can only change by "turning into" a set of three different nulls (two positive and one negative), through a process known as *pitchfork bifurcation*; any other change would alter the topological degree in the region surrounding the original null point.

Accounting for bifurcations modifies our expectation above about the null points in the Earth's magnetosphere. Introducing a neglected effect, such as the solar wind, might cause a bifurcation in the topology, resulting in a magnetosphere with either zero or four null points, for example. It is possible with a little more effort to construct the separatrix surfaces for each possibility. Each configuration would persist under a wide range of changes, up to another bifurcation.

A common simplified model of solar coronal magnetic fields, shown in Fig. 4.5, has obvious similarities to the axisymmetric case from Fig. 4.3 (where we might envision looking down on a dipolar active region embedded in a larger-scale background field). A set of low-lying coronal loops are anchored to positive and negative photospheric regions. Overlying field lines of the large-scale solar magnetic field are separated from these by separatrices originating in a magnetic X-point. The two sets of field lines play roles identical to the equatorial field and IMF, respectively, in the magnetospheric example. In both cases, when the two contributions are exactly antiparallel the coronal model has a symmetry, and a continuum of X-points forms. The symmetry in this model is broken when the infinite bands of flux are replaced by finite patches. As in the magnetospheric example, the continuum of null points fractures into a pair of null points connected by a null–null line.

4.1.3 Current sheets

The most common discontinuities found in magnetic fields occur at surfaces, such as shocks. Any discontinuity

$$[\mathbf{B}] = \lim_{\epsilon \to 0}\{\mathbf{B}(\mathbf{x} + \epsilon \hat{\mathbf{n}}) - \mathbf{B}(\mathbf{x} - \epsilon \hat{\mathbf{n}})\}. \tag{4.10}$$

must be perpendicular to the surface normal $\hat{\mathbf{n}}$ (i.e. parallel to the surface), to avoid a divergence in the magnetic field (Griffiths, 1999). The discontinuity leads to a surface current density $\mathbf{K} = \hat{\mathbf{n}} \times [\mathbf{B}]/\mu_0$. Any field component normal to the surface produces a large force density (formally infinite) of the kind characteristic of both fast and slow MHD shocks. In such cases the field lines are abruptly bent, but not topologically changed. Without a normal component the surface is termed a tangential discontinuity (TD), the only type of discontinuity which can be in equilibrium (Cowley *et al.*, 1997).

It is possible for the edge of a TD to include null points which are locally Y-shaped. Such a null cannot be characterized by its derivative Jacobian matrix M_{ij}, since derivatives are not defined at the discontinuity. If the TD occurs on a smooth sheet, however, the Y-type null will locally resemble one from a field which is otherwise current-free (first studied by Green (1965) and Syrovatskii (1971)), shown in Fig. 4.6. In a three-dimensional version of such a TD (Longcope and Cowley, 1996), field lines following the edge of the TD will either diverge away from or converge toward the null point in the direction erstwhile ignored (making it a positive or negative Y-null, respectively). A pair of Y-type null points at the edges of a TD of finite breadth, as in Fig. 4.6, have the same topological degree (Greene, 1993) as a single regular null point; the degree is either positive or negative. In both two and three dimensions it is possible to create the null pair continuously by deforming a regular null point (Syrovatskii, 1971; Longcope and Cowley, 1996).

The two-dimensional current sheet equilibrium proposed by Green (1965) and Syrovatskii (1971) is current-free everywhere except in the current sheet. Such a structure can be described by a potential $A(x, y)$ which is the real (or imaginary) part of a complex potential that is analytic except at a branch cut defining the current sheet. This powerful technique has been used to construct equilibria resembling realistic coronal current sheets (Priest and Raadu, 1975; Tur and Priest, 1976; Hu and Low, 1982). General formulations developed by Aly and Amari (1989) and by Titov (1992) permit the construction and evolution of equilibria of arbitrary complexity, containing numerous current sheets.

A magnetic discontinuity is, of course, a mathematical idealization. In reality, the magnetic field will change its direction by $\Delta\mathbf{B}$ over some finite distance δ and the current density will reach a large but finite value $|\mathbf{j}| \sim |\Delta\mathbf{B}|/(\mu_0\delta)$. Typically a non-ideal effect, for example resistivity, which prevents $\delta \downarrow 0$. Such an effect,

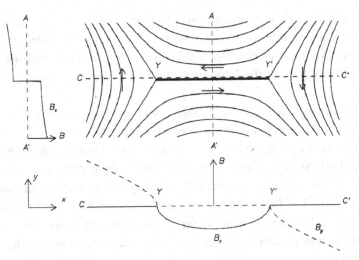

Fig. 4.6. A tangential magnetic discontinuity (TD) of the Green–Syrovatskii type in a two-dimensional magnetic field. The upper right panel shows magnetic field lines as solid curves, and indicates by arrows the direction of the field. The two Y-type nulls, labeled Y and Y', lie at either end of the current sheet (bold line). Two broken lines, labeled AA' and CC', are cuts along which the magnetic field is plotted to the left and below the field, respectively. Along the vertical cut (AA', left), the horizontal magnetic field B_x (solid line), suffers a discontinuity at the current sheet ($y = 0$). The vertical field B_y is zero over the entire cut. The horizontal cut (CC', lower line) passes just above the current sheet, YY'. The horizontal field B_x (solid line), is negative within the current sheet and vanishes outside it. The vertical field B_y (broken line) vanishes everywhere over the sheet (it is a TD) and is non-zero outside it. Note that both components vanish at the null points Y and Y'.

along with the self-consistent plasma dynamics, determines how thin the layer will actually become (see Chapter 5). Like so many other mathematical idealizations, the TD is valuable when appropriately applied. On scales much larger than δ there will be little evidential difference between a layer which is thin and a genuine discontinuity.

4.2 Regions of different topology

4.2.1 Open and closed fields

Most space physics studies are concerned with particular volumes of magnetized plasma such as the magnetosphere or the solar corona. Certain field lines may remain entirely within the volume of interest, either forming closed curves or wandering ergodically. All other field lines cross the external boundary of the volume either once or twice; the points at which they cross are called *footpoints*.

Field lines with only one footpoint are referred to as *open*; they often extend from the boundary to infinity. Those with two footpoints are called *closed*. Figure 4.3 shows examples of both open field lines, anchored to either the northern or southern hemisphere of the Earth, and closed field lines, overlying the Earth's equator.

If a field contains both open and closed field line regions then there must be an interface of some sort between these regions. The interface itself must consist of field lines, but these can be neither open nor closed (lest they belong to one or the other region). Such a surface is by definition a separatrix; an example is the fan surface originating at a magnetic null point, as in Fig. 4.4. The fan surfaces in the right-hand panel of Fig. 4.4 separate the closed field lines girdling the Earth's equatorial regions from the open field lines anchored near its poles; the separatrices separate the open field lines from field lines not anchored to the Earth at all (i.e. purely interplanetary magnetic field).

In similar models, often used for the solar corona, the lower boundary is the photospheric surface. Such models generally include closed field lines, for example in active regions (i.e. two sunspots of opposite magnetic polarity). They also include open field extending from the photosphere out into interplanetary space. Coronal holes, which appear darker in EUV and X-ray images, are believed to correspond to regions of open magnetic field lines (in fact, a frequently valid but imprecise equivalence is often seen in the literature between the terms "coronal hole" and "open field region"). These often surround the north and south solar poles but also often occur at lower, even equatorial, latitudes.

The outward plasma flow which is the solar wind tends to drag coronal field lines with it ever outward. Thus beyond some distance from the Sun one expects all field lines to be open field lines.[†] Measurements of the halo electron population reveal that the real solar wind also includes some closed field lines (Gosling *et al.*, 1987). These are often attributed to transient events where bundles of formerly closed magnetic field lines are in the process of being "opened".

Omitting transient events, and thereby closed field lines, leaves a steady-state heliospheric field consisting entirely of open field lines. These field lines can be categorized as either *outward* or *inward* according to the sign of B_r at their footpoints. The boundary between sectors of outward and inward field lines is a TD known as the *heliospheric current sheet (HCS)*. In reality the HCS is not a genuine discontinuity but consists of a complex dynamic structure, perhaps including many bundles of closed field lines or even U-shaped field with no footpoints at all. In analogy with the caveat at the end of Section 4.1 following the TD discussion,

---

[†] Here it should be noted that magnetic field lines extending all the way to the heliopause (the edge of the solar wind) are effectively "open" in all practical senses (enumerated in Section 4.1.1) in which a field line is physically meaningful.

the HCS is an idealization of the actual situation; but it has proven to be a useful idealization.

The most common quantitative model of the global corona is the potential-field source-surface (PFSS) model first introduced by Schatten *et al.* (1969) and Altschuler and Newkirk (1969) and refined by subsequent investigators. The magnetic field is taken to be current-free, $\mathbf{B} = -\nabla\chi$, in the region between the photosphere and an outer shell at $r = R_S$ called the source surface (see Fig. 4.7). The source-free, or solenoidal, condition $\nabla \cdot \mathbf{B} = 0$ means that the scalar potential must be harmonic, $\nabla^2\chi = 0$, within the range $R_\odot < r < R_S$. The upper boundary, $r = R_S$, represents the distance beyond which all field is open. In order to achieve this, the PFSS model requires that the field be purely radial there ($B_\theta = B_\phi = 0$) and thus that $\chi(R_S, \theta, \phi) = 0$. The general solution to Laplace's equation satisfying this boundary condition is

$$\chi(r, \theta, \phi) = \sum_{\ell=1}^{\infty} \sum_{m=0}^{\ell} \left[\left(\frac{R_S}{r}\right)^{\ell+1} - \left(\frac{r}{R_S}\right)^{\ell} \right]$$
$$\times P_\ell^m(\cos\theta) [g_\ell^m \cos(m\phi) + h_\ell^m \sin(m\phi)], \tag{4.11}$$

where $P_\ell^m(x)$ is the associated Legendre polynomial.[†] The coefficients g_ℓ^m and h_ℓ^m are fixed using measurements of the photospheric magnetic field.

A simplified heliospheric field model, consisting of only open field, can be matched to the PFSS model of the corona. A field line which is outward (inward) at the source surface at $r = R_S$ has $B_S = B_r(R_S) > 0$ ($B_S < 0$). These sectors are separated by one or more curves along which $B_S = 0$; they therefore represent the base of the HCS (of which there may be more than one). At radii approaching the source surface from below, Eq. (4.11) becomes strongly dominated by its lowest poles, $\ell = 1$ and $\ell = 2$, which correspond to the photosphere's overall dipole and quadrupole; one consequence of its low-order nature is that $B_S(\theta, \phi) = -\partial\chi/\partial r|_{r=R_S}$ tends to be very smooth, vanishing along a single closed curve dividing the $r = R_S$ sphere in two. In this commonly occurring situation there is a single HCS separating an inwardly directed hemisphere from an outwardly directed hemisphere. Occasionally this topology will change via bifurcation, giving rise to a slightly more complex situation with, for example, two or more distinct outward or inward sectors.

Since both B_θ and B_ϕ vanish everywhere over the source surface, a sector boundary, $B_r = 0$, is a curve of genuine magnetic null points. Such a one-dimensional continuum of nulls is one of the constructions expressly discounted in Section 4.1.2

[†] In practice the expansion includes a pre-factor $\sim (R_S/R_\odot)^\ell$, and $P_\ell^m(x)$ is defined using the Schmidt normalization in order to keep all coefficients of roughly comparable magnitude. Equation (4.11) is intended only to indicate the basic form of the potential.

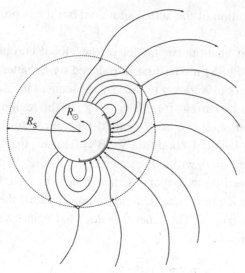

Fig. 4.7. Schematic depiction of a potential-field source-surface (PFSS) model as viewed from above the Sun's north pole (the sense of rotation is indicated by the semicircular arrow). The broken circle shows the source surface at $r = R_S$. The field is made purely radial at this surface by setting $B_\theta = B_\phi = 0$. Between the source surface and the solar surface, $r = R_\odot$, the magnetic field is potential: $\mathbf{B} = -\nabla\chi$. Field lines are anchored to the photosphere in the negative region (the shaded segment) and in the positive regions. Outside the source surface the field is swept back in a spiral by the solar rotation. Two null points (X-points, actually lines of null points) are shown as open circles on the source surface. The upward and downward separatrices are shown extending from these circles. The sector boundaries between heliospheric fields of opposite radial polarity are shown as the bolder darker curves.

since it is non-generic. It owes its appearance now to the definition of the source surface as a surface on which B_θ and B_ϕ vanish simultaneously (in general circumstances, these functions would vanish on separate surfaces, which would generically intersect transversally only along curves). Each null point on this curve is neither positive nor negative; all are X-type nulls whose neutral direction (along the eigenvector corresponding to the $\lambda_2 = 0$ eigenvalue) is parallel to the sector boundary curve.

Each null point has two distinct separatrix curves extending downward into the current-free corona: one forward and one backward (see Figs. 4.7 and 4.8). For each closed sector-boundary curve on the source surface, the forward separatrix curves form a continuous separatrix surface mapping downward to negative-polarity photospheric regions; this corresponds to the bold broken line in the right-hand panel of Fig. 4.8. The backward separatrices form independent separatrix surfaces mapping to positive regions. These surfaces are the separatrices between open and closed magnetic field line regions. Their footpoints trace photospheric curves, which are

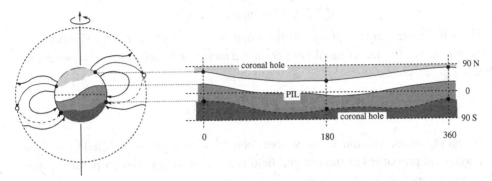

Fig. 4.8. A schematic depiction of coronal hole boundaries defined by the source surface model. The left-hand panel is a meridional cut, and the right-hand panel shows the full solar surface ($r = R_\odot$) plotted as the sine of the latitude vs. the longitude. The bold solid curves are the footprints (intersections with the solar surface) of the upward and downward separatrices respectively; the black curve is the polarity-inversion line (PIL). The shaded regions are, from top to bottom, the outward coronal hole (light grey), the positive closed-flux region (white), the negative closed-flux region (medium grey), and the inward coronal hole (dark grey). The same shading in the left-hand panel indicates how these regions might appear on the disk. (Reproduced from Longcope, 2005.)

the theoretical manifestations of coronal hole boundaries in the PFSS model. The coronal field becomes increasingly complex at decreasing radii and can also contain isolated coronal null points at radii $R_\odot < r < R_S$. Thus the separatrices leaving the smooth sector boundary may map to fairly complex coronal hole boundaries.

4.2.2 *Topology of closed fields (active regions)*

Studies concerned with a single active region, or even complexes of active regions, are generally restricted to closed field lines. The photospheric surface, $z = 0$, is the only boundary and each field line has two footpoints on it. One footpoint lies in the positive polarity region (i.e. where $B_z(x, y, 0) > 0$) and the other in the negative polarity region. In many models the function $B_z(x, y, 0)$ vanishes along smooth photospheric curves called *polarity inversion lines* (PILs).[†] In a distinct class of coronal models, positive and negative regions, representing sunspots or magnetic flux elements, are separated by extended areas where $B_z(x, y, 0) = 0$, intended to represent regions of very weak or intermixed field. We demonstrate below that the topologies of the coronal field for a photospheric field with PILs and for one with regions of $B_z = 0$ will be different.

[†] These structures are also sometimes referred to as *neutral lines*. This same term is used to describe a locus of nulls ($|\mathbf{B}| = 0$) in a two-dimensional model. We prefer the term PIL since it emphasizes that only one component of the field, the radial, vanishes there: typically the field is horizontal. The high degree of intermittency of the magnetic field at the solar surface means that large-scale PILs only emerge in moderate-resolution magnetic maps.

4.2.2.1 Field line mappings

We will denote generic points in the positive and negative polarity regions as $\mathbf{x}_\pm = (x_\pm, y_\pm, 0)$. The coronal field defines a *mapping* between these two sections of the photospheric surface,

$$\mathbf{x}_- = \mathbf{X}(\mathbf{x}_+), \qquad \mathbf{x}_+ = \mathbf{X}(\mathbf{x}_-), \tag{4.12}$$

by making an association $\mathbf{x}_+ \leftrightarrow \mathbf{x}_-$ between the footpoints of each field line. The topological properties of the coronal field are thus related to the properties of this mapping function $\mathbf{X}(\mathbf{x})$. Here the term "topological" refers to characteristics preserved under continuous changes in footpoints. Topological boundaries therefore correspond to discontinuities in the mapping function $\mathbf{X}(\mathbf{x})$.

There are broadly speaking two reasons why mapping discontinuities are thought to be physically significant. The first is that magnetic reconnection (see Chapter 5) is a local process that results in topological changes in field lines. This occurs most naturally at boundaries where topologically distinct field lines are physically close enough to be involved in the same local process. Thus magnetic reconnection often occurs at topological boundaries such as separatrices or separators. These same features are associated with discontinuities in the mapping $\mathbf{X}(\mathbf{x})$.

The second reason to focus on mapping discontinuities is that they are natural places for current sheets (tangential discontinuities) to form. Horizontal plasma motions of the photosphere will change the mapping function $\mathbf{X}(\mathbf{x})$ owing to the frozen-in-field-lines theorem (see Section 4.1.1). The coronal field will also change, possibly in some very interesting and complicated way, but the change in the mapping will not depend on this coronal evolution. An oft-used modeling strategy is to derive, or at least constrain, the coronal field using the mapping. A particularly popular application of this strategy is in inferring the presence of intense current densities from discontinuities in the mapping function.

If the photospheric flow field is smooth then it will not introduce new discontinuities into the mapping function (Longcope, 2005). It would seem reasonable to assume that where the footpoint mapping is continuous, the coronal magnetic field producing it is also continuous – there are no TDs within it. In a seminal paper, Parker (1972) suggested that in fact this assumption is invalid and that TDs do form routinely in the coronal field where the footpoint mapping is continuous. This so-called *Parker problem* has been the subject of intensive study since it was first posed and remains unresolved to this day (Antiochos, 1987; Jensen, 1989; Van Ballegooijen, 1988; Longcope and Strauss, 1994; Parker, 1994, 2004, Craig and Sneyd, 2005; Low, 2006b). The contrary statement is, however, broadly accepted: *where the footpoint mapping is discontinuous, there may be discontinuities in the coronal field* (but not necessarily). Thus mapping discontinuities are used as indicators of

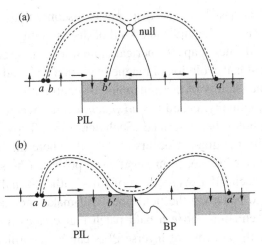

Fig. 4.9. Two-dimensional illustrations of field line mappings exhibiting the two possible types of mapping discontinuity. In each case neighboring footpoints *a* and *b* map to points *a'* and *b'* separated by a considerable distance. Each time the photospheric field is quadrupolar with three polarity-inversion lines (PILs), indicated by the vertical lines. The negative (downward) photospheric regions are shaded, and the vertical and horizontal arrows show the sense of the photospheric field. (a) A field with a coronal null point. Although it is a two-dimensional illustration we take the null to be negative, with the spines indicated by bold solid lines and the fan field by the thinner solid lines. Footpoints *a* and *b* map from opposite sides of the fan surface to points near each of the spine field footpoints. (b) A bald patch where a coronal field line (solid) grazes the photospheric surface, crossing in the inverse sense, from negative to positive, as indicated by the horizontal arrow. (Reproduced from Longcope, 2005.)

locations where current sheets, or at the least very intense current layers, will form naturally in response to the dynamics of the coronal magnetic field.

4.2.2.2 Separatrices

Where then will the footpoint mapping be discontinuous? Figure 4.9 illustrates the two types of mapping discontinuity possible when the coronal field itself is continuous. In each case, two neighboring footpoints in the positive polarity region (labeled *a* and *b*) map to points in the negative polarity (*a'* and *b'*) separated by a significant distance. Even if the points *a* and *b* can be brought arbitrarily close together, *a'* and *b'* would remain a finite distance apart. There is thus a discontinuity in the mapping at the point between them indicated by the solid curves emanating from between *a* and *b* in each panel.

The first kind of discontinuity (Fig. 4.9(a)) occurs when there is a null point in the coronal magnetic field. The separatrix from this null point, a fan surface indicated by the bold solid lines, maps to a curve in the photosphere (in the negative-polarity regions, if the null is positive). The mapping is discontinuous across this entire curve. There is also a more complicated kind of discontinuity at the two points to which the spine field lines map. Numerical simulations have confirmed that the natural three-dimensional evolution of magnetic plasma will lead to intense current concentrations in the vicinity of these mapping discontinuities.

The second discontinuity, called a *bald patch*, occurs where the coronal field line "grazes" the photospheric surface (Seehafer, 1986; Titov *et al.*, 1993). As shown in Fig. 4.9(b), the grazing occurs at the PIL where the horizontal field is directed in the "inverse" sense, from negative to positive (the sense is indicated by an arrow above each PIL). At sections of PILs with "normal" horizontal field (i.e. from positive to negative) the field lines will make short Ω-shaped curves. It is the inverse orientation above a bald patch that creates the U-shape of the grazing. In practice the portions of inverse PILs can be identified as those where $(\mathbf{B} \cdot \nabla)B_z|_{z=0} > 0$.

Models in which polarity regions are separated by field-free areas, rather than PILs, will not have bald patches. In these models, often called *magnetic charge topology* (MCT) models, fan surfaces are the only separatrices. Many of the null points from which these fan surfaces originate are located at the photospheric level ($z = 0$) within the field-free areas. A connectivity matrix identifying how much flux from each positive region connects to each negative region (or vice versa) provides a simplified version of the photospheric mapping function. The matrix can be assembled using the skeleton from all the null points (Longcope and Klapper, 2002; Beveridge and Longcope, 2005).

4.2.2.3 *Quasi-separatrix layers*

At places where the photospheric mapping suffers a step discontinuity, terms in the derivative matrix (i.e. the Jacobian) $\mathcal{D}_\pm(\mathbf{x}_\pm)$, where

$$
\mathcal{D}_+(\mathbf{x}_+) \equiv \begin{pmatrix} \dfrac{\partial X_-}{\partial x_+} & \dfrac{\partial X_-}{\partial y_+} \\[2mm] \dfrac{\partial Y_-}{\partial x_+} & \dfrac{\partial Y_-}{\partial y_+} \end{pmatrix}, \qquad \mathcal{D}_-(\mathbf{x}_-) \equiv \begin{pmatrix} \dfrac{\partial X_+}{\partial x_-} & \dfrac{\partial X_+}{\partial y_-} \\[2mm] \dfrac{\partial Y_+}{\partial x_-} & \dfrac{\partial Y_+}{\partial y_-} \end{pmatrix},
$$

$$(4.13)$$

will diverge in the manner of a Dirac δ-function. The divergence will occur along a curve of photospheric points which form the separatrix surfaces of the coronal field, when extended upward. As explained above this could be either a fan surface of a null point or a bald patch separatrix.

It was proposed by several authors (Van Ballegooijen, 1985; Longcope and Strauss, 1994; Priest and Démoulin, 1995; Démoulin *et al.*, 1996) that derivative matrices $\mathcal{D}_{\pm}$ which were extremely large could have almost the same practical significance as those that were formally infinite. The locations at which the derivatives are deemed to be sufficiently large, in some specified sense, are called *quasi-separatrix layers* (QSLs). To identify QSLs one must settle on a specific method of quantifying the magnitude of the matrix, i.e. either $\mathcal{D}_{+}$ or $\mathcal{D}_{-}$. Several different methods based on such quantities as the largest eigenvalue or the matrix norm have been proposed (Priest and Démoulin, 1995; Démoulin *et al.*, 1996; Titov *et al.*, 2002). Each method produces a scalar function over the entire photospheric surface. The function is dimensionless and typically of order unity for well-behaved mappings such as translations or rigid rotations. Points where the scalar function exceeds some fraction, say one-tenth, of the maximum value are designated as QSLs. This criterion is satisfied within an area, rather than a curve; this area, when extended upward into the corona, becomes a flattened volume (i.e. a *layer*) rather than a surface.

Let us conclude by describing how one might locate QSLs in a magnetic field whose components are stored on a three-dimensional computational grid. The mapping functions $\mathbf{X}(\mathbf{x})$ are computed on a two-dimensional grid of photospheric points, which might be finer than the one on which $\mathbf{B}$ is represented. Beginning at a grid point $\mathbf{x}_{+,ij}$ where $B_z > 0$, a field line is traced, by solving Eq. (4.2), until it returns to $z = 0$ at a point $\mathbf{x}_-$. The final position vector $\mathbf{X} = \mathbf{x}_-$ is assigned to the initial point $\mathbf{x}_{+,ij}$. The process is repeated for every photospheric grid point in the positive region and then for every one in the negative region. The derivative matrix $\mathcal{D}_{\pm}$ can then be approximated by finite differences or computed by integrating the matrix $\partial B_i / \partial x_j$ along the field line (Longcope and Strauss, 1994). This method yields the matrices $\mathcal{D}_{\pm}$, from which the values of the scalar quantity described above can be found. Contour plots of this quantity or its logarithm will show the QSLs.

A relationship between QSLs and separatrices has been demonstrated for continuous fields which can be approximated by discrete, isolated, photospheric flux regions. The QSLs in the former correspond to the genuine separatrices in the latter (Titov *et al.*, 2002).

4.3 Magnetic helicity

Using the magnetic helicity provides a simple alternative to characterizing every magnetic field line in a given volume. The basic quantity is defined within a volume $\mathcal{V}$ which completely encloses all field. That is to say, $B_n = \hat{\mathbf{n}} \cdot \mathbf{B} = 0$, on all boundaries ∂V, so all field lines are either closed curves or ergodic. The magnetic helicity (see also Eq. 3.72) in this case is given by the volume integral

$$H_m = \int_V \mathbf{A} \cdot \mathbf{B}\, d^3 x, \qquad (4.14)$$

where $\mathbf{A(x)}$ is the vector potential for the magnetic field: $\mathbf{B} = \nabla \times \mathbf{A}$. The vector potential suffers from a well-known ambiguity because the gradient of an arbitrary scalar field can be added to it without changing $\mathbf{B(x)}$. As long as $\hat{\mathbf{n}} \cdot \mathbf{B} = 0$ everywhere on ∂V such a change, known as a gauge transformation, will not change the value of H_m. (This is easily verified using integration by parts and the fact that $\nabla \cdot \mathbf{B} = 0$). This *gauge invariance* is reassuring – there is no physical significance associated with a gauge transformation, so any quantity that is altered by a gauge change is of dubious physical significance.

What makes the magnetic helicity particularly useful is that it does not change even when plasma motions change $\mathbf{B(x)}$, provided that they do so according to the ideal induction equation (3.24). The vector potential of the evolving field will satisfy

$$\frac{\partial \mathbf{A}}{\partial t} = \mathbf{v} \times \mathbf{B}, \qquad (4.15)$$

whose curl is easily seen to return Eq. (3.24). Time-differentiating the helicity integral (4.14) and using both Eqs. (3.24) and (4.15) confirms that $dH_m/dt = 0$.

What is more remarkable still is that when non-ideal effects such as resistivity invalidate the assumption of ideal induction, the helicity is still conserved to a greater degree than, say, the magnetic energy (Berger, 1988). That is to say, resistivity converts magnetic energy to heat but appears to do little to the magnetic helicity. The basic reason for this is that, comparing the two quadratic integrals, the energy involves one more derivative of $\mathbf{A}$ than does the helicity. Their dissipation rates, under resistivity, also differ by a single derivative. This means that the fine scale structure, which is equivalent to large wave numbers in the Fourier transform of $\mathbf{A}$, affects the dissipation of energy more than the dissipation of helicity.

4.3.1 Topological interpretation of helicity

The magnetic helicity turns out to be intimately related to the topology of field lines. The relationship is most easily demonstrated for magnetic fields of a very specialized form. Assume that all the magnetic field in volume V is confined to two very thin closed flux tubes (see Section 6.4), as depicted in Fig. 4.10; outside these

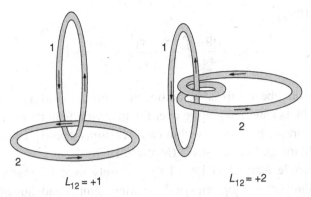

$L_{12} = +1$

$L_{12} = +2$

Fig. 4.10. The linking number between a pair of closed curves labeled 1 and 2. For clarity the curves are shown with finite width. Each curve is traversed in the direction indicated by the arrow. The pair on the left is linked with linking number $L_{12} = +1$; the pair on the right with $L_{12} = +2$.

tubes $\mathbf{B} = \mathbf{0}$. The tubes have fluxes Φ_1 and Φ_2 respectively, in the forms of field lines running circumferentially, i.e. parallel to their central axes. In this special case the magnetic helicity within the volume $\mathcal{V}$ can be written as

$$H_\mathrm{m} = 2L_{12}\Phi_1\Phi_2, \tag{4.16}$$

where L_{12} is an integer called *Gauss's linking number*. The linking number expresses the number of times curve 1 links curve 2 in the right-handed sense, as illustrated in Fig. 4.10. Unlinked curves have linking number $L_{12} = 0$; curves linked once in the left-handed sense have $L_{12} = -1$.

A pair of linked tubes can never be unlinked using just continuous motions and deformations. This fact is internalized by most people at an early age and explains their astonishment when a magician appears to violate it by pulling apart two seemingly solid and linked rings. A related but less appreciated fact, expressed by the frozen-in-field-line theorem, is that ideal plasma motions can bend and deform field lines but never break them or destroy them. This means that such motions cannot change L_{12}, Φ_1, or Φ_2. It therefore follows naturally that such motion will not change the magnetic helicity.

It was demonstrated by Gauss that the linking number of two parameterized curves, $\mathbf{r}_1(\ell_1)$ and $\mathbf{r}_2(\ell_2)$, is given by the double integral

$$L_{12} = \frac{1}{4\pi} \oint_1 \oint_2 \frac{\mathbf{r}_1 - \mathbf{r}_2}{|\mathbf{r}_1 - \mathbf{r}_2|^3} \cdot \frac{d\mathbf{r}_1}{d\ell_1} \times \frac{d\mathbf{r}_2}{d\ell_2} \, d\ell_2 d\ell_1. \tag{4.17}$$

To understand why such a formidable expression occurs in terms of the magnetic helicity, write the contribution to $\mathbf{A}$ from tube 1 using the Biot–Savart law (Griffiths, 1999):

$$\mathbf{A}(\mathbf{x}) = \frac{\Phi_1}{4\pi} \oint_1 \frac{\mathbf{x} - \mathbf{r}_1}{|\mathbf{x} - \mathbf{r}_1|^3} \times \frac{d\mathbf{r}_1}{d\ell_1} \, d\ell_1. \tag{4.18}$$

Valid for $\mathbf{x}$ outside tube 1, this is the dominant contribution to $\mathbf{A}(\mathbf{x})$ inside tube 2. Using it in Eq. (4.14) and integrating over the magnetic field there, and obtaining an analogous expression for $\mathbf{A}(\mathbf{x})$ within tube 1, returns a product equal to $2\Phi_1\Phi_2$ times the double integral in Eq. (4.17) (Moffatt, 1978).

While the double integral in Eq. (4.17) is rarely used in practice, it clearly exhibits several important properties of the linking number and thus of the helicity. The value of the double integral is not changed by exchanging the indices (so that $L_{21} = L_{12}$) or by reversing the senses of both curves: $d\mathbf{r}_1/d\ell_1 \rightarrow -d\mathbf{r}_1/d\ell_1$ and $d\mathbf{r}_2/d\ell_2 \rightarrow -d\mathbf{r}_2/d\ell_2$. The integral will change sign (but not magnitude) if both curves are mirror-reflected: $z_1 \rightarrow -z_1$ and $z_2 \rightarrow -z_2$. Owing to this last property we can refer to H_m as a *pseudo scalar*, but it really means that the helicity quantifies the preferred sense of "handedness" in the magnetic field – which is also called its *chirality*.

To generalize the linking idea to a more realistic field we consider a volume filled with N thin, closed, isolated flux tubes. A series of steps like those just described allow the helicity to be written as

$$H_m = \sum_{i=1}^{N} \sum_{j \neq i} L_{ij} \Phi_i \Phi_j. \tag{4.19}$$

(Setting $N = 2$ and using the fact that $L_{12} = L_{21}$ quickly returns Eq. (4.16).) In this more complex situation, the particular value $H_m = 0$ can mean either that all flux tubes are mutually unlinked or that they are interlinked but in right-handed and left-handed senses in equal numbers (weighted according to their flux). Equation (4.19) can be further generalized to a continuous magnetic field, provided that the latter has no ergodic field lines, by replacing the sums with integrals (it becomes unnecessary to exclude self-linking from the inner integral, since this occurs on a field line of infinitesimal flux). The integral version is still equivalent to the original equation, Eq. (4.14), as was the two-flux-tube case. Equation (4.14) applies to every field, even those with ergodic field lines; it is the only version regularly used.

4.3.2 Twist and writhe

There are many instances in space physics of magnetic field confined to thin flux tubes (see Chapter 6). For example, solar active regions are believed to form when flux tubes of $\Phi \sim 10^{20}$–10^{22} Mx (1 Mx = 1 G cm^2) rise buoyantly to the surface from the dynamo layer in the tachocline. Transient structures in the solar wind, called magnetic clouds, are believed to be flux ropes ejected from the corona. Small transient events observed on the day-side of the magnetosphere, called flux transfer events (FTEs, Russell and Elphic, 1978) are believed to be flux tubes created by reconnection at the magnetopause. It is therefore worth characterizing the magnetic helicity of a generic flux tube.

We assume here that the magnetic field within a volume $\mathcal{V}$ is entirely confined to a single thin closed tube whose axis is the curve $\mathbf{r}(\ell)$, parameterized by the arc length. The unit vector tangent to the axis is

$$\hat{\mathbf{t}} = \frac{d\mathbf{r}}{d\ell}. \tag{4.20}$$

The magnetic field within the tube is largely parallel to the axis, $B_t\hat{\mathbf{t}}$, but also includes an azimuthal component twisting about the axis. As a result of the azimuthal component, there is an axial current density within the tube[†]

$$\mu_0 J_t = \hat{\mathbf{t}} \cdot \nabla \times \mathbf{B} \simeq 2q(\ell)B_t, \tag{4.21}$$

where $q(\ell)$ is the local pitch of the magnetic field lines about the axis at position ℓ. If the axis were completely straight, $\hat{\mathbf{t}} = \hat{\mathbf{x}}$ say, neighboring field lines would wrap around it once over a length $\Delta x = 2\pi/|q|$. If $q > 0$ then the field lines wrap in a right-handed helix.

We apply the helicity integral (4.14) to the structure described above and assume that the cross section of the tube is small compared with the scale of its axis. The resulting helicity can be decomposed into two terms (Berger and Field, 1984; Moffatt and Ricca, 1992),

$$H_{\mathrm{m}} = \Phi^2(Tw + Wr), \tag{4.22}$$

called the *twist* and the *writhe*, respectively, each of which depends on different properties of the tube. The easier to understand, the twist,

$$Tw \equiv \frac{1}{2\pi} \oint q(\ell)\,d\ell, \tag{4.23}$$

can be interpreted as the total number of times that the neighboring field lines wrap around the axis. It should be noted, however, that Tw need not be an integer even if all the field lines within the tube close on themselves. Furthermore, considering a straight axis, $\hat{\mathbf{t}} = \hat{\mathbf{x}}$, hides the actual difficulty of tracking the angles made by field lines near an axis when that axis itself is deformed.

[†] Since $\mathbf{B} = 0$ outside the tube, it easy to show that the tube carries no net current. There must, therefore, be a return current at the outer surface of the tube. There will also be an even larger azimuthal surface current to confine the axial field B_t. Neither surface current plays any role in the helicity of the tube.

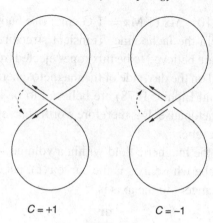

$$C = +1 \qquad\qquad\qquad C = -1$$

Fig. 4.11. Illustration of the crossing number C for a closed curve. The curve on the left has one positive crossing. The broken-line pointer shows how the upper segment must be rotated to align it with the lower one: in the positive (counterclockwise) direction. The right-hand curve has $C = -1$.

The writhe term depends only on the shape of the axis $\mathbf{r}(\ell)$ but is more cumbersome to deal with. It can be formally expressed as a double integral

$$Wr = \frac{1}{4\pi} \oint \oint \frac{\mathbf{r} - \mathbf{r}'}{|\mathbf{r} - \mathbf{r}'|^3} \cdot \frac{d\mathbf{r}}{d\ell} \times \frac{d\mathbf{r}'}{d\ell'} \, d\ell \, d\ell' \qquad (4.24)$$

where $\mathbf{r}' = \mathbf{r}(\ell')$. This equation bears an obvious similarity to the linking number, Eq. (4.17), except that here both integrals are performed over the same curve – which is the *only* curve. The integral will not diverge as long as the curve is smooth enough that $\hat{\mathbf{t}}(\ell) \times \hat{\mathbf{t}}(\ell') \to 0$ sufficiently fast as $\ell' \to \ell$.

This double integral provides some insight into the nature of writhe. It is (cf. (4.18)) invariant under a reversal in the direction of integration but will change sign (not magnitude) upon a mirror reflection $z(\ell) \to -z(\ell)$. If the curve were entirely confined to a plane then $\mathbf{r} - \mathbf{r}'$, $\hat{\mathbf{t}}$, and $\hat{\mathbf{t}}'$ would all lie in that same plane and their triple product would vanish. It follows that planar curves have no writhe.

4.3.2.1 Computing the writhe

There are several useful and enlightening ways to compute the writhe of a particular curve, but using integral (4.24) is not among them. One method involves counting the number of times the curve appears to cross itself when viewed from a particular direction. Figure 4.11 shows two different figure-8 curves, viewed face on. In each case, the curve appears to intersect itself at the central point, but we maintain that the curve does not *actually* touch itself there – the intersection is only *apparent*. The figure on the left crosses once in a sense we designate as positive, since the upper part (from our viewpoint) would have to be rotated in the positive (counterclockwise)

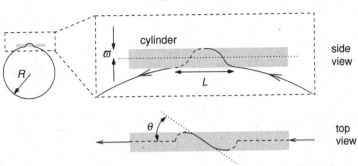

Fig. 4.12. A simplified model of an active region flux tube deformed by emergence and Coriolis effects. The deformed section wraps around a cylinder of radius ϖ (shaded).

direction to align it with the lower piece (as indicated by the dashed arc). The figure-8 on the right has a single *negative* crossing.

The signed crossing number C of a curve is found by subtracting the number of negative crossings from the number of positive crossings. The left- and right-hand curves in Fig. 4.11 have $C = +1$ and $C = -1$ respectively. In general the crossing number will depend on the angle from which the curve is viewed, but the writhe is always independent of this angle. To calculate the writhe of a space curve one averages the signed crossing numbers from viewing angles uniformly distributed over all 4π steradians:

$$Wr = \langle C \rangle. \tag{4.25}$$

If the figure-8 curves in Fig. 4.11 lie almost flat (but not so flat that the curves self-intersect) then the same crossing number will be evident from all directions above the front of the page. A mental transformation should convince you that when viewed from behind the crossing will have exactly the same sign – in spite of the change in roles played by the upper and lower parts. Only when viewed edge-on will the curve appear not to cross itself. Provided that the curve is sufficiently flat, these edge-on views will subtend a negligible fraction of the possible directions and so $Wr = +1$ for the left-hand curve and $Wr = -1$ for the right-hand curve.

We conclude by demonstrating the above technique on a curve more relevant to space physics: a simplified model of an active-region flux tube. Let us begin with a flux tube that is a circle of radius R within the Sun (say the radius of the tachocline). To mimic active region emergence and deflection by Coriolis forces, we deform a section of the tube around an imaginary cylinder of radius $\varpi \ll R$ (see Fig. 4.12). The deformed section was initially of length L, and the cylinder around which it wraps is a tangent to the circle of radius R (see the left-hand part

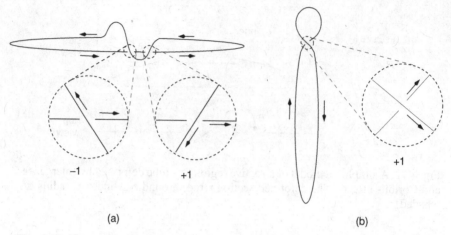

Fig. 4.13. The writhed axis from Fig. 4.12 viewed from two different angles. The entire curve is shown and the cylindrical deformation is exaggerated for clarity. (a) A view slightly different from "top view". Two crossings are enhanced as insets to show that they are of opposite signs. (b) A view very close to the axis of the cylinder. Only one crossing (positive) occurs, also enhanced as an inset.

of the upper panel). Thus the apex of the tube ends up at radius $r = R + 2\varpi$. The curve wraps one complete turn about the cylinder in the right-handed sense. When viewed from the top, the apex of the tube is tilted by $\theta = 2\pi\varpi/L$ with respect to the plane of the original circle. Clearly the deformation creates a non-planar curve (except in the limit $\varpi \to 0$) which might therefore have non-zero writhe.[†]

To compute the writhe of the curve we have just constructed we view it from various angles. In views close to the one called "top view" in Fig. 4.12, the deformed upper portion appears to cross the far side of the circle, as shown in Fig. 4.13(a) (in deducing the writhe we must consider the entire curve). From this viewpoint there are two crossings of opposite sign, yielding a signed crossing number $C = 0$. Any view near to the "side view" of Fig. 4.12 will show no crossings at all. Only when viewed from very close to the cylinder axis will the curve appear to cross itself, with $C = +1$, as shown in Fig. 4.13(b). These crossings appear only when viewed from angles to the cylinder axis within an angle $\theta = 2\pi\varpi/L$. (These are viewing angles within the arc labeled θ in Fig. 4.12). There is, in fact, a half-cone (half-angle θ) within which that crossing will appear. Viewed from the rear within the corresponding half cone, there will also be a single positive crossing. A cone

---

[†] There are corners at the edge where the helix abuts the circle. Such discontinuities violate the conditions under which integral (4.24) converges, so they must be smoothed. The specific manner of smoothing will affect the final answer. We prefer to ignore this technical difficulty in the interests of clarity. However, the value of Wr that we find does depend on how this issue is actually resolved.

of half-angle $\theta \ll 1$ subtends a solid angle equal to $\pi\theta^2$ steradians, so the two half-cones, front and rear, contribute $C = +1$ within the solid angle $\pi\theta^2$, while all other angles have $C = 0$. Averaging over all 4π steradians of viewing angles then gives

$$Wr \simeq \frac{1}{4\pi}\pi\theta^2 = \pi^2\frac{\varpi^2}{L^2}, \qquad (4.26)$$

as the writhe of the deformed circle.

The curve above is highly idealized but still serves to illustrate several important and general facts about writhe. First, the writhe is not a topological quantity: it changes as the curve is continuously deformed. Even for one complete helical turn, the writhe is not an integer; it depends on the amplitude of the helical distortion – here it is proportional to the amplitude squared. A helical distortion of right-handed sense introduces *positive* writhe; its mirror reflection, a left-handed distortion, would naturally be negative. Finally, repeating the steps for tubes which were initially some planar curve other than a circle will yield the same writhe, provided that the deformed portion is small enough; the writhe characterizes the degree of deformation out of the plane.

4.3.2.2 *Utility of twist and writhe*

The tools of twist and writhe prove useful in understanding many aspects of magnetic topology in space physics. For example, the magnetic fields in many active regions are found to carry current within their flux tubes (Pevtsov *et al.*, 1994). Since internal current contributes to twist but not to writhe, we infer that a flux tube which emerges to create an active region must have some twist, Tw (Leka *et al.*, 1996). Large samples of active regions show a significant tendency for regions in the northern hemisphere to have a left-handed twist ($Tw < 0$) while those in the south have a right-handed twist (Pevtsov *et al.*, 1995; Zirker *et al.*, 1997).

It is also known that the Coriolis effect creates a deformation in a flux tube's axis as it rises through the solar convective envelope. The rising plasma expands as it enters layers of lower pressure. To conserve angular momentum while its moment of inertia is increasing, the tube's apex rotates more slowly. Viewed from the rotating frame of the Sun, this appears as a *retrograde* rotation of the tube's rising apex. This sense of apparent rotation will bring the leading (western) leg closer to the equator and drive the following leg away from the equator (Fan *et al.*, 1994). Sunspot pairs have long been known to exhibit this sense of tilt, referred to as *Joy's law* (Hale *et al.*, 1919; Howard, 1991). A flux tube in the northern hemisphere will be tilted in the sense shown in Fig. 4.12: it will have a positive writhe.

To understand how a right-handed deformation, $Wr > 0$, of a northern flux tube is consistent with the observation of a left-handed twist, we must return to the magnetic helicity. The tube rises through a highly conductive plasma, so its evolution is almost certainly governed by the ideal induction equation. If the tube began as a planar curve with no twist then its total helicity would be $H_m = 0$. This value would be preserved even as the apex was deformed by the Coriolis effect. The tube would therefore have $Tw = -Wr$, which would be negative in the northern hemisphere.

A second example concerns the eruption of a twisted flux rope from the corona. Some models of solar filaments hold that the magnetic field surrounding them (the so-called filament channel) forms a twisted flux rope. When the twist is sufficiently great it can serve as a source of free energy which can drive flows to decrease the twist magnitude. This spontaneous plasma flow, known as a *kink instability*, is well documented in laboratory plasmas. The flow is ideal and hence will not change the helicity of the rope: $H_m = \Phi^2(Tw + Wr)$. Therefore a decrease in the magnitude of the twist must be compensated by an offsetting change in the writhe. The axis of the flux rope about a filament is initially straight: $Wr \simeq 0$. The spontaneous exchange of twist for writhe, the kink instability, will cause a helical deformation of the axis with the same handedness as the twist. This instability is reminiscent of the kinking of an elastic band when it is excessively twisted.

4.3.3 Relative helicity

The requirement that $\hat{\mathbf{n}} \cdot \mathbf{B} = B_n = 0$ on all boundaries makes the helicity equation (4.14) inapplicable to most volumes of interest in space physics. In order to extend this tool to situations where the field is normal to the boundaries, Berger and Field (1984) introduced the *relative helicity*. The helicity of a field $\mathbf{B}(\mathbf{x})$, inside a volume $\mathcal{V}$, is defined relative to a reference field $\mathbf{B}_0(\mathbf{x})$ which must have the same distribution of normal field on all boundaries:

$$(\mathbf{B} - \mathbf{B}_0) \cdot \hat{\mathbf{n}} = 0, \qquad \mathbf{x} \in \partial\mathcal{V}. \tag{4.27}$$

The relative helicity is then defined by the integral

$$H_R = \int_{\mathcal{V}} \mathbf{A} \cdot \mathbf{B} \, d^3x - \int_{\mathcal{V}} \mathbf{A}_0 \cdot \mathbf{B}_0 \, d^3x, \tag{4.28}$$

where $\mathbf{A}_0$ is the vector potential generating the reference field. This vector potential is required to match the tangential components of $\mathbf{A}$ at surfaces. This condition, $(\mathbf{A} - \mathbf{A}_0) \times \hat{\mathbf{n}} = 0$ on $\partial\mathcal{V}$, can be satisfied by performing the appropriate gauge transformations on $\mathbf{A}_0$. Provided this is done, H_R is invariant under gauge transformations as long as they do not violate the boundary condition on $\mathbf{A} - \mathbf{A}_0$.

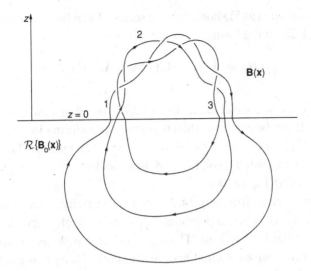

Fig. 4.14. An illustration of how a reference field defines a topological helicity, i.e. the relative helicity. The field $\mathbf{B}(\mathbf{x})$ is defined in the corona ($z > 0$) but its field lines have footpoints in the photosphere ($z = 0$). Applying the mirror-reflection operator, $\mathcal{R}$, to the reference field $\mathbf{B}_0(\mathbf{x})$ gives a field with which to continue the field lines below the photosphere ($z < 0$). This permits field lines to form closed curves, for which topological quantities such as the linking number are well defined. In this figure $L_{13} = -1$.

The need for a reference field is evident when one considers the topological interpretation of magnetic helicity. Equation (4.17) shows that helicity can be related to the linking number between pairs of field lines, a topological invariant. Two field lines with footpoints on the boundary ∂V are not, however, closed curves, so there is no unambiguous way of deciding whether they link one another. A reference field is a means of extending the field beyond the volume of interest, so that the field lines do not have footpoints. In the common situation where V is the coronal half-space $z > 0$, the field is extended by matching it to a mirror reflection of $\mathbf{B}_0$, as illustrated in Fig. 4.14. Applying Eq. (4.14) to this extended field and recalling the sign change under mirror reflection yields Eq. (4.28).

A common choice of reference field is the potential field, $\mathbf{B}_0 = -\nabla \chi$, since this is unique when all its normal components are prescribed.[†] The potential must be harmonic, $\nabla^2 \chi = 0$, and satisfy Neumann conditions at all boundaries, $\partial \chi / \partial n = -B_n$. This choice has the appealing property that for cases where Eq. (4.14) might be used, i.e. when $B_n = 0$ on all boundaries, the potential field is identically zero and the relative helicity, Eq. (4.28), gives the same value.

[†] Strictly speaking this is only true in simply connected volumes. Finn and Antonsen (1985) showed that in toroidal volumes (of great interest in laboratory plasma physics) it is also necessary to specify the toroidal flux of the reference field.

Finn and Antonsen (1985) showed that a series of algebraic steps could be used to rewrite Eq. (4.28) in the form

$$H_R = \int_V (\mathbf{B} - \mathbf{B}_0) \cdot (\mathbf{A} + \mathbf{A}_0)\, d^3x. \tag{4.29}$$

While mathematically equivalent to Eq. (4.28), the Finn–Antonsen equation is often preferred. It can be easily verified that gauge transformations to *either* vector potential, $\mathbf{A}$ or $\mathbf{A}_0$, will leave Eq. (4.29) unchanged. The tangential components of $\mathbf{A}_0$ therefore need not match those of $\mathbf{A}$ on the boundary when using Eq. (4.29), as they must when using Eq. (4.28).

It is particularly clear from Eq. (4.29) that the helicity of the reference field, relative to itself, is zero. When a potential field is used as reference then the relative helicity of a potential field is zero. The potential field is the field with minimum magnetic energy for a given normal field distribution. Using it as a reference field means that a non-vanishing relative helicity implies an energy above the minimum, i.e. non-zero free energy.

The potential field is not, however, the only reference field in use. Low (2006b) proposed a relative helicity which uses as a reference one component of the Chandrasekhar–Kendall decomposition of $\mathbf{B}(\mathbf{x})$. Moreover, even the term "potential field" can be variously interpreted as a field confined to a computational box or one extending to infinity in the lateral direction. One should always remain alert to the dependence on the reference field implicit in the relative helicity.

4.3.4 Helicity flux

The relative helicity has proven very useful in theoretical models; however, it is not possible to measure it directly since it integrates the magnetic field over an entire volume. It is much easier to measure its time rate of change, and this is now routinely done for the magnetic helicity of the solar corona (van Driel-Gesztelyi *et al.*, 2003).

If the ideal induction equation (3.24) holds within a volume V then the magnetic helicity, defined by (4.14), will not change regardless of the velocity field present. The relative helicity, however, can change if the normal field on the boundary changes. Provided that these changes also satisfy the ideal induction equation for a velocity at the boundary, the helicity change is (Berger and Field, 1984)

$$\frac{dH_R}{dt} = 2\int_{\partial V} (\mathbf{v} \cdot \mathbf{A}_0) B_n\, da \;-\; 2\int_{\partial V} (\mathbf{B} \cdot \mathbf{A}_0) v_n\, da, \tag{4.30}$$

where $v_n = \hat{\mathbf{n}} \cdot \mathbf{v}$ is the *outward-directed* normal velocity. The two terms are frequently interpreted respectively as the winding of the field and the advection of helicity out of the volume. Intuitively instructive as this may be, one must bear in mind that owing to the inherent gauge ambiguity of $\mathbf{A}$ the values of the individual terms are not well defined, even though their sum is. For the same reason it is not possible to interpret either integrand as the surface density of the helicity flux (Pariat *et al.*, 2005).

For the particular case of a coronal magnetic field occupying the half-space $z > 0$, and subject to purely horizontal flows, the helicity change is

$$\frac{dH_R}{dt} = -2 \int_{z=0} (\mathbf{v} \cdot \mathbf{A}_0) B_z \, dx \, dy. \tag{4.31}$$

Note that the outward normal points down, $\hat{\mathbf{n}} = -\hat{\mathbf{z}}$, so that, while it might appear that the term involving horizontal flow has changed its overall sign, in fact it has not. The potential field is a natural choice for the reference field since it is determined from B_z alone and can be written using a Green function:

$$\mathbf{A}_0(\mathbf{x}) = \frac{1}{2\pi} \int_{z'=0} \frac{\hat{\mathbf{z}} \times (\mathbf{x} - \mathbf{x}')}{|\mathbf{x} - \mathbf{x}'|^2} B_z(\mathbf{x}') \, dx' \, dy', \quad z = 0. \tag{4.32}$$

If maps of the vertical magnetic field $B_z(x, y, 0)$ and of the photospheric velocity $\mathbf{v}(x, y, 0)$ used are substituted into Eqs. (4.31) and (4.32) this yields an observational measure of the helicity change. The first map may be constructed from a magnetogram. The vertical is approximately the line of sight near the center of the solar disk, so the more common longitudinal magnetograms may be used there. Away from disk center, B_z may be constructed from the three components measured by vector magnetograms (Stenflo, 1994). Velocity maps are generally made by cross-correlating small areas from different times; this is called local correlation tracking (November and Simon, 1988). These maps have been combined to measure the helicity flux (and its integral, the helicity change) for numerous solar active regions over long intervals (Chae, 2001; Démoulin *et al.*, 2002; Kusano *et al.*, 2002; Nindos *et al.*, 2003; van Driel-Gesztelyi *et al.*, 2003).

Some understanding of helicity can be achieved by considering the simplified case where the photospheric field, $B_z(x, y, 0)$, is localized to circular unipolar patches, perhaps representing the cross sections of flux tubes. Each patch is assumed to undergo horizontal motion consisting of translation and rigid rotation at angular velocity ω_i. In this case Eq. (4.31) takes the form

$$\frac{dH_R}{dt} = -\frac{1}{2\pi} \sum_i \Phi_i^2 \omega_i - \frac{1}{2\pi} \sum_i \sum_{j \neq i} \Phi_i \Phi_j \frac{d\theta_{ij}}{dt}, \tag{4.33}$$

where Φ_i is the signed vertical flux in patch i and θ_{ij} is the polar angle of the separation between the centers of patches i and j.

The first term on the right of Eq. (4.33) is sometimes called the *spinning* contribution (Longcope *et al.*, 2007) since it quantifies the injection of helicity due to the spinning of each patch. A patch of either sign spinning in a counterclockwise sense ($\omega_i > 0$) will inject negative helicity. This is natural since this sense of spinning would create a left-handed twist ($Tw < 0$) in a flux tube anchored to it. One complete rotation, $\omega_i \Delta t = 2\pi$, will change the helicity by $\Delta H_R = -\Phi_i^2$. There have been many observations (Brown *et al.*, 2003, for example) of sunspots spinning at rates as high as 3 degrees per hour ($\omega_i \sim 2 \times 10^{-3}$ rad/s); this motion adds helicity to the active region corona above it.

The second term on the right of Eq. (4.33), the *braiding helicity flux*, quantifies the helicity change occurring as patches move around one another. An opposing pair moving around one another in a clockwise sense ($d\theta_{ij}/dt > 0$) will inject positive helicity. If each has equal flux and spins at the same rate and sense at which they rotate then their spinning and braiding contributions will cancel. The cancellation makes sense, because a flux tube connecting the patches would be subject to nothing more than rigid rotation. That does not change the internal geometry of the field and thus should not change the helicity. Therefore, in order to inject helicity, the two patches must spin and rotate differently. The full braiding term in Eq. (4.33) actually includes a much wider range of contributions, including contributions from the relative motion of like-signed sources. In the end it does, however, quantify the helicity change arising from the restricted range of motions considered.

5

Magnetic reconnection

TERRY G. FORBES

5.1 Preamble

The widespread interest in reconnection results from the fact that it is a fundamental process that occurs in magnetized plasmas whenever the connectivity of the field lines changes in time. Reconnection is most commonly associated with geomagnetic and solar activity because such changes in field line connectivity can be directly observed, but there are many other, less well-known, applications ranging from meteorites and comet tails to accretion disks and galactic jets. Reconnection is also found in laboratory devices that have been built to study the feasibility of controlled thermonuclear reactors, as well as in several experiments that have been specifically designed to study reconnection as a basic plasma process. Those aspects of magnetic reconnection that depend primarily on the topology of the magnetic field tend to be of universal application. However, aspects that depend on the detailed characteristics of the plasma itself, such as its temperature and density, tend to be restricted to the specific application where such characteristics occur. Thus, there is no universal theory that can be applied to all situations.

5.2 Basic concepts

The term *magnetic reconnection* was introduced by Dungey (1953a), who was interested in the problem of particle acceleration in the Earth's magnetosphere. Earlier studies (Giovanelli, 1946; Hoyle, 1949) had considered the acceleration of particles at magnetic neutral points in the presence of an electric field produced by plasma convection, but these studies did not include the magnetic field produced by the current associated with the motion of the particle. Using the framework of non-ideal magnetohydrodynamics (MHD), Dungey argued that this current would take the form of a thin sheet in which the diffusion of the magnetic field would necessarily dominate. Furthermore, this diffusion would cause field lines passing

113

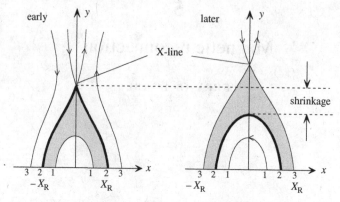

Fig. 5.1. The magnetic field line topology in the reconnection model of flare loops at two different times. The shaded area is the X-ray loop system, while the numbers at the base mark the footpoints of individual field lines. The outermost edges of the flare ribbons are located at X_R and $-X_R$.

through the current sheet to change their connectivity to one another. This process was described as field line "disconnection" followed by "reconnection".

5.2.1 Definitions

Magnetic reconnection can be defined quite generally as a change in the connectivity of field lines in time. This definition is based on the supposition that field lines can be uniquely identified and tracked in time. In an ideal plasma, i.e. one which satisfies the frozen-flux condition (see Section 3.2.3, the motion of an individual field line can be defined as being identical to the plasma motion in the direction perpendicular to the magnetic field. However, reconnection necessarily requires field lines to violate the frozen-flux condition so that they can change their connectivity, and when this happens it is not always possible to define the velocity or connectivity of a field line uniquely (Priest *et al.*, 2003). A tokomak plasma is an example of where problems can occur, although even here one can still talk about reconnection provided that there are well-organized flux surfaces. Generally reconnection is well defined in three dimensions if all the field lines map to surfaces, or localized regions, where the frozen-flux condition holds, because the connectivity of the field lines at any point in time can be uniquely established. Although it is possible to define a generalized reconnection process that does not require an X-type field line topology (see Section 5.3 below), all known applications to date do involve such a topology.

Before considering the full complications of reconnection in three dimensions, let us start by considering reconnection in two dimensions. Figure 5.1 shows a simple configuration derived from the standard two-dimensional model of a flare-loop system. The magnetic field is shown at two different times. The panel on the left shows a time after the flare has occurred when the field is starting to relax. The panel on the right shows a later time. In each panel the horizontal line at the bottom corresponds to the solar surface. In the model it is assumed that the surface can be represented by an infinitely conducting and stationary plate. The feet of all the field lines are anchored (i.e. frozen) to this plate, so that a field line mapping to a given point on the surface maintains its identity for all times. As the field lines relax, they move toward the X-line in matched pairs and undergo reconnection. This reconnection process causes the region of closed field lines to grow with time. Newly closed field lines have a cusp shape while field lines lower down have a rounded shape, and the newly closed field lines pull away from the X-line as it moves upwards. This downward motion is usually referred to as "shrinkage" in the solar context, while in the Earth's geomagnetic tail it is referred to as "dipolariza-tion". In the solar model the outermost edge of the flare-loop system lies along the field lines connecting the surface to the X-line. These lines constitute a separatrix as defined in the previous chapter, and in the model the footpoints of the separatrix correspond to the outer edge of the flare ribbons.

The field lines that have not yet reconnected can be described as "open" field lines, meaning that they extend upwards to infinity or to at least a very large distance. The field lines forming the loops below the X-line are said to be closed, meaning that they have a finite length and have both feet attached to the surface. In Fig. 5.1 the rate of reconnection is just the rate at which magnetic flux is transferred from the open region to the closed region. Quantitatively, we can express the rate of change of this flux as

$$\Phi_B = \frac{d}{dt} \int_0^{y_0(t)} B_x(0, y) dy = \frac{d}{dt} \int_0^{X_R(t)} B_y(x, 0) dx \qquad (5.1)$$

where y_0 is the location of the X-line on the y-axis. The time rate of change of this flux can be related to the electric field using Faraday's law (Eq. 10.9). The final result can be expressed as

$$\frac{d\Phi_B}{dt} = E_0 - E(X_R, 0) \qquad (5.2)$$

where E_0 is the electric field at the X-line and $E(X_R, 0)$ is the convective electric field $\mathbf{v} \times \mathbf{B}$ at the surface, at the footpoint of the field line that maps to the X-line (the separatrix). For solar flares the electric field, $E(X_R, 0)$, is negligible compared with the electric field at the X-line, so the rate of reconnection is related very simply to the apparent motion of the footpoint of the separatrix:

$$E_0 = B_y(X_0, 0) \dot{X}_0. \qquad (5.3)$$

The electric field E_0 at the X-line prescribes the rate at which magnetic flux is transferred from the open to the closed region. Since the flux crossing the y-axis below the X-line is the same as the flux between $x = 0$ and the separatrix footpoint at X_0, the electric field can be expressed simply as the vector product of the normal magnetic field at the separatrix footpoint and the apparent velocity of the footpoint across the surface.

For reconnection in the Earth's magnetotail, $E(X_R, 0)$ is not negligible and its value must be measured to determine the reconnection rate. Such measurements can be made from the ground, using radar signals reflected off the ionosphere to determine the convective flow and the corresponding convective electric field (see Blanchard *et al.*, 1996).

5.2.2 *MHD theory versus kinetic theory*

Reconnection can occur in a broad range of environments. Some, such as the convection zone in the outer layer of the solar interior, are completely dominated by binary particle collisions. However, in many cases of interest the plasma is nearly collisionless, i.e. the mean free path for binary collisions between particles is much greater than the characteristic scale length of the system. The Earth's magnetosphere is a classic example of such a collisionless system. Because MHD does not explicitly treat individual particle motions, one may jump to the false conclusion that it is of little use in collisionless plasmas. In reality, MHD and its two-fluid variants usually describe the global behavior of collisionless plasmas very well. For example, beyond ten solar radii the solar wind is a completely collisionless plasma, yet MHD models describe its global velocity, temperature, and density quite well, including such time-dependent aspects as shock disturbances, stream interactions, and turbulence.

Thus ideal MHD, and even non-ideal MHD, can successfully describe the average bulk properties of collisionless plasmas. Ideal (dissipationless) MHD embodies conservation principles, such as mass, momentum, and energy conservation, which are universal for both collisional and collisionless systems (Parker, 1996). Resistive MHD, which assumes that dissipation is a local property of the plasma, can be valid if wave–particle or wave–wave interaction prevent long-range interactions in the plasma.

Despite the past successes of MHD in explaining the dynamics of collisionless plasmas, however, there is good reason to be cautious when applying it to reconnection. By definition, reconnection cannot occur in ideal MHD because it depends on the diffusion of field lines through the plasma, often in small-scale structures such as current sheets. In this situation kinetic theory is essential for calculating the effective resistivity.

To determine a realistic resistivity for a collisionless plasma requires consideration of the generalized Ohm's law. For a fully ionized plasma it can be written as

$$\mathbf{E} = -\mathbf{v} \times \mathbf{B} + \frac{\mathbf{j}}{\sigma} + \frac{m_e}{ne^2}\left[\frac{\partial \mathbf{j}}{\partial t} + \nabla \cdot (\mathbf{vj} + \mathbf{jv})\right] - \frac{\mathbf{j} \times \mathbf{B}}{ne} - \frac{\nabla \mathbf{p}_e}{ne}, \quad (5.4)$$

where $\mathbf{vj}$ and $\mathbf{jv}$ are dyadic tensors and $\mathbf{p}_e$ is the electron stress tensor (Rossi and Olbert, 1970). The first term on the right-hand side of this equation is the convective electric field, while the second term is the field associated with Ohmic dissipation caused by electron–ion collisions. The conductivity σ is the inverse of the electrical resistivity η_e. The next three terms describe the effects of electron inertia and the next to the last term expresses the Hall effect. Ion inertia can be considered negligible because the large mass of the ions means that they do not contribute significantly to a change in the current density. Finally, the last term includes the electron gyroviscosity, which is considered by many to be important at a magnetic null (Strauss, 1986; Dungey, 1994). For a partially ionized plasma, collisions between charged particles and neutrals lead to additional terms associated with ambipolar diffusion.

Although all the terms on the right-hand side of the generalized Ohm's law (5.4), other than the first, allow field lines to slip through the plasma, they do not all produce dissipation. For example, the inertial terms do not cause the entropy of the plasma to increase. Thus, even though one may speak of inertial effects as creating an effective resistivity, this resistivity does not necessarily lead to dissipation.

Which terms are important in a particular situation depends not only upon the plasma parameters but also upon the length and time scales for variations of these parameters (Elliott, 1993; Sturrock, 1994). For magnetic reconnection, we normally want to know which non-ideal terms are likely to be significant within the current sheet where the frozen-flux condition is violated. Since each non-ideal (i.e. diffusion) term in the generalized Ohm's law contains either a spatial or temporal gradient, we can estimate the significance of any particular term by computing the gradient scale length L_0 required to make the term as large as the value of the convective electric field $\mathbf{v} \times \mathbf{B}$ outside the diffusion region.

Consider, for example, the three inertial terms $[\partial \mathbf{j}/\partial t + \nabla \cdot (\mathbf{vj} + \mathbf{jv})]$ on the right-hand side of the generalized Ohm's law. If we assume that $\nabla \approx 1/L_0$, $|\mathbf{j}| \approx B_0/(\mu L_0)$, and $\partial/\partial t \approx V_0/L_0$, say, where L_0 is a typical length scale and V_0 a typical velocity, then these three terms will be of the same order as the convective electric field if

$$\frac{m_e}{ne^2}\frac{V_0 B_0}{\mu L_0^2} \approx V_0 B_0.$$

In other words, in order for the inertial terms to be important in a current sheet, its thickness L_{inertia} should be given by

$$L_{\text{inertia}} \approx c \left(\frac{m_e \epsilon_0}{ne^2} \right)^{1/2} \approx \lambda_e, \tag{5.5}$$

where

$$\lambda_e = \frac{c}{\omega_{\text{pe}}} = \left(\frac{m_e}{ne^2 \mu} \right)^{1/2} = 5.30 \times 10^6 \, n^{-1/2} \tag{5.6}$$

is the electron inertial length or skin depth, $c = (\epsilon_0 \mu)^{1/2}$ is the speed of light, and $\omega_{\text{pe}} = [(e^2 n_e)/(\epsilon_0 m_e)]^{1/2}$ is the electron plasma frequency.

Similarly, for the Hall term $\mathbf{j} \times \mathbf{B}/(ne)$,

$$\frac{B_0^2}{ne\mu L_0} \approx V_0 B_0$$

or

$$L_{\text{Hall}} \approx \frac{c}{M_A} \left(\frac{\tilde{\mu} m_p \epsilon_0}{ne^2} \right)^{1/2} \approx \frac{\lambda_i}{M_A}, \tag{5.7}$$

where

$$\lambda_i = \frac{c}{\omega_{\text{pi}}} = \left(\frac{\tilde{\mu} m_p}{ne^2 \mu} \right)^{1/2} = 2.27 \times 10^8 \left(\frac{\tilde{\mu}}{n} \right)^{1/2} \tag{5.8}$$

is the ion inertial length or skin-depth. The Alfvén Mach number equals $M_A = V_0/V_A = V_0 B_0 (\mu \tilde{\mu} m_p n)^{-1/2}$, with $\tilde{\mu} = \overline{m}/m_p$ the mean atomic weight, $\omega_{\text{pi}} = [(q_i^2 n_i)/(\epsilon_0 m_i)]^{1/2}$ the ion plasma frequency, and V_A the Alfvén speed.

For the electron stress term $\nabla \mathbf{p}_e/(ne)$ we can write

$$\frac{nk_B T}{neL_0} \approx V_0 B_0$$

if we assume that $|\mathbf{p}_e| \approx nk_B T_e$ and $T_e \approx T_i \approx T$. Solving for L_0 leads to

$$L_{\text{stress}} \approx \frac{\beta^{1/2}}{M} R_{\text{gi}}, \tag{5.9}$$

where

$$\beta = nk_B T \left(\frac{2\mu}{B_0^2} \right) = 3.47 \times 10^{-29} \frac{nT}{B_0^2} \tag{5.10}$$

and

$$R_{\text{gi}} = \frac{(k_B T_i m_p \tilde{\mu})^{1/2}}{e B_0} = 9.49 \times 10^{-7} \frac{(T_i \tilde{\mu})^{1/2}}{B_0} \tag{5.11}$$

are the plasma β-parameter and the ion gyroradius, respectively.

Finally, for the collision term $\mathbf{j}/\sigma$,

$$\frac{B_0}{\mu\sigma L_0} \approx M v_A B_0,$$

where the product $(\mu\sigma)^{-1}$ is also the magnetic diffusivity η. Using Spitzer's formula for the collisional resistivity η_e of a plasma (see Eq. 3.19) we obtain

$$\eta_e = \frac{1}{\sigma} = \frac{(k_B m_e T_e)^{1/2}}{n e^2 \lambda_{mfp}}, \tag{5.12}$$

where

$$\lambda_{mfp} = 3(2\pi)^{3/2}\frac{(k_B T_e \epsilon_0)^2}{n e^4 \ln\Lambda} = 1.07 \times 10^9 \frac{T_e^2}{n \ln\Lambda} \tag{5.13}$$

is the mean free path for electron–ion collisions (Schmidt, 1966). Combining these expressions with those for the electron and ion inertial lengths we obtain

$$L_{collision} \approx \frac{\beta^{1/2}}{M}\frac{\lambda_e \lambda_i}{\lambda_{mfp}}. \tag{5.14}$$

Note that the length scale $L_{collision}$ of the spatial variations required to achieve significant field-line diffusion is inversely proportional to the mean free path λ_{mfp}. As λ_{mfp} increases, the diffusion caused by collisions becomes less effective, and increasingly sharper gradients are required to maintain the size of the dissipation term $\mathbf{j}/\sigma$.

Tables 5.1 and 5.2 list various plasma parameters along with the characteristic scale lengths for four different regions where reconnection is thought to occur. The parameter L_e is the global (external) scale size of the region, and the fundamental quantities from which the other parameters are derived are the density n, temperature T, and magnetic field B. For convenience, we assume that the Alfvén Mach number M_A is unity and that the electron and ion temperatures are roughly equal. The most extreme plasma environments listed in Table 5.1 occur in the magnetosphere, which is completely collisionless, and in the solar interior, which is highly collisional.

In addition to the parameters discussed above, Table 5.1 also lists the value of the *Debye length*

$$\lambda_D = \left(\frac{\epsilon_0 k_B T_e}{n e^2}\right)^{1/2} = 69.0 \left(\frac{T_e}{n}\right)^{1/2}. \tag{5.15}$$

The number of particles within a Debye sphere (i.e. $4\pi n\lambda_D^3/3$) must be larger than unity in order for the generalized Ohm's law to hold. Otherwise, the collective behavior which characterizes a plasma breaks down. The number of particles in a

Table 5.1. *Comparison of plasma parameters in different environments (in MKS units, i.e. length scales in* m, *n in* m^{-3}, *T in* K, *B in tesla, electric fields in* $V\,m^{-1}$*)*

Parameter	Laboratory experiments[a]	Terrestrial magnetosphere[b]	Solar corona[c]	Solar interior[d]
L_e	10^{-1}	10^7	10^8	10^7
n	10^{20}	10^5	10^{15}	10^{29}
T	10^5	10^7	10^6	10^6
B	10^{-1}	10^{-8}	10^{-2}	10^1
λ_D	10^{-6}	10^3	10^{-3}	10^{-10}
R_{gi}	10^{-3}	10^5	10^{-1}	10^{-4}
λ_i	10^{-2}	10^6	10^1	10^{-6}
$\ln \Lambda$	11	33	19	3
λ_{mfp}	10^{-2}	10^{16}	10^4	10^{-9}
β	10^{-2}	10^{-1}	10^{-4}	10^4
$L_u\,(\approx R_m)$	10^3	10^{14}	10^{14}	10^{10}
E_D	10^3	10^{-13}	10^{-2}	10^{11}
$E_A\,(= v_A B)$	10^4	10^{-2}	10^5	10^4
$E_{SP}\,(= E_A/\sqrt{R_{me}})$	10^2	10^{-9}	10^{-3}	10^{-2}

[a] MRX at Princeton Plasma Physics Laboratory.
[b] Plasma sheet.
[c] Above a solar active region.
[d] Base of the solar convection zone.

Table 5.2. *Diffusion lengths (in meters) from the generalized Ohm's law*

Characteristic length	Laboratory experiments[a]	Terrestrial magnetosphere[b]	Solar corona[c]	Solar interior[d]
$L_{inertia}(\lambda_e)$	10^{-4}	10^4	10^{-1}	10^{-8}
$L_{Hall}(\lambda_i)$	10^{-2}	10^6	10^1	10^{-6}
L_{stress}	10^{-3}	10^5	10^{-3}	10^{-2}
$L_{collision}$	10^{-4}	10^{-7}	10^{-7}	10^{-3}

[a] MRX at Princeton Plasma Physics Laboratory.
[b] Plasma sheet.
[c] Above a solar active region.
[d] Base of the solar convection zone.

Debye sphere for the environments shown in Table 5.1 ranges from 10^{14} for the magnetosphere to only about four for the solar interior at the base of the convection zone. Also shown in the table is the Lundquist number L_u, which is the same as the magnetic Reynolds number R_m when the flow and Alfvén speeds are the same. For a collisional plasma, the Lundquist number based on L_e can be expressed as

$$L_u = \frac{L_e v_A}{\eta} = \frac{L_e T_e^{3/2} B_0}{(\tilde{\mu}n)^{1/2} \ln \Lambda} 2.07 \times 10^8. \qquad (5.16)$$

Here η is the magnetic diffusivity, which is defined as $1/(\mu\sigma) = \eta_e/\mu$, where σ is the conductivity and η_e is the electrical resistivity. In the expression on the far right, η has been replaced by Spitzer's formula for the electrical resistivity of a collisional plasma (Priest and Forbes, 2000).

The characteristic scale lengths in Table 5.1 provide an indication of which terms in the generalized Ohm's law are likely to be important for the reconnection of current sheets. As with MHD shocks and turbulence, the large-scale dynamics of the flow causes the current sheet to thin until it reaches a length scale where field line diffusion is effective. Thus, in principle, the term with the largest characteristic length scale in Table 5.2 is the one that will be most important. Since the Hall term has the largest length in every environment except the solar interior, one might conclude that it is generally the most important. However, this conclusion does not take into consideration the fact that the Hall term tends to zero in the region of a magnetic null point or sheet. The Hall term on its own does not contribute directly to reconnection, since it freezes the magnetic field to the electron flow. To know whether a particular term is really as important as is suggested by its relative length scale requires a complete analysis of the kinetic dynamics, which is a rather formidable task. An excessively small length scale, however, does indicate that any process associated with that term is unlikely to be important. Therefore, on this basis we can conclude that collisional diffusion is not important in the terrestrial magnetosphere or the solar corona and that the electron inertial terms and the Hall term are not important in the solar interior.

Although the collision length scale $L_{collision}$ is equally small in both the magnetosphere and the corona, the general importance of the collisions in these two regions is quite different. In the magnetosphere the collision mean free path λ_{mfp} is nine orders of magnitude larger than the global scale size L, but in the corona it is four orders of magnitude smaller than the global scale. Thus, we can be confident that collisional transport theory applies to large-scale structures in the corona even though it is not applicable within thin current sheets or dissipation layers. By contrast, in the magnetosphere collisions are so few that collisional transport theory does not apply at any scale.

Another important issue concerning the applicability of collisional theory is the strength of the electric field, in a frame moving with the plasma. If this field exceeds the *Dreicer electric field*, defined by

$$E_{\mathrm{D}} = \frac{e \ln \Lambda}{4\pi \epsilon_0 \lambda_{\mathrm{D}}^2} = \frac{e^3 \ln \Lambda \, n}{4\pi \epsilon_0^2 k_{\mathrm{B}} T_{\mathrm{e}}} = 3.02 \times 10^{-13} \frac{n \ln \Lambda}{T_{\mathrm{e}}}, \qquad (5.17)$$

runaway acceleration of electrons will occur. The most likely location for the production of runaway electrons in a reconnection process is in a thin current sheet that forms at the null point. This field could be as large as the convective electric field based on the Alfvén speed, that is

$$E_{\mathrm{A}} = v_{\mathrm{A}} B_0 = 2.18 \times 10^{16} \frac{B_0^2}{(\tilde{\mu} n)^{1/2}},$$

or as low as the Sweet–Parker electric field

$$E_{\mathrm{SP}} = \frac{E_{\mathrm{A}}}{R_{\mathrm{m}}^{1/2}},$$

where R_{m} is the magnetic Reynolds number based on the inflow Alfvén speed (i.e. the inflow Lundquist number). As shown in Table 5.1, the Dreicer field in the magnetosphere is much smaller than E_{A} or E_{SP}, so runaway electrons will always be generated by reconnection there. However, in the solar interior the Dreicer field is so large that the runaway acceleration of electrons never occurs. In the intermediate regimes of the laboratory and the solar corona the Dreicer field lies between E_{A} and E_{SP}, so perhaps runaway electrons are only produced when very fast reconnection occurs.

Even in completely collisionless environments such as the Earth's magnetosphere, it is still sometimes possible to express the relation between electric field and current density in terms of an anomalous resistivity. For example, Lyons and Speiser (1985) showed that the electron inertial terms in the generalized Ohm's law lead to an anomalous resistivity

$$\frac{1}{\sigma^*} = \frac{\pi B_z}{2ne},$$

where B_z is the field normal to the current sheet. This resistivity is derived solely from a consideration of the particle orbits, and in the magnetotail current sheet it may be larger than any anomalous resistivity due to wave–particle interactions. A typical example of the latter is the anomalous resistivity due to ion acoustic waves (Priest, 1982; Benz, 1993).

5.3 Reconnection in two dimensions

In this section we outline basic two-dimensional approaches to magnetic reconnection, using on resistive MHD theory. More detailed information can be found in

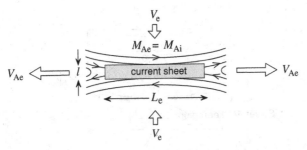

Fig. 5.2. The Sweet–Parker field configuration. Plasma (open arrows) flows into the upper and lower sides of a current sheet of length L_e but must exit through the narrow tips of the sheet, of width l. Because the field is assumed to be uniform in the inflow region, the external Alfvén Mach number $M_{Ae} = v_e/v_{Ae}$ at a large distance is the same as the internal Alfvén Mach number M_{Ai} at the midpoint edge of the current sheet.

the original papers and in recent books by Biskamp (2000) and Priest and Forbes (2000).

5.3.1 Steady-state models

A few years after Dungey's introduction of the concept of reconnection, Sweet (1958) and Parker (1957) developed the first quantitative model. In order to make the analysis as analytically tractable as possible, they focused on the problem of two-dimensional steady-state reconnection in an incompressible plasma. They assumed that reconnection occurs in a current sheet whose length is set by the global scale L_e of the field, as shown in Fig. 5.2. Under these conditions they determined that the speed of the plasma flowing into the current sheet is approximately

$$v_e = v_{Ae} L_u^{-1/2} \tag{5.18}$$

where $L_u = \mu_0 L_e v_{Ae}/\eta$ is the Lundquist number and η is the magnetic diffusivity. The global scale length is L_e, and $v_{Ae} = B_e/\sqrt{\mu_0 \rho_e}$ is the Alfvén speed in the inflow region. Sometimes L_u is referred to as the magnetic Reynolds number, although the latter term is normally used for a similar number based on a typical flow speed rather than the Alfvén speed. The outflow speed of the plasma from the current sheet is V_{Ae} and does not depend on the value of L_u. The reconnection rate in two dimensions is measured by the electric field at the reconnection site. This electric field is perpendicular to the plane of Fig. 5.2 and prescribes the rate at which magnetic flux is transported from one topological domain to another. In two-dimensional (2D) steady-state models this electric field is uniform in space. Therefore the Alfvén Mach number, $M_{Ae} = v_e/v_{Ae}$, provides a quantitative measure of the reconnection

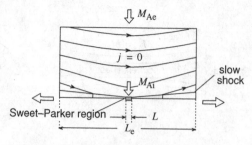

Fig. 5.3. Petschek's field configuration. Here the length L of the Sweet–Parker current sheet is much shorter than the global scale length L_e, and the magnetic field in the inflow is non-uniform. Two pairs of standing slow-mode shocks extend outwards from the central current sheet. Petschek's model assumes that the current density in the inflow region is zero and that there are no external sources of field at large distance.

rate relative to the characteristic electric field $v_{Ae} B_e$. In terms of this number, the Sweet–Parker reconnection rate is just $M_{Ae} = L_u^{-1/2}$.

In astrophysical and space plasmas L_u is very large ($L_u \gg 10^6$), so Sweet–Parker reconnection is usually too slow to account for phenomena such as geomagnetic substorms or solar flares. Petschek (1964) proposed a model with an increased rate of reconnection resulting from the use of a current-sheet length greatly reduced from that of the Sweet–Parker model. He did this by encasing their current sheet, in an exterior field with global scale length L_e. He also introduced two pairs of standing slow-mode shocks radiating outwards from the tip of the current sheet, as shown in Fig. 5.3. In Petschek's solution most of the energy conversion comes from these shocks, which accelerate and heat the plasma to form two hot outflow jets.

Petschek also assumed that the magnetic field in the inflow region was current-free and that there were no sources of field at large distances. These assumptions, together with the trapezoidal shape of the inflow region created by the slow shocks, lead to a logarithmic decrease in the magnetic field as the inflowing plasma approaches the Sweet–Parker current sheet. This variation in the field leads in turn to Petschek's formula for the maximum reconnection rate, namely

$$M_{Ae[Max]} = \pi / (8 \ln L_u) \qquad (5.19)$$

where L_u is again the Lundquist number and M_{Ae} is the Alfvén Mach number in the region far upstream of the current sheet, as shown in Fig. 5.3. Because of its logarithmic dependence on L_u, the Petschek reconnection rate is many orders of magnitude greater than the Sweet–Parker rate, and for most space and laboratory applications Petschek's formula predicts that $M_{Ae} \approx 10^{-1}$ to 10^{-2}.

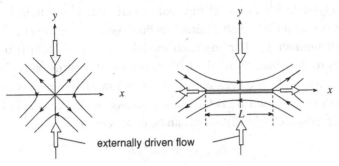

externally driven flow

Fig. 5.4. Syrovatskii's field configuration. Unlike Petschek's configuration, this has external sources which produce an X-type configuration even when local sources of current are absent. The application of external driving to an X-type configuration (left-hand panel) creates a current sheet (right-hand panel) whose length L depends on the temporal history of the driving and the rate at which reconnection operates. The fastest reconnection rate occurs when L is equal to the external scale length L_e.

Petschek's model uses the Sweet–Parker model to describe the flow of plasma and fields in the diffusion region. Because the Sweet–Parker model only gives average properties for this region, such as its length and thickness, no detailed matching is possible between the flows in the diffusion region and the flows in the external region outside. This lack of detailed matching is sometimes misunderstood to mean that there is no matching at all (Biskamp, 2000), but in fact the average properties of the diffusion region are rigorously matched to the external region to the extent that the Sweet–Parker model allows (Vasyliūnas, 1975b).

It is not always appreciated that Petschek's reconnection model is a particular solution of the MHD equations which applies only when special conditions are met. First, it requires that the flows into the reconnection region arise spontaneously without external forcing (Forbes, 2001). In general, driving the plasma externally creates a significant current density in the inflow region, and this violates Petschek's assumption that the inflow field is approximately potential. Second, Petschek's solution also requires that there be no external source of field in the inflow region. In other words, the field in question must be just the field produced by the currents in the diffusion region and the slow shocks. In many applications of interest neither of these conditions is met.

An alternative approach to reconnection in current sheets was pioneered by Green (1965) and Syrovatskii (1971), who considered what happens when a weak flow impinges on an X-line in a strongly magnetized plasma, as indicated in Fig. 5.4. The imposed flow creates a current sheet which achieves a steady state when the rate of field line diffusion through the sheet matches the speed of the flow. A quantitative model of this process was published by Somov (1992).

For a steady-state MHD model the spatial variation of the field in the inflow region is the key quantity which determines how the reconnection rate scales with the Lundquist number L_u. For any such model, the electric field is uniform and perpendicular to the plane of the field. Thus outside the diffusion region we have $E_o = -v_y B_x$, where E_o is a constant, v_y is the inflow along the axis of symmetry (the y-axis in Fig. 5.4), and B_x is the corresponding field. Thus the inflow Alfvén Mach number, M_{Ae}, at large distances can be expressed as

$$M_{Ae} = M_{Ai} B_i^2 / B_e^2 \tag{5.20}$$

where M_{Ai} is the Alfvén Mach number at the current sheet, B_i is the magnetic field at the edge of the current sheet, and B_e is the magnetic field at a large distance.

In Syrovatskii's model the field along the inflow axis of symmetry is

$$B_x = B_i (1 + y^2/L^2)^{1/2} \tag{5.21}$$

where B_i is the field at the current sheet, y is the coordinate along the inflow axis, and L is the length of the current sheet. Combining (5.21) with (5.20) yields

$$M_{Ae} = M_{Ai} \frac{1}{1 + L_e^2/L^2}, \tag{5.22}$$

which has its maximum value when $L = L_e$. Thus the maximum reconnection rate in Syrovatskii's model scales as $L_u^{-1/2}$, the same as for the Sweet–Parker model.

By comparison, the field in Petschek's model along the inflow axis varies approximately as

$$B_x = B_i \frac{1 - (4/\pi) M_{Ae} \ln(L_e/y)}{1 - (4/\pi) M_{Ae} \ln(L_e/l)} \tag{5.23}$$

where l is the current sheet thickness. (This expression for the field is only a rough estimate; the actual variation in the region $y < L$ (Vasyliūnas, 1975b; Priest and Forbes, 2000) is more complex.) Evaluating B_x at $y = L_e$ and substituting the result into Eq. (5.20) gives

$$M_{Ai} = M_{Ae}[1 - (4/\pi) M_{Ae} \ln(L_e/l)]^{-2}. \tag{5.24}$$

The Sweet–Parker theory can be used to eliminate L_e/l, so as to obtain an expression for M_{Ae} as a function of L_u. This expression has a maximum value, given by Eq. (5.19).

The spatial variation in the field in the inflow region is given for the Syrovatskii and the Petschek models in Fig. 5.5. Although both fields increase with distance away from the current sheet, the rate at which they increase is markedly different. At large distances the rate of increase in the Syrovatskii model is dominated by the external field, whose variation is fixed and independent of the reconnection rate. By contrast, the variation in the Petschek model is closely coupled to the reconnection

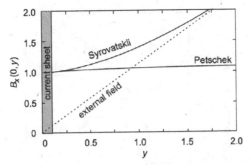

Fig. 5.5. The variation in the magnetic field, in the inflow region of Petschek's and Syrovatskii's models, along the axis of symmetry (the y-axis). At large distances the variation in Syrovatskii's model is determined by external field sources at infinity. The magnetic fields are normalized to their values at the edge of the current sheet, and the distance y is normalized to the length L of the current sheet. For the Petschek curve, $M_{Ae} = 0.02$, $l = 0.1L$, and $L_e = 2L$.

rate, disappearing altogether when the rate goes to zero. This is one of the main reasons why the two models give such different predictions for the reconnection rate. It also explains why the numerical simulation by Biskamp (1986) of the evolution shown in Fig. 5.4 produces a Sweet–Parker rather than a Petschek-type scaling.

Even in circumstances where Petschek's model would be expected to apply it apparently does not. Several numerical simulations (Biskamp, 1986; Scholer, 1989) were carried out in an attempt to verify the steady-state solution found by Petschek (1964), but none of these simulations has been able to replicate the scaling results that are predicted by Petschek's solution if the resistivity is kept uniform and constant. Only when a non-uniform, localized, resistivity model (Ugai, 1988; Yan et al., 1992) is used does the Petschek configuration appear. The fact that the resistivity apparently needs to be non-uniform does not contradict Petschek's model because the model makes no explicit assumption about whether the resistivity is uniform or not. It is equally valid for both cases because it assumes only that the region where resistivity is important is localized. The numerical experiments carried out to date imply that the diffusion region can be localized only by enhancing the resistivity near the X-line. Whether there might be other ways to localize the diffusion region (e.g. assuming a non-uniform viscosity) remains unknown.

Although Petschek assumed that the current density j in the inflow region was zero to first order, it is not actually necessary to make such an assumption to obtain a solution. More generally, j can be non-zero to first order in the expansion of the inflow equations, so that the inflow magnetic field is no longer determined by solving Laplace's equation ($\nabla^2 A = 0$) for the vector potential A but by solving Poisson's equation ($\nabla^2 A = -\mu_0 j$) instead (Priest and Forbes, 1986). The relaxation of the assumption that j is zero introduces an additional degree of freedom,

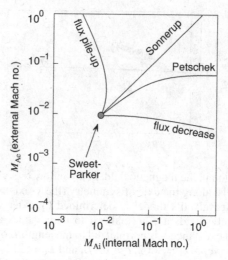

Fig. 5.6. External Alfvén Mach number M_{Ae} vs. the internal Alfvén Mach number M_{Ai}, for the family of solutions obtained by Priest and Forbes (1986). These solutions were obtained by an expansion in terms of the inflow Alfvén Mach number for small variations of the field around the uniform inflow field assumed in the Sweet–Parker model. Solutions with the labeled characteristics are obtained for different choices of the parameters describing the distant boundary conditions.

so that there is now a family of solutions (Fig. 5.6). These solutions can be summarized in terms of the relation between M_{Ai}, the internal Alfvén Mach number at the entrance to the diffusion region, and M_{Ae}, the Alfvén Mach number at the exterior inflow boundary (Fig. 5.3). The relation is

$$\frac{M_{Ae}^{1/2}}{M_{Ai}^{1/2}} = 1 - \frac{4}{\pi} M_{Ae} (1 - b) \left[0.834 - \ln \tan \left(\frac{\pi}{4} L_u^{-1} M_{Ae}^{-1/2} M_{Ai}^{-3/2} \right) \right]$$

(5.25)

where b is a parameter whose value corresponds to the assumptions made about the inflow boundary conditions at $y = L_e$. The relation (5.25) is plotted in Fig. 5.6 for $S = 100$ for various values of b. When $b = 0$ Petschek's solution is obtained, and when $b = 1$ a solution equivalent to that of Sonnerup (1970) is obtained. When $b < 1$, the solution somewhat resembles Syrovatskii's solution in that the magnetic field increases markedly with distance away from the current sheet. For these solutions the maximum reconnection rate is the same as for the Sweet–Parker model.

As b increases beyond unity, a flux-pile-up regime occurs where the magnetic field increases as the diffusion region current sheet is approached (Fig. 5.7). For a very strong flux pile-up with $b \gg 1$, the flow approaches the MHD stagnation-point-flow solution found by Parker (1973) and Sonnerup and Priest (1975). The stagnation-point flow appears to be very fast since formally there is no limit to

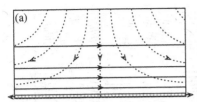 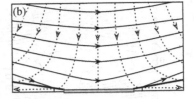

Fig. 5.7. (a) The stagnation-point-flow solution obtained by Parker (1973) for magnetic field annihilation at a current sheet, and (b) the closely related flux-pile-up solution obtained by Priest and Forbes (1986).

the value of M_{Ae}. However, large values of M_{Ae} require large variations in the gas pressure, which are not possible unless the plasma β is very much greater than unity. For low-β plasmas the amount of pile-up is limited, and when this limitation is taken into consideration the reconnection rate is found to scale at the relatively slow Sweet–Parker rate (Litvinenko et al., 1996; Priest, 1996; Litvinenko and Craig, 1999). For a low-β plasma the fastest rate occurs for $b = 0$, which is Petschek's solution.

Vasyliūnas (1975b) was the first to point out that the differences between reconnection solutions are often related to the behavior of the gas and magnetic pressures in the inflow region. The inflow can be characterized as undergoing a compression or an expansion depending on whether the gas pressure increases or decreases as the plasma flows in towards the X-point. These compressions or expansions can be characterized further as being of the fast-mode type or the slow-mode type, depending on whether the magnetic pressure changes in the same sense as the gas pressure (fast-mode type) or in the opposite sense (slow-mode type). For the family of solutions above one finds that in Petschek's solution ($b = 0$) the gas pressure is uniform to second order in the expansion parameter M_{Ae} so that, to this order, the plasma is neither compressed nor rarefied as it flows towards the X-point. For solutions with $b < 0$ the plasma undergoes a slow-mode compression, while for $b > 1$ it undergoes a slow-mode expansion. Between $b = 0$ and $b = 1$ the solutions have a hybrid character; slow-mode and fast-mode expansions exist in different regions of the inflow. It is also possible to have hybrid solutions in which slow-mode expansions and slow-mode compressions co-exist in different regions of the inflow (Strachan and Priest, 1994), although this does not happen for the above family of solutions.

5.3.2 Two-dimensional time-dependent reconnection

So far, we have only considered the development of steady-state models of reconnection. However, for many phenomena (such as flares and geomagnetic substorms)

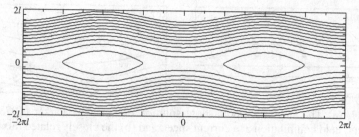

Fig. 5.8. Typical magnetic island structure resulting from the tearing instability of a plane current sheet. The contour lines show magnetic flux surfaces of the magnetic field (Forbes, 2006).

reconnection occurs on such short time scales that a steady or quasi-steady state does not exist. For example, in high-speed coronal mass ejections (with velocities greater than about 1000 km/s) the current sheet created by the ejection grows in length at a speed which is of the order of, or in excess of, the ambient Alfvén wave speed (Lin and Forbes, 2000). Steady-state reconnection is primarily confined to situations where the external field and the flows vary slowly compared with the characteristic Alfvén time scale of the system.

Time-dependent reconnection was first considered by Dungey (1958), who noted that motions in the vicinity of an X-line in a strongly magnetized plasma can lead to the very rapid formation of a current sheet. The first explicit solution demonstrating this possibility was published by Imshennik and Syrovatskii (1967). They found that if the gas pressure is negligible and the resistivity is small then, during the initial formation of the sheet, the electric field E at the X-line grows at a rate proportional to $(t_c - t)^{4/3}$ where t_c is about twice the Alfvén time scale of the system. At the time t_c the electric field becomes infinite, but the assumptions underlying the solution break down before this time is reached. Although there have been several analyses and extensions of this solution (Priest and Forbes, 2000), little effort has been made to apply this theory to highly dynamic phenomena such as flares.

Most theoretical effort on time-dependent reconnection has concentrated on the *tearing instability* (Furth et al., 1963), illustrated in Fig. 5.8. This is a non-ideal instability in which the magnetic reconnection of field lines plays a central role. Tearing has been invoked in some flare models as a mechanism for releasing magnetic energy (Heyvaerts et al., 1977) but, considered as a possible flare mechanism, it suffers from the fact that resistive tearing is relatively slow (Steinolfson and Van Hoven, 1984). The onset of most flares occurs over a time period on the order of the Alfvén time scale in the corona, but the time scale of the tearing mode is much slower, being a combination of the Alfvén time scale and the much slower resistive time scale. The actual growth rate depends on the wavelength of the perturbation.

For a simple current sheet the growth rate is zero if the wave number k is such that $kl > 1$, where l is the width of the sheet. This means that a current sheet which is shorter than 2π times its width is stable to tearing. For $L_u^{1/4}kl < 1$, small perturbations grow exponentially on a time scale of

$$\tau_{tm} = (kl)^{2/5} L_u^{3/5} \tag{5.26}$$

where τ_{tm} is the growth period of the tearing mode and L_u is the Lundquist number based on the current sheet thickness and the external Alfvén speed. The fastest growing mode occurs when $kl \approx L_u^{-1/4}$ and the corresponding growth rate time scale is $L_u^{1/2}$. For values of kl less than $L_u^{-1/4}$, this expression is not valid.

The threshold condition $kl < 1$ for tearing is a consequence of the fact that the field lines of the initial current sheet resist being bent. An initial perturbation which has a relatively short wavelength tends to straighten out before significant reconnection can occur but, as the wavelength of the perturbation increases, there is more time for diffusion to act. This diffusion occurs in a thin layer in the center of the sheet having a thickness of order $L_u^{-2/5}l$ for the shortest wavelength mode and $L_u^{-1/4}l$ for the fastest growing mode.

The growth and stability of the tearing mode can be affected by many factors, such as the geometry of the sheet, line-tying, the presence of a guide field (a magnetic field component in the direction of the current), externally driven flows, the mechanisms of magnetic diffusivity, and so on. A discussion of such effects can be found in Galeev (1979), Priest and Forbes (2000), and Wesson (1987).

Although a large body of literature is devoted to 2D reconnection, there are still some fundamental questions that remain unanswered. For example, what is the rate of reconnection in a rapidly driven system where the current sheet grows at a rate of the order of the ambient Alfvén speed? This situation occurs in large coronal mass ejections whose speed typically exceeds the local Alfvén wave speed. Also, the effects of strong radiation, thermal conduction, and partial ionization have been only partially explored.

5.4 Reconnection in three dimensions

The addition of a third dimension introduces new reconnection properties and behaviors that are fundamentally different from those in two dimensions. Perhaps the most significant of these is that reconnection can occur in configurations where all field lines have the same topology. That is, the mappings of all field lines of the system are topologically identical. In such configuration the concept of a moving field line becomes ambiguous (Priest *et al.*, 2003).

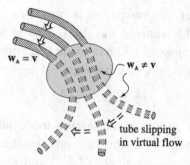

Fig. 5.9. Schematic diagram of the ambiguity that occurs in the definition of field line velocity in three dimensions. The wide broken lines indicate the projection of magnetic flux tubes onto the two-dimensional plane (from Priest *et al.*, 2003).

Figure 5.9 illustrates the situation. Here it is supposed that an ideal conductor, having zero electrical resistivity, surrounds a non-ideal region where reconnection occurs. We suppose for simplicity that there is no fluid flow in the ideal region. Because of the reconnection occurring within the non-ideal region, the mapping of the field lines from their entry point to their exit point changes with time. In three dimensions the change in the connectivity of the field lines can occur in a distributed way, so that any given field line is continually slipping, changing its point of connection to the boundary. Consequently, if one makes a movie of the field line motion by tracing out the field lines at their incoming boundary point, these field lines will appear to move not only within the non-ideal region, but also within the ideal region lying beyond the other side of the non-ideal region. This "virtual" motion in the ideal region occurs even though the fluid in the exterior region is at rest. Thus the virtual motion contradicts the normal requirement that the field lines in a stationary ideal conductor should be at rest. This duality in the definition of the field line velocity is also evident if one simply considers using the exit point of the field line, rather than its entrance point, to trace out the field lines. In this case the field lines on the outgoing side will appear to be at rest while the field lines on the incoming side will appear to move (Priest *et al.*, 2003). Such behavior cannot occur in two dimensions.

5.4.1 *Definitions revisited*

In two dimensions, linear null points fall into two types, namely X-points and O-points, where the neighboring magnetic field lines have X-type or O-type topology, respectively (see the previous chapter). The notion of magnetic reconnection is then fairly straightforward. Reconnection occurs at an X-point, where two pairs of separatrices meet; during the process of reconnection pairs of magnetic field lines are advected towards the X-point, where they are disconnected and then reconnected. The mapping of magnetic field lines from one boundary to another is discontinuous and, therefore, during the process of reconnection there is a sudden change in the magnetic connectivity of plasma elements owing to the presence of a localized diffusion region where ideal MHD breaks down.

These features of two-dimensional reconnection carry over into three dimensions, provided that there is a well-defined topology. Such a topology will exist if there are appropriate null-point pairs or localized point sources on the boundary (see Chapter 4). However, in many cases of interest an exact topology does not exist. This is especially the case in the solar corona, where reconnection is thought often to occur when null points are not present (Démoulin *et al.*, 1997). To cope with such situations, Schindler *et al.* (1988) proposed that the definition of reconnection should be extended to include all effects of local non-idealness that generate a component ($E_{\parallel}$) of the electric field along a magnetic field. In other words if a localized region exists where

$$\int E_{\parallel} \, ds \neq 0 \qquad (5.27)$$

then reconnection is occurring. Some authors (see Priest and Forbes, 2000) have argued that such a definition is too general, since it includes examples of magnetic diffusion or slippage (such as in double layers or shock waves) that have not been traditionally included in the concept of reconnection.

If nevertheless, we accept the idea that reconnection occurs whenever there is an electric field parallel to the magnetic field then the question arises of how the rate of reconnection is to be defined. In the two-dimensional example of Section 5.3.1, the rate is given by the electric field at the X-line. In the absence of an exact topology and in the presence of a distributed parallel electric field, it is no longer obvious how one should define the rate. However, for the three-dimensional example of Fig. 5.10, such a rate must exist since there is a definite value of flux that is changing its connectivity as a function of time. Hesse *et al.* (2005) showed that the reconnection rate, i.e. the rate at which magnetic flux is newly connected, is given by the maximum value of the electric potential obtained by integrating the parallel electric field along the magnetic field line.

5.4.2 Reconnection in fields without distinct topological mappings

In this section we provide a kinematic example of how the generalized definition of reconnection can be applied to a three-dimensional configuration that has no exact topology. (The term "kinematic" refers to the fact that the model does not incorporate the continuity, momentum, and energy equations as would an MHD

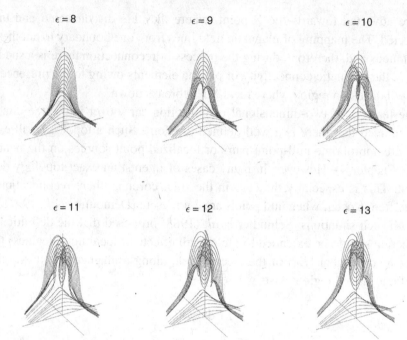

$\epsilon = 8$ $\epsilon = 9$ $\epsilon = 10$

$\epsilon = 11$ $\epsilon = 12$ $\epsilon = 13$

Fig. 5.10. Evolution of the magnetic field in the Hesse *et al.* (2005) model for line-tied reconnection in the corona.

model.) The magnetic field model for this example is (from Hesse *et al.*, 2005)

$$B_x = -1 - \epsilon \frac{1 - (y/L_y)^2}{1 + (y/L_y)^2} \frac{1}{1 + (z/L_z)^2}, \tag{5.28}$$

$$B_y = \frac{1}{4}x, \tag{5.29}$$

$$B_z = \frac{1}{5}. \tag{5.30}$$

Here L_y and L_z are arbitrary scale lengths, which are both set to 5 in the model. This magnetic field does not vanish anywhere, and therefore it does not have the separatrix surfaces or separator line that would exist if null points were present. The effects of reconnection on the magnetic field are simulated by varying the parameter ϵ which determines the amount of flux that has been reconnected at a given time. Figure 5.10 displays an overview of the magnetic field structure for different values of ϵ. The starting value was $\epsilon = 0$, which corresponds to a sheared magnetic arcade. Increasing ϵ leads to the formation of a flux rope that lies above the initial arcade.

From Faraday's equation (Eq. 10.9) we can calculate the inductive electric field associated with the changing magnetic field of the model. By integrating the component of this field parallel to the magnetic field, we obtain the electric potential shown in Fig. 5.11. This potential is not necessarily the total potential; it represents

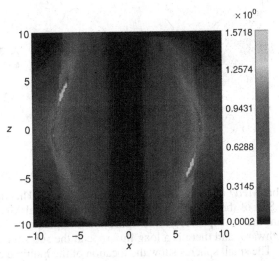

Fig. 5.11. Mapping onto the $z = 0$ surface of the net electric potential produced by induction in the Hesse *et al.* (2005) model for $\epsilon = 10$ in Eq. (5.28).

only the potential caused by the motion of the magnetic field. For this reason it is sometimes referred to as a pseudo-potential (see Hesse *et al.*, 2005). The resulting distribution of the potential on the surface is reminiscent of the chromospheric ribbons produced by solar flares in decaying active regions where the geometry of the magnetic field is similar to that used in the model. At any given time the rate of reconnection is given by the peak value of the inductive potential.

5.4.3 *Numerical simulations of three-dimensional reconnection*

A large number of fully three-dimensional (3D) MHD simulations have been carried out in which reconnection occurs. However, in many of these simulations the focus has been not on the reconnection process itself but on some other phenomenon or structure in which reconnection plays an important, but perhaps secondary, role. In the last few years there has been an increasing interest in determining the exact nature of the reconnection process occurring in these simulations and in carrying out new simulations specifically designed to determine the nature of 3D reconnection. Figure 5.12 shows the magnetic configuration that occurs in a global simulation of the interaction between the solar wind and the terrestrial magnetic field (Dorelli *et al.*, 2007) for a southward orientation of the interplanetary magnetic field. For a northward magnetic field, the magnetic topology is relatively simple, consisting of a two magnetic nulls with a single separator line between them. However, for the southward case shown in the figure, a more complex situation occurs with many nulls (indicated by the small spheres) and many separator lines connecting these nulls.

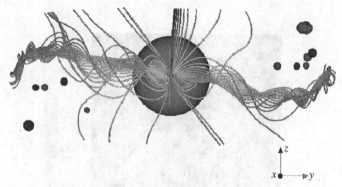

Fig. 5.12. Global MHD modeling of Earth's magnetosphere. This figure shows a view from the Sun of the dayside magnetopause from a global MHD simulation of the terrestrial magnetosphere. In this figure the interplanetary magnetic field is oriented southward, and there is a long flux rope at the sub-solar region of the magnetopause. The small spheres show the location of the multiple neutral points that form. (Courtesy of J. Dorelli.)

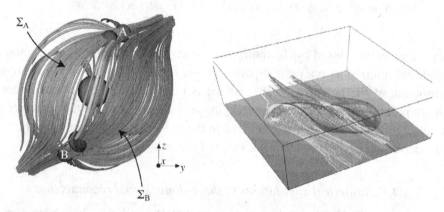

Fig. 5.13. Magnetohydrodynamic simulations of separator reconnection. (Left) A view from the Sun of the Earth's dayside magnetopause for a purely northward IMF, with the Earth's dipole tilted in the GSE yz-plane by $45°$ (compare with the bottom panel in Fig. 4.3 showing a potential field model in the case where the Earth's dipole field is aligned with the z-axis). A separator line (see Fig. 4.2 for the detailed geometry of such a null–null line) extends across the magnetopause and terminates in the polar cusp neutral points that are indicated by the spheres (from Dorelli *et al.*, 2007). (Right) A simulated solar coronal magnetic skeleton. Two separatrix surfaces intersect to form three separator lines (from Maclean *et al.*, 2006; Hayes *et al.*, 2007).

Idealized simulations are sometimes carried out in order to understand better the complex interactions that occur in more realistic global simulations. Figure 5.13 shows two examples of what is known as separator-type reconnection (Priest and Forbes, 2000). This type of reconnection is common to both the terrestrial magnetosphere (left-hand panel) and the solar corona (right-hand panel). The example in the left-hand panel is an idealized version of the reconnection process in the Earth's magnetosphere (Dorelli *et al.*, 2007), while the example in the right-hand panel is an idealized version of the reconnection process in compact (i.e. non-eruptive) solar flares (Maclean *et al.*, 2006). The topologies of the magnetic fields shown in both cases are complicated and inherently three dimensional, but many aspects of this 3D complexity are common to both situations.

Three-dimensional numerical simulations have also been carried out to test various 3D analytical models. An example of this type of simulation is the one carried out by Heerikhuisen and Craig (2004) to test a 3D analytical solution obtained by Craig *et al.* (1995). This was is an exact analytical solution of the incompressible resistive MHD equations, and it is closely related to the stagnation-point-flow (Sonnerup and Priest, 1975) and flux-pile-up solutions (Priest and Forbes, 2000) mentioned earlier in this chapter. There has been considerable discussion over the last 30 years about the physical significance of these types of solution. Because the mathematical expression for the reconnection rate in these solutions is independent of the electrical conductivity of the plasma, some authors (e.g. Sonnerup and Priest, 1975) have argued that they correspond to fast reconnection. Other authors (Litvinenko *et al.*, 1996), however, argued that these solutions correspond to very slow reconnection because of the assumption of incompressibility. The numerical simulations have shown that the latter conclusion is correct. Nevertheless, reconnection of this type has been observed to occur in global MHD simulations of the magnetosphere when the interplanetary magnetic field is northward (see Dorelli *et al.*, 2007).

Finally, several 3D numerical simulations have been carried out in conjunction with laboratory experiments specifically designed to study the kinetic aspect of the reconnection process. A recent review of these experiments was given by Yamada (2007).

5.5 Topics for future research

Many aspects of magnetic reconnection have yet to be explored. Even long-studied topics such as steady-state two-dimensional reconnection are not fully understood. Many questions remain about how time-dependent reconnection works in impulsively driven phenomena such as solar flares and geomagnetic substorms. For example, during the impulsive phase of eruptive solar flares the current sheet where reconnection occurs can grows at a rate that exceeds the Alfvén time scale of the system. This rapid growth means that no steady-state reconnection theory applies during the impulsive phase, and there are virtually no theories that predict how the reconnection rate scales with the plasma resistivity in such a situation.

Despite growing evidence that the reconnection process in both solar flares and the terrestrial magnetosphere may be turbulent (McKenzie, 2000; Nakamura *et al.*, 2002), only a few studies have addressed the issue of turbulent reconnection (e.g. Ichimaru, 1975; Lazarian and Vishniac, 1999). The occurrence of plasma turbulence in a highly structure environment poses a severe challenge to large-scale numerical simulations, so progress in this area may be slow for some time to come.

As we emphasized in the early part of this chapter, it seems likely that kinetic effects play an important role in determining how reconnection works both in the terrestrial magnetosphere and in the solar corona. The low rate of particle collisions in the magnetosphere and the super-Dreicer electric fields that are generated in the corona rule out any models that use transport coefficients derived from standard collisional theory. The fact that collisional effects are important does not, in general, rule out the application of MHD theory and simulations in either environment. As Parker (1996) pointed out, ideal MHD theory does not require the plasma to be collisional; it only requires that the electric field in the rest frame of the plasma be small compared with the convective electric field. The principal limitation of ideal MHD theory, insofar as reconnection is concerned, is that reconnection is inherently non-ideal. Thus, some sort of non-ideal treatment is required of the current sheet region where reconnection occurs. In the magnetosphere and the corona this non-idealness requires a consideration of the generalized Ohm's law or its equivalent. The only region where collisional theory is likely to be valid within the current sheet is the solar convection zone, where particle–particle collisions dominate. It seems likely that as numerical simulations of the convection zone become more sophisticated and realistic, interest in how reconnection works in this region will increase.

6

Structures of the magnetic field

MARK B. MOLDWIN, GEORGE L. SISCOE, AND
CAROLUS J. SCHRIJVER

6.1 Preamble

One goal this book pursues is to identify phenomena throughout the heliosphere that can be said to be universal. An excellent example of a universal subject is the morphology of magnetically defined structures. The heliosphere is full of distinctive magnetic forms that recur in widely different places. What they are and why this is so are questions we take up here. As a first stab at distinguishing magnetically defined structures, we put them into three groups: current sheets, of which the heliospheric current sheet (HCS) is the largest example; flux tubes, with sunspots as a prototype; and cells, in which we include cavities such as magnetospheres. These three classes make up the common forms of heliophysical magnetic structure that exist on MHD time and distance scales (we are not concerned here with kinetic-scale structures that inhabit the dissipation range of turbulence). Our tasks are to explain why these magnetic structures arise naturally and to describe examples of each. It is important to note that these structures are idealized mental constructs to approximate a real world; we are trying to describe a continuum in black-and-white terms. We should realize that current sheets are not mathematical planes, flux tubes are surrounded by other fields, and cells are leaky in the real world. Nevertheless, these abstractions should help us to think and communicate.

That magnetically defined structures in the heliosphere have been seen to take common forms has led people to recognize over time that there actually exists such a thing as the magnetic organization of cosmic matter (as described, for example, in the NRC report "Plasma physics of the local cosmos", 2001, after which this volume is named). The concept of magnetically organized matter helps to define heliophysics as a unique field of study (cf. Section 1.2) of interesting plasma phenomena in the universe. Parker's monumental *Cosmical Magnetic Fields* (1979) paints a picture of space constantly animated by magnetic field born deep in the convection layer of the Sun, pursuing necessarily a life of turbulent, sometimes violent

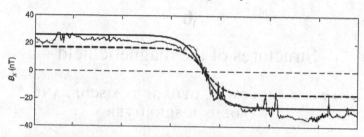

Fig. 6.1. The Harris current sheet formula fitted to a crossing of the Cluster spacecraft of a quiet current sheet in Earth's magnetotail. Two curves result from taking either the field or the current as the independent variable. (From Thompson *et al.*, 2005.)

dissipation in a hopeless quest to reach equilibrium, while structuring itself into sheets, tubes, and cells.

6.2 Current sheets in cosmic plasmas

Our focus here is mainly on current sheets in the form of tangential disconti- nuities (see Fig. 4.6) or rotational discontinuities that evolve into tangential-like discontinuities (e.g. De Keyser *et al.*, 1998). Tangential discontinuities, it may be recalled from Chapter 5, are non-propagating surfaces, across which no magnetic flux passes as the magnetic field changes direction or strength or both, while the total pressure (magnetic plus thermal see Section 6.4) is continuous. Tangential discontinuities form the boundaries of magnetic tubes and cells where dynamo- created fields eventually end up and spend the rest of their lives. Note that shock waves are also current sheets, but their role in heliophysics is to mediate the inter- actions between super-Alfvénic plasma streams or other streams or bodies such as planets. Thus shock waves from a separate topic and will be treated in Vol. II of this series.

The classical physical–mathematical model of a current sheet is the one- dimensional Harris sheet (Harris, 1962) in which the field strength varies as the hyperbolic tangent across the sheet, from a constant value on one side to the same constant value on the other side. The direction of the field at the two sides is arbitrary (except that the field must be everywhere coplanar with the sheet), but usually the field is assumed to reverse direction across the sheet. A plasma pressure is then specified such that the total pressure is constant through the sheet (it varies as the inverse square of the hyperbolic secant). The thickness can be arbitrary as long as one-dimensionality is maintained as an approximation. As Fig. 6.1 shows, the Harris-sheet model sometimes gives a good fit to the current sheet in Earth's quiet magnetotail (at other times the fit is not as good; for example, the sheet can be bifurcated, see Thompson *et al.*, 2005 and references therein).

In his *Spontaneous Current Sheets in Magnetic Fields* (1994), Parker develops at length the theme that current sheets (tangential discontinuities) inevitably form in naturally occurring turbulent plasmas (i.e. plasmas without special symmetries that preclude current-sheet formation). He showed that the hydromagnetic equations treat magnetic field lines formally like light rays passing through a refracting medium. Such a medium, if turbulent, causes flux surfaces to develop holes into which fields not formerly adjacent intrude from both sides and come into contact, creating a surface of discontinuity. Parker used the result to propose a means by which the corona might be heated through current sheet dissipation (by reconnection and nanoflares) and consequent wave generation (see Chapter 8). Another seemingly unavoidable way for current sheets to form in the corona is through the expansion of magnetic flux tubes (discussed below) that poke out of the photosphere, where, surrounded by plasma with a higher gas and lower magnetic pressure, they are kept relatively isolated. These rise into the corona where the plasma pressure is low. The flux tubes expand until at some altitude they press against each other forming a beehive pattern of flux tubes separated by current sheets (Cranmer and Van Ballegooijen, 2005). One cannot measure current sheets in the corona either in situ (of course) or remotely, but possibly some are advected out with the solar wind and make up some part of a steady traffic of magnetic discontinuities that spacecraft beyond the Earth's magnetosphere have found to be always present (Ness *et al.*, 1966; Siscoe *et al.*, 1968). If so, these current sheets constitute part of the energy-containing range of interplanetary turbulence, i.e. the part that retains the memory of a solar source (Chapter 7).

Interplanetary space is a honeycomb of outwardly advecting current sheets. Most are probably endproducts of the in-transit decay of Alfvénic turbulence, but a significant number could have been formed in the corona and possibly retain information pertinent to coronal heating. Some of these have actually been caught in the act of dissipating magnetic energy through reconnection, at 1 AU (Gosling *et al.*, 2005, 2007).

In an early statistical survey of discontinuous changes in field direction in the range 30° to 180°, Burlaga and Ness (1969) found that such discontinuities swept past the observing spacecraft at an average rate of about one per hour. Most of the discontinuities were found at the low end of the range of changes in field direction considered by the survey. Vasquez *et al.* (2007) later showed that the number of current sheets continues to increase for smaller directional changes and that including discontinuities down to a 3° change brings the frequency of encounters up to about ten per hour. Early studies found solar wind current sheets to be consistent with tangential discontinuities (e.g. Burlaga and Ness, 1969) but, as more examples were studied and the method of minimum variance analysis was used to determine the direction of the normal to the current sheet

(Sonnerup and Cahill, 1967), rotational discontinuities (recognized by a measurable normal component of the magnetic field) were judged to dominate (Smith, 1973; Neugebauer *et al.*, 1984; Lepping and Behannon, 1986; Söding *et al.*, 2001). When multi-point measurements were finally able to be used to determine the direction of the normal to the current sheet, because of the increase in the number of spacecraft, the number of rotational discontinuities that could be definitely identified on the basis of a measurable normal field component dropped to less than the number of tangential discontinuities that could be definitely identified on the basis of a change in field strength (Horbury *et al.*, 2001). Finally, with the advent of measurements from the closely spaced Cluster spacecraft quartet, the number of definitely identified rotational discontinuities dropped to zero (Knetter *et al.*, 2004).

In each of these studies, however, most discontinuities were unclassifiable as either rotational or tangential, being instead consistent with both, i.e. having a small change in field strength (rotational-like) and also a small normal component (tangential-like); this caused Neugebauer (2006) to conclude that the majority of solar wind discontinuities are unassignable from current measurements. But she noted that one aspect of the observations breaks the symmetry of unassignability: changes in the components of the magnetic field consistently correlate with changes in the components of velocity, which is not a property of tangential discontinuities. Tangential discontinuities would leave them uncorrelated. It is, however, a property of rotational discontinuities, but the observed correlation coefficient is significantly less than theory predicts. Moreover, why, if they are rotational discontinuities, are their normal components unmeasurably small, making them appear as a tangential discontinuity? Thus, the situation is left unresolved by the correlation of field and velocity components.

Vasquez *et al.* (2007a, b) proposed to cut the Gordian knot by granting most-likely status to the only current-sheet-formation scenario that can account for all observed properties of the large, unassignable, group of discontinuities: a nonlinear, dissipative, cascade of Alfvénic turbulence into planar current sheets with normals perpendicular to the ambient field (and therefore tangential-like) yet retaining much of the correlation between field and velocity components that characterizes the parent sea of Alfvénic turbulence. This scenario assigns the bulk of solar wind current sheets to the status of noise, but there remain the definitely assignable tangential discontinuities, based on a significant change in field strength, which make up about 20% of the population. This is the about-one-per-hour group, which could indeed constitute a coronal signal. Their interarrival spacing is consistent with the scale of expanded photospheric flux tubes and with the changes in the anisotropy directions of solar energetic particles presumably following flux tubes from the Sun (Bartley *et al.*, 1966; McCracken and Ness, 1966).

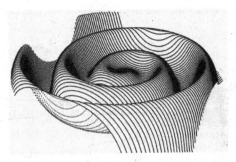

Fig. 6.2. The heliospheric current sheet modeled for a 15° tilt of the solar magnetic equator. For a slow wind at 400 km/s, the figure is 25 AU across, extending to beyond Saturn's orbit; for a fast wind at 800 km/s, the figure is 12.5 AU across, extending only to just beyond Jupiter's orbit (see Chapter 8 for the properties of the solar wind and the details of the "Parker spiral" of which this figure shows one aspect). (From Jokipii and Thomas, 1981.)

The current sheets discussed above are temporary inhabitants of the heliosphere, ending their stay by heating the corona or by riding the solar wind into interstellar space. In contrast, the heliospheric current sheet (HCS) is a semi-permanent resident. This mother of all current sheets for heliophysics (see e.g. Fig. 9.3) forms the border between solar wind streams from evolving coronal holes of opposite magnetic polarity and generally on opposite hemispheres. Its plane approximates the symmetry plane of the dipole component of the Sun's magnetic field. Across the HCS, therefore, the magnetic field reverses direction, in contrast with the transient discontinuities discussed above across which the change in field direction is usually small. As a proxy for the solar dipole's symmetry plane, the HCS circles the Sun completely, originating several solar radii above the Sun, where the streams from the two polar coronal holes come together and presumably extend to the termination shock if not beyond, a distance around 100 AU. The HCS is typically tilted with respect to the Sun's rotational equator, approaching it in orientation only at the minimum of the solar activity cycle. In its tilted configuration the HCS is shaped into a wavy surface (Fig. 6.2) by the combined action of solar rotation and radial solar wind outflow; this same combination curves the interplanetary magnetic field (IMF) into Parker spirals (Chapter 9), which HCS waves mimic. As the solar activity progresses through its maximum phase, during which the Sun's dipole magnetic moment changes sign, the HCS tilts, warps, and at times fragments into more than a single structure (cf. Fig. 8.1).

The HCS is observed always to be encased in a sheath of high-density slow-moving solar wind, with which it forms an integral unit known variously as the heliospheric plasma sheet, the streamer belt, or simply the slow wind. But it could also be called the trans-heliospheric highway since it carries a lively traffic of magnetic "freighters" (plasmoids, a plasma confined by a helically twisted flux

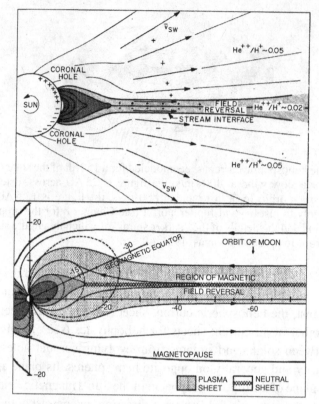

Fig. 6.3. (Upper panel) The heliospheric current sheet, labeled "field reversal", and its plasma sheet (Gosling *et al.*, 1981) and (lower panel) the magnetospheric current sheet, here the "neutral sheet", and its plasma sheet (Ness, 1969).

rope) transporting plasma and magnetic field out of the heliosphere. The plasmoids (defined rather loosely as "plasma-magnetic entities") range in size from small "blobs" to enormous coronal mass ejections (CMEs, see e.g. Wang *et al.*, 1998; Crooker *et al.*, 2004). The traffic in magnetic flux arises from episodic magnetic reconnection occurring within and around arches of closed magnetic fields in the corona that lie at the inner edge of this "highway" (Fig. 6.3).

The role of the heliospheric current sheet as an interface between magnetic fields of opposite radial polarity and as a highway for plasmoid transport makes it the prototype of current sheets that play the same dual role inside planetary magnetospheres. In the magnetospheric case, the solar wind snags magnetic field lines from the planet's two poles on the sunward side and stretches them anti-sunward to form the characteristic two-lobe magnetospheric tail across which the magnetosphere's analog of the heliospheric current sheet separates the two lobes. The magnetospheric current sheet is thus an extension of the planet's magnetic equator, in analogy to the HCS and the Sun's magnetic equator. In the planetary case, the equatorial current sheet may be deformed by the tilt of the dipole axis toward or away from the Sun, which Fig. 6.3 depicts, as happens in the case of the Sun as it goes through the sunspot cycle. An important aspect of the analogy in all cases is the traffic that occurs of plasmoids along the current sheet away from the parent body, including CMEs of all sizes in the heliosphere, substorm plasmoids in Earth's magnetosphere, and rotationally driven plasmoids of internally generated plasma in the magnetospheres of Jupiter and Saturn (Chapter 13 and Vol. II).

6.3 Magnetic flux tubes

Flux tubes define the basic geometry of magnetic fields because $\nabla \cdot \mathbf{B} = 0$. This means that a magnetic field line can neither begin nor end; it either closes on itself or goes to infinity. Therefore, a tube generated by field lines through the circumference of a sufficiently small circle centered on an arbitrary field line forms either a torus or infinite magnetic "spaghetti". Thus, the volume of space occupied by any magnetic field configuration is in fact made up not of field lines, which have no volume, but of volume-occupying flux tubes (the TRACE extreme ultraviolet image of the solar corona in Fig. 8.9 makes the point visible). Unlike a field line, a flux tube is a physically meaningful magnetic entity since the flux per unit area is the magnetic field strength. In cosmic plasmas, flux tubes are more than a physically meaningful way to partition a magnetic field under study; they are isolated entities governed by a Newtonian equation of motion, while preserving their flux and (in some cases) the mass of its entrained plasma. Our goal in this section is to describe qualitatively the several ways in which isolated flux tubes arise and to give examples of their occurrence.

To start, we follow a greatly compressed version of the relevant discussion given in Parker (1979, Section 21.4). Flux tubes are born at the base of the convection zone, where dynamo action generates most magnetic flux in the solar system. The mother load of magnetic flux produced here – possibly aided by the Raleigh–Taylor instability – spawns flux tubes that rise buoyantly to break the surface and create bipolar pairs of magnetic flux bundles, shredded into ensembles of smaller flux tubes that poke through the photosphere. Throughout the solar convective envelope they are shuffled around by convective motions, but that process is only visible at the solar surface (Chapter 8). Thus, the law $\nabla \cdot \mathbf{B} = 0$, dynamo action at depth, Raleigh–Taylor shredding, buoyant rising, and convective shuffling are the topological and dynamical elements that lead to the complex pattern of magnetic flux tubes of varying sizes seen in the photosphere.

At the photospheric level and higher, magnetic reconnection enters as the dominant flux-tube-generating mechanism, taking over from the Raleigh–Taylor

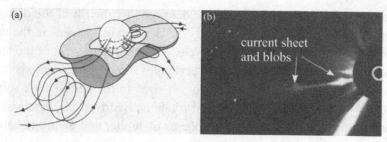

Fig. 6.4. (a) The relation between a CME flux tube (lower left) and the heliospheric current sheet (Crooker *et al.*, 1993) and (b) an image of the smaller kind of coronal ejection showing "blobs" (Lin *et al.*, 2005).

instability at depth. Reconnection joins the legs of magnetic arches pushed together by photospheric flows. Higher up, reconnection can operate in the neck formed by the ballooning of the apex of a flux-tube arch. In both cases a flux tube is formed which has a flux surface as its underside, i.e. in cross section the field lines seen in projection are topological circles closing above the photosphere. The field thereby becomes enclosed by a flux surface forming a defined body (ignoring the ends which might be still rooted in the photosphere), on which surface stresses can act and set the body in motion.

The body thus defined is a flux tube, and the surface stresses acting on it include buoyancy, diamagnetic repulsion, and the Reynolds stress of the jet emanating from the reconnection site under it, where it budded off the magnetic arch. The stresses act so as to expel the flux tube from the Sun and will do so unless the tube is too massive, too strongly rooted at the ends, or too firmly tethered by the overlying magnetic field (see Vol. II, also e.g. Pneuman, 1984; Forbes and Isenberg, 1991; Antiochos *et al.*, 1999; Siscoe *et al.*, 2006). The conditions necessary for expulsion are so commonly met that definite counterparts at 1 AU of coronal mass ejections (CMEs) are frequently encountered, and blobs seen in coronagraph images may also have been observed in the form of small flux tubes (Moldwin *et al.*, 2000; Fig. 6.4). At any time, the heliosphere contains several hundred fully fledged coronal mass ejections in transit to the termination shock and also countless smaller blobs. Conditions favorable for flux-tube formation and ejection most commonly occur under the heliospheric current sheet, which, as already noted, is where heliospheric flux-tube traffic peaks.

6.4 Definition of a flux tube

A flux tube is the volume enclosed by a set of field lines that intersect a simple closed curve. The frozen-in flux condition of ideal MHD describes a parcel of plasma threaded by magnetic field lines as a conserved entity, whose motion can be followed (Section 3.2.3.1). This has important and far-reaching implications and makes understanding the structure of flux tubes important.

In the absence of gravity and viscosity, Eq. (3.2) yields a simplified MHD equation of motion:

$$\rho \frac{d\mathbf{v}}{dt} = -\nabla p + \nabla \cdot \mathcal{T}, \tag{6.1}$$

where

$$\mathcal{T}_{ij} = \frac{B_i \, B_j - \delta_{ij} \, B^2/2}{\mu_0}. \tag{6.2}$$

This quantity is called the Maxwell stress tensor. Suppose, for the purpose of illustration, that the magnetic field is approximately uniform and is directed along the z-axis. In this case, the above equation of motion reduces to

$$\rho \frac{d\mathbf{v}}{dt} = -\nabla \cdot \mathbf{p}, \tag{6.3}$$

where

$$\mathbf{p} = \begin{pmatrix} p + B^2/(2\mu_0) & 0 & 0 \\ 0 & p + B^2/(2\mu_0) & 0 \\ 0 & 0 & p - B^2/(2\mu_0) \end{pmatrix}. \tag{6.4}$$

Note that the magnetic field increases the plasma pressure p by an amount $B^2/(2\mu_0)$ in directions perpendicular to the magnetic field and decreases the plasma pressure by the same amount in the parallel direction. Thus, the magnetic field gives rise to a magnetic pressure $B^2/(2\mu_0)$ acting perpendicular to the field lines and a magnetic tension $B^2/(2\mu_0)$ acting along the field lines. Since the plasma is tied to the magnetic field lines (i.e. the field lines are frozen in), it follows that magnetic field lines embedded in an MHD plasma act rather like mutually repulsive rubber bands.

An implication of the frozen-in condition is that a flux tube can change shape but the topology of the field cannot change. As seen in Chapters 4 and 5, by adding small amounts of dissipation ideal MHD is violated and reconnection becomes possible, with dramatic consequences for plasma stability.

Another implication of the frozen-in flux condition is that the configuration of the magnetic field can determine the configuration of the plasma. In other words, the magnetic field pressure can confine the plasma. This has implications for laboratory magnetic fusion experiments and gives rise to a variety of magnetic field structures in space and astrophysical plasmas. Also, flux tubes can have heat, mass, and energetic particles flowing along them. Though they are often considered isolated structures they are obviously embedded in a surrounding magnetized plasma that often has a different origin. This can give rise to current sheets between the flux tubes, which through magnetic reconnection allow the exchange of mass, momentum, energy, magnetic helicity, and flux.

In a magnetized plasma, steady-state conditions occur when the plasma is in pressure equilibrium with its surroundings. From the discussion above, it is clear that both the plasma and magnetic pressures need to be considered. So, one way of examining the stability of a particular configuration is to examine the total pressure across the structure. A stable magnetic configuration can be a magnetic bottle, where the plasma pressure inside is balanced by the magnetic pressure of the surrounding flux tubes. This is the basis behind confined plasma devices such as tokamaks. Examining Eq. (6.1) shows that a flux tube in equilibrium will satisfy the condition

$$\nabla p - \mathbf{j} \times \mathbf{B} = \mathbf{0}. \tag{6.5}$$

For small plasma β (the ratio of plasma pressure and magnetic pressure; see Eq. 3.11), the pressure gradient term is negligible and Eq. (6.5) reduces to

$$\mathbf{j} \times \mathbf{B} = \mathbf{0}, \tag{6.6}$$

This is called force-free equilibrium. As in deriving Eq. (6.1), using Ampère's law (4.1) we can rewrite this equation in two parts:

$$\mathbf{j} \times \mathbf{B} = (\mathbf{B} \cdot \nabla)\frac{\mathbf{B}}{\mu_0} - \nabla\left(\frac{B^2}{2\mu_0}\right). \tag{6.7}$$

The first term is the magnetic tension force and the second term is the magnetic pressure gradient force. The magnetic tension force acts along the inward normal of a field line if it is curved (it acts so as to relax the tension in the curved field line). The magnetic pressure gradient force is completely analogous to the plasma pressure gradient force.

6.4.1 Modeling of magnetic flux ropes: force-free fields

A simple model of a flux tube that satisfies the *force-free* equilibrium condition is often called the constant-α model (Burlaga, 1988), because if the Lorentz force $\mathbf{j} \times \mathbf{B}$ vanishes then the non-trivial solution is that $\mathbf{j}$ and $\mathbf{B}$ are parallel. Therefore we can write

$$\nabla \times \mathbf{B} = \alpha \mathbf{B}, \tag{6.8}$$

where in general α is spatially dependent. The scalar function $\alpha(\mathbf{r})$ is not completely arbitrary since $\mathbf{B}$ must satisfy the condition that the field is source-free, i.e. $\nabla \cdot \mathbf{B} \equiv 0$. In addition, currents cannot have a divergence, so that $\nabla \cdot (\nabla \times \mathbf{B}) = 0$. So, using Eq. (6.8), we obtain

$$\nabla \cdot (\nabla \times \mathbf{B}) = \nabla \cdot (\alpha \mathbf{B})$$
$$= \alpha \nabla \cdot \mathbf{B} + \mathbf{B} \cdot \nabla \alpha.$$

Hence

$$\mathbf{B} \cdot \nabla \alpha = 0, \tag{6.9}$$

so that α must be constant along a field line. An example of a force-free magnetic configuration is a simple flux tube that has a twist. A twisted flux tube is called a flux rope.

Many different flux-rope models have been developed to explain space plasma observations in a number of settings. A cylindrically symmetric force-free model is widely used because of its simple analytic form and often very good fit to observations (Lepping *et al.*, 1990). One model uses Bessel function solutions to provide the magnetic field configuration. To demonstrate that observations are consistent with these flux rope models, simulated spacecraft trajectories are "flown" through the model flux rope and a comparison is made with observations. Lepping *et al.* (1990) developed an iterative method of varying the free parameters (the core field strength, the twist, and the distance from the spacecraft trajectory to the rope's axis, i.e. the impact parameter) to obtain a best fit to observations.

There has been much effort to model non-force-free and non-cylindrically-symmetric flux rope configurations (e.g. Mulligan and Russell, 2001). This has mainly been driven by interplanetary coronal mass ejection and magnetic cloud observations that clearly show the expansion and evolution of a structure as it passes a spacecraft. In addition, in the Earth's magnetotail the vertical extent of the flux rope is significantly smaller than the downtail extent, indicating that the cross section is elliptical rather than circular (Moldwin and Hughes, 1991). Many of these models do a better job at fitting the data by introducing the additional free parameters discussed above.

6.5 Definition of a flux rope

A universal object in the plasma universe is the magnetic flux tube. These are the "elementary particles" of the magnetized plasma realm. They obey Newtonian mechanics and the MHD equations of motion treat them as constructs that can be "followed" by accounting for all the forces acting on the magnetized plasma. Flux tubes are a useful construct in laboratory magnetic fusion experiments to astrophysical plasmas to the heliophysical domain discussed in this chapter. Because of the presence of a magnetic field component parallel to the current in a current sheet, multiple reconnection most often makes twisted flux tubes, called flux ropes. These flux ropes are observed on the Sun, in the solar wind, at the magnetopause, in the magnetotails of the magnetized planets, and above the ionopause of Venus. They have a wide range of scale sizes and play an important role in transporting mass, momentum, and magnetic flux throughout the magnetosphere.

6.5.1 Solar flux ropes

Many theories for the formation of coronal flux ropes have been developed (e.g. Rust and Kumar, 1994; Priest *et al.*, 1996; Kuperus, 1996; Kuijpers, 1997; MacKay *et al.*, 1997; Zirker *et al.*, 1997; Rust, 2001). Solar observations indicate that flux ropes can form gradually within the solar corona over a period of days to weeks. The region from the photosphere to the lower corona is highly dynamic, and there are several processes that can affect the formation of flux ropes, including: (i) the emergence of twisted magnetic fields from the convection zone inside the Sun (e.g. Pevtsov *et al.*, 2003); (ii) the distortion of coronal fields by shear flows acting on the photospheric footpoints of the coronal field lines (DeVore, 2000; Chae, 2001); (iii) the concentration of axial magnetic fields at a polarity inversion line in the photosphere and the subsequent formation of a coronal flux rope by magnetic reconnection associated with photospheric flux cancellation (Van Ballegooijen and Martens, 1989); (iv) the interactions between neighboring active regions (Martens and Zwaan, 2001; Lin and Van Ballegooijen, 1995); and (v) the relaxation of the corona to a minimum-energy state while conserving its magnetic helicity (Nandy *et al.*, 2003). The observed coronal flux ropes may be formed by a combination of the above processes (Van Ballegooijen *et al.*, 2007). There currently are many different viable flux-rope generation models. As in many fields of science, there is probably more than one way to generate a similar structure; therefore, multiple mechanisms may be responsible for flux rope generation on the Sun.

6.5.2 CMEs and interplanetary CMEs

Magnetic reconnection is thought to play an integral role in the energization of solar plasma and the reconfiguration of solar magnetic fields (Chapter 5). Magnetic reconnection in the solar corona (e.g. Kopp and Pneuman, 1976; Hiei *et al.*, 1993; see also the review by Forbes *et al.*, 2006) has also been suggested as being responsible for producing and ejecting magnetic flux ropes, also known as magnetic clouds when they are in interplanetary space (Gosling, 1990, 1993). This model suggests that reconnection occurs on the sunward side of expanding coronal loops to form a flux-rope structure (Fig. 6.5). These magnetic structures have been named magnetic clouds and have been shown to be a subset of coronal mass ejections

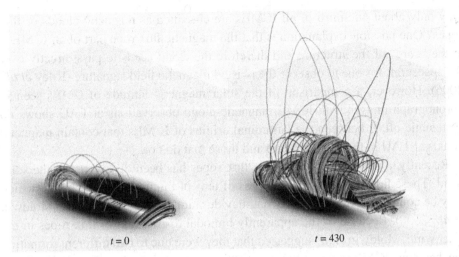

Fig. 6.5. Model results of the formation of a twisted flux rope from a magnetic flux eruption from the solar surface. (From Amari *et al.*, 2000.)

(CMEs; see e.g. Burlaga *et al.*, 1982; Klein and Burlaga, 1982; Gosling, 1990). Magnetic clouds, as defined by Klein and Burlaga (1982), are regions with a radial dimension of ~ 0.25 AU at 1 AU in which the magnetic field strength is high and the magnetic field direction changes appreciably owing to the rotation of one component of **B** (approximately) within a plane. Magnetic clouds are also inferred to be expanding at 1 AU owing to the higher pressure observed inside the structure compared with that in the surrounding medium.

Coronal mass ejections often show flux-rope structures in coronagraph images. Flux-rope signatures seen in situ in the solar wind have been called interplanetary CMEs (or ICMEs). Not all events identified as ICMEs have all these signatures (e.g. Zwickl *et al.*, 1983). Gosling *et al.* (1987) argued that the most robust signature of an ICME at 1 AU is the presence of counter-streaming suprathermal electrons. However, studies of fast ICMEs driving interplanetary shocks have found that other signatures are just as good, if not better, than the presence of counter-streaming electrons in identifying ICMEs. Richardson and Cane (1993) found that plasma proton temperature depressions were the best signature of fast CMEs, while Zwickl *et al.* (1983) found that low-variance, enhanced, magnetic fields (e.g. Tranquille *et al.*, 1987) were equally as good as counter-streaming electron signatures.

There are several databases of coronal mass ejections (e.g. Cane and Richardson, 2003; Jian *et al.*, 2006), which use different identification criteria. Though most events are common to all the lists, the 10% or so that are unique to one list or another indicates the wide variety of plasma and magnetic field signatures that are collectively called an ICME. One question that comes out of these studies is why only about one-third of all ICMEs are classified as magnetic clouds or flux ropes? One possible explanation is that the magnetic flux rope part of an ICME is at the "core" of the structure and therefore the cloud needs to pass directly over the spacecraft for one to observe the twisted magnetic field signature (Riley *et al.*, 2006). However, a comparison of the solar magnetic latitude of CMEs seen in coronograph images with ICME magnetic-cloud observations at 1 AU shows no systematic offset between the latitudinal origins of ICMEs that contain magnetic clouds at 1 AU in the ecliptic plane and those that do not.

Recently, a new class of magnetic flux ropes has been discovered in the solar wind. These flux ropes have scale sizes of tens of minutes to a few hours, which may be compared with the nearly one-day duration of magnetic clouds (Moldwin *et al.*, 2000). Because of the apparently bimodal distribution of flux ropes in the solar wind, Moldwin *et al.* suggested that they were due to two different formation mechanisms: ICMEs are due to large-scale reconnection in the solar corona, while small-scale flux ropes may be due to magnetic reconnection across the heliospheric current sheet.

6.5.3 *Flux ropes in the Earth's magnetotail*

Like the heliospheric current sheet (HCS) that serves as an interface between the magnetic fields from the north and south magnetic poles of the Sun, the magneto-sphere's tail current sheet separates the opposite polarities of the Earth's magnetic field. The current sheet is embedded in the plasma sheet that divides the magnetotail into north and south magnetic lobes characterized by near empty flux tubes that have opposite polarities. The north lobe's magnetic field points toward the Earth and the south lobe's magnetic field points away from the Earth (see e.g. the bottom panel in Fig. 6.3). Like the heliosphere on the large scale, the magnetosphere is shaped like a windsock, the magnetospheric cavity extending far behind the Earth. This is due to the solar wind momentum and pressure confining and pushing the magnetotail away from the Sun, analogously to a comet's ion tail being "blown" away from the Sun by the solar wind.

Also like the HCS, the Earth's current sheet can be considered as a trans-magnetosphere highway, since it also carries plasma and magnetic flux within plasmoids that are ejected downstream. These plasmoids have twisted flux-rope topologies and are created by multiple magnetic reconnection across the Earth's current sheet.

These flux ropes have vertical scale sizes comparable with the size of the plasma sheet, but they have a wide range of downtail dimensions. Frequently flux rope plas-moids have diameters of many tens of R_E, though the average is $16.7 \pm 13.0\ R_E$ (Moldwin and Hughes, 1992b). Ieda *et al.* (1998), using geotail observations,

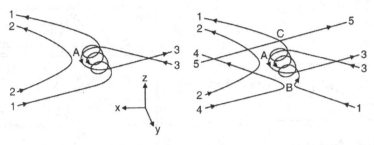

Fig. 6.6. Reconnection of plasma-sheet field lines with a B_y-component leads to the formation and subsequent ejection of a twisted flux tube or flux rope (Hughes and Sibeck, 1987).

estimated that plasmoids have dimensions that average $10R_E$ (the downtail length) $\times$ $40R_E$ (the cross-tail width) $\times 10R_E$ (the height). Plasmoids are strongly correlated with geomagnetic substorms (Moldwin and Hughes, 1993) and are often associated with high-speed tailward flow (Moldwin and Hughes, 1992b). The near-Earth neutral line model (Hones, 1976) predicts that plasmoids are formed by magnetic reconnection in the near or mid-tail current sheet, which completely severs the plasma sheet away from the Earth, leading to the rapid propagation of the plasmoid downtail. These plasmoids are three-dimensional twisted flux ropes (e.g. Moldwin and Hughes, 1991, 1992b) and are given their twist by the penetration of the IMF B_y component across the Earth's magnetotail. The formation of these flux ropes across the current sheet is similar to the formation of coronal mass ejection events (Gosling *et al.*, 1995). Figure 6.6 depicts schematically how multiple reconnection across a magnetotail current sheet that has a B_y or cross-tail field component gives rise to a twisted flux-rope topology. Instead of creating closed loops the reconnection process connects fields that have been displaced in the *y*-direction, leading to a twisted flux rope.

Observationally, flux ropes are identified by a south-then-north (or vice versa) bipolar magnetic signature as their internal field, coiled in the sense of earthward in the north and tailward in the south (or vice versa), moves tailward over a spacecraft. The left-hand panel of Fig. 6.7 shows the characteristic signature of a plasmoid flux rope observed in the deep magnetosphere. Note the bipolar signature in the B_z (north–south) component, which is coincident with the strong "core" field observed in the B_y and B_x components.

All flux-rope plasmoids observed beyond $100R_E$ downtail have solar-wind-like or higher velocities moving downtail. However, earthward of $100R_E$, flux-rope plasmoids are often observed moving earthward with reversed polarities (i.e. south-then-north). These earthward-moving flux-rope plasmoids are thought to be due to remnant or "fossil" flux ropes formed by magnetic reconnection that did not extend all the way through the plasma sheet into the lobe. This would "trap" the flux rope

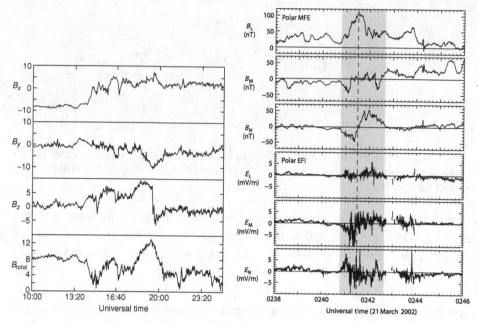

Fig. 6.7. (Left) The magnetic field on 28 January 1983 observed by ISEE-3 deep in the Earth's magnetotail, showing the presence of a magnetic flux rope plasmoid. (Right) Polar magnetic and electric field observations for a flux transfer event or FTE. (From Le *et al.*, 2008.)

on earthward convection paths, giving rise to south–north bipolar signatures and slowly earthward propagating structures (see e.g. Moldwin and Hughes, 1994).

Because of the large vertical extent of plasmoid flux ropes and the magnetosheath pressure confining the lobes, a passing plasmoid perturbs the surrounding lobe magnetic field in a characteristic manner, which is similar to that of a snake swallowing an egg. Bipolar signatures coincident with a compression of the magnetic field are observed in the lobe and are called traveling compression regions (TCRs; see e.g. Slavin *et al.*, 1984). Observations of a TCR by IMP-8 in the midtail and of a plasmoid flux'rope by ISEE-3 in the deep tail sequentially after a substorm provide the most definitive indication of the relationship between TCRs and plasmoid flux ropes (e.g. Moldwin and Hughes, 1992a).

6.5.4 *Flux ropes at the magnetopause*

Twisted flux tubes are often observed at the Earth's magnetopause and are called flux transfer events (FTEs; they are part of the so-called Dungey cycle, see Section 10.4.3). They are the result of multiple magnetic reconnections across the dayside magnetopause. Figure 6.8 shows a schematic of the dayside magnetopause after reconnection has created a flux-transfer event. Flux ropes on the magnetopause

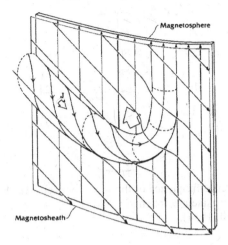

Fig. 6.8. A flux transfer event at the Earth's magnetopause (from Russell and Elphic, 1979).

are formed by reconnection between the southward IMF and the northward geomagnetic field (see the review by Farrugia *et al.*, 1988).

Whereas solar wind magnetic field observations are often given in the geocentric–solar-ecliptic (GSE: *x*, Earth–Sun line; *z*, ecliptic north pole) coordinate frame and magnetotail magnetic field observations are given in the geocentric–solar–magnetospheric (GSM: *x*, Earth–Sun line *z*, projection of magnetic dipole axis onto GSE *yz*-plane) coordinate frame, observations at the magnetopause are usually placed in magnetopause normal coordinates. These are often called boundary normal coordinates and are often abbreviated as LMN coordinates, where N is the normal to the magnetopause and L points northward along the magnetopause, so that the NL-plane contains the GSM *z*-axis, while M points westward and completes the right-handed coordinate system. Magnetopause crossings are often described in these normal coordinates on the assumption that the magnetopause is a tangential discontinuity, the fields on either side of such a boundary are tangential to the surface. Therefore the cross product of the two directions of the magnetic field on either side of the boundary is the normal to that plane. If the magnetic field does connect across the magnetopause, a normal can still be found by identifying the direction of the field that is the most nearly constant (e.g. Sonnerup and Cahill, 1967). The right-hand panel of Fig. 6.7 shows an FTE observed by the Polar spacecraft. The magnetic and electric field observations are shown in magnetopause normal coordinates (Le *et al.*, 2008). Note that the bipolar turning in the normal component is coincident with the large "core" field signature in the L component.

These magnetic signatures can be seen up to an Earth radius from the actual magnetopause crossing, owing to the perturbation, due to the event, of the

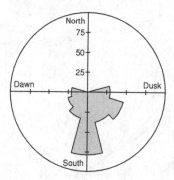

Fig. 6.9. The occurrence frequency of flux-transfer events (FTEs) as a function of the angle of the interplanetary magnetic field (IMF) in the *yz* GSM plane (from Berchem and Russell, 1984).

surrounding magnetosheath or boundary layer. This is similar to the TCR signature seen in the magnetotail lobe due to the passing plasmoid flux rope. The region around the flux transfer event is often called the field line draping region and has characteristic field and associated plasma flows (Farrugia *et al.*, 1987).

The association of FTEs and sporadic or patchy magnetic reconnection is borne out by many observed signatures, i.e. those that can most easily be explained by temporal changes in the outflow from the reconnection site during a southward IMF. The strong correlation of FTE signatures below the cusp with a southward IMF also points to the relationship between FTEs and magnetic reconnection. Figure 6.9 shows this correlation between the IMF north or south direction and FTE occurrence (Berchem and Russell, 1984).

Much progress has been made in understanding the often complex field, plasma flow, and particle signatures observed with FTE by using numerical simulations (e.g. Raeder, 2006; Omidi and Sibeck, 2007). These results indicate that FTEs are due to multiple reconnection on the dayside magnetopause, and furthermore the Raeder (2006) results predict that there should be a seasonal dependence on the occurrence frequency of FTEs, owing to the dipole tilt effect on FTE generation. Figure 6.10 gives for comparison results from high-resolution simulations without dipole tilt (left-hand panel), in which no FTEs occur, and the results for a 34° tilt relative to the oncoming solar wind, in which an FTE develops.

6.6 Flux ropes at other planets

Flux ropes, either in the form of plasmoids in the magnetotail or as FTEs at the magnetopause, have also been found in the magnetospheres of Mercury (Russell and Walker, 1985) and Jupiter (Walker and Russell, 1991). Limited observations of the Mercury dayside magnetopause showed that FTEs there have short durations, of only a few seconds, although they appear to have a higher occurrence frequency

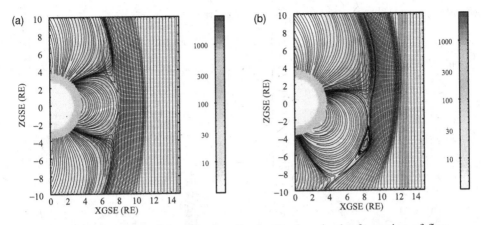

Fig. 6.10. Illustration of the role that dipole tilt plays in the formation of flux-transfer events. (a) No dipole tilt; (b) a dipole tilt of 34°. Multiple X-lines lead to the formation of flux-transfer events. The gray scale represents the plasma pressure. See also the color-plate section. (From Raeder, 2006.)

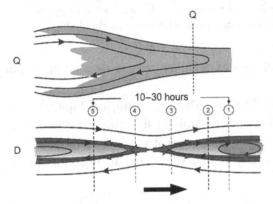

Fig. 6.11. Schematics of the configuration of the Jovian magnetotail for quiet (Q) and disturbed (D) times. The numbered lines indicate different field and flow observations from the Galileo spacecraft during disturbed intervals (from Kronberg *et al.*, 2005).

than at Earth. The MESSENGER mission should provide new observations about the frequency and scale size of Hermian FTE.

Observations of the Jovian magnetotail with the Galileo spacecraft suggest that periodic processes similar to substorms occur with typical time scales of several days. However, instead of magnetic flux building up in the magnetotail owing to magnetic reconnection on the dayside, as at Earth, the Jovian magnetosphere substorm is driven by the plasma loading of fast rotating magnetic flux tubes (Kronberg *et al.*, 2005; see Fig. 6.11). The idea is that during the "loading" phase, the Jovian plasma sheet is in a stable configuration with the plasma corotating. However, owing to centripetal forces the plasma sheet is thinned, leading to

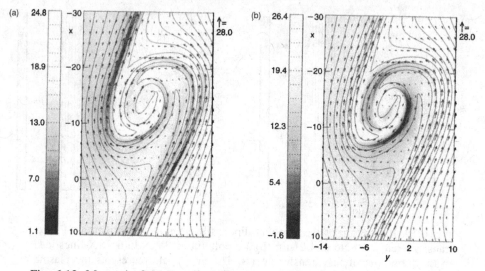

Fig. 6.12. Magnetic field lines and vectors of Kelvin–Helmholtz vortices produced on the magnetopause in a two-dimensional MHD simulation. The gray scale shows the B_z component (a) in the simulation plane and (b) projected onto magnetospheric coordinates (from Otto and Fairfield, 2000).

reconnection, which creates radial inward and outward plasma flows and the ejection of plasmoids.

The comparison of FTEs at Mercury, Earth, and Jupiter has been used to understand the physical factors controlling the temporal and spatial scales of FTEs. Zeleny and Kuznetsova (1988) suggested that the physical dimension of the magnetopause and the strength of the solar wind driver control these parameters.

6.6.1 Flux ropes and the Kelvin–Helmholtz instability

Flux ropes were discovered in the Venus ionosphere using data from Pioneer Venus Orbiter (Russell and Elphic, 1979). This may be somewhat surprising considering that Venus does not have an intrinsic magnetic field and therefore not a true large-scale magnetosphere. However, Venus has a very dense atmosphere and ionosphere that interact with the solar wind to make an induced magnetosphere (Chapter 13; see Fig. 13.12). Currents flowing in this ionosphere, at the ionopause, help to stand-off the solar wind, and the dynamics of this interaction can create small-scale magnetic flux ropes. For low solar wind dynamic pressure conditions, the ionopause forms at a high altitude where collision processes can be neglected. During these conditions the ionopause appears as a tangential discontinuity. Therefore a velocity shear across this interface can lead to the development of a Kelvin–Helmholtz (K–H) instability (Terada *et al.*, 2002); see Fig. 6.12. The K–H instability has been invoked as the origin of the Venus flux ropes (Wolff *et al.*, 1980; Thomas and Winske, 1991) although the conditions at the Venus ionopause, particularly the inefficiency of the downward transport against buoyancy after rope generation (e.g. Luhmann and Cravens, 1991), are not explained well in this interpretation.

The K–H instability has also been invoked explain the formation of twisted flux tubes at the Earth's magnetopause (Otto and Fairfield, 2000; Fig. 6.12). Though reconnection and the K–H instability can transport momentum and energy across the magnetopause boundary, since the K–H instability is ideal it cannot transport mass. However, there have been a number of observations of rotating and rapidly changing magnetic fields at the magnetopause during northward IMF conditions that suggest that the K–H instability is active at the magnetopause and can create twisted flux tubes (e.g. Fairfield *et al.*, 2000).

6.7 Magnetic cells

Magnetic fields tied to gravitating bodies will expand to fill all space unless prevented from doing so. Examples include the magnetic fields of the Sun, Mercury, Earth, Jupiter, the jovian satellite Ganymede, Saturn, Uranus, and Neptune. In every case, however, the magnetic field's expansionist ambition is checked by some other magnetic-field-bearing plasma expanding from somewhere else. Each magnetic field is therefore encased within a definable volume, which we refer to as a cell. In the Sun's case, the cell is the heliosphere. In the other cases mentioned the cells are the planetary magnetospheres. We will add other types later. The cellular structure that results is not like a honeycomb in a beehive, which the flux-tube structure in the corona and solar wind resembles. Rather, it is like a set of Russian nesting dolls; one cell is encased within another. For example, Ganymede's magnetic field is enveloped by Jupiter's magnetosphere, which is enveloped by the heliosphere, which is enveloped by the plasma of the local interstellar medium, etc. One can imagine the nesting to continue at least up to the scale size of the galaxy. Staying within the heliosphere, the scale sizes of the objects already mentioned cover seven orders of magnitude, as Fig. 6.13 illustrates.

Planetary magnetospheres are the archetypes of magnetic cellular structure within the heliosphere. Since planetary magnetospheres form the subject of Chapters 10 and 13, it is not our intention to review these examples here. The magnetosphere chapters indeed discuss structures and processes that are common to the magnetospheric type of cellular magnetic structure, but the real focus is on how these structures and processes differ among the whole set of known magnetospheres. They are divided into kinds according to whether the solar wind interacts mainly with the planetary magnetic field or with the planetary ionosphere or according to whether magnetospheric processes are driven mainly by external or internal

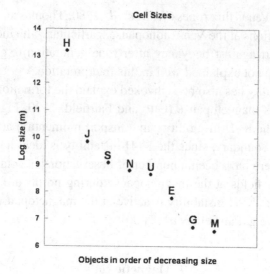

Fig. 6.13. Estimates of scale sizes of heliophysical examples of magnetic cells, ordered by magnitude (taken from Chapters 10 and 13): H, heliosphere; J, Jupiter; S, Saturn; N, Neptune; U, Uranus; E, Earth; G, Ganymede; M, Mercury.).

sources. Scaling laws are presented that in principle allow one to estimate the parameters that characterize the shared magnetospheric structures and processes for an arbitrary instance of the type. For example, expected properties of the Earth's paleomagnetosphere – the magnetosphere that existed when Earth had a different dipole moment, as during magnetic reversals – have been determined using such scaling laws (e.g. Vogt and Glassmeier, 2001).

Instead of emphasizing the differences between heliospheric instances of cellular magnetic structures, here we focus on how as a collection of disparate objects they perform a common role in heliophysics. Heliophysical magnetic cells arise from the conflict between independently created magnetized plasmas trying to expand into each other's space. Each case is a macroscale battle for territory at the boundary between competing plasmas. The boundary constantly shifts to maintain a balance of force as one side or the other gains or loses strength. Think of the magnetopause moving in response to a variable solar wind. The battlefield, which usually contains at least one shock front, strong spatial gradients, and steep velocity shears, becomes through the microscale and macroscale instabilities thereby engendered a prolific producer of energetic particles, MHD and plasma waves, turbulence, and electromagnetic emissions. At the actual rampart that separates the invading and defending magnetic fields, magnetic reconnection acts as a catapult that allows the invader to overwhelm the defender and extend the attack deep into the defending cell, down to the ionosphere and throughout the tail in the case of the magnetosphere. Such invasions create trapped radiation belts, auroras, substorms, and storms. In general, these byproducts of the conflict – energetic particles, waves, turbulence, EM emissions, trapped radiation, storms, etc. – dissipate or carry away a small fraction of the flux of energy directed against the attacked magnetic cell, as, for example, the solar wind energy flux is incident on a planetary magneto-sphere. But from the viewpoint of both physics and space weather, this generally small fraction of the incident energy flux creates much of all that is interesting in space.

Heliophysical cells, therefore, are generalized factories that convert into the energetic phenomena of heliophysics the energy that expanding magnetized plas-mas direct against each other as they fight for territory. Viewed in this way, the term "heliophysical cell" is not a mere inert synonym for a magnetosphere. Instead it embraces all definable volumes occupied by plasmas deriving from independent sources and engaged in energy-converting battles over space. The defining essen-tial of a heliophysical cell is that it is a volumetric source for the conversion of the directed energy of expanding magnetized plasmas into energetic heliophys-ical phenomena. The heliosphere itself, defined by the surface where the solar wind pushes against the local interstellar medium, creating anomalous cosmic rays at its termination shock front (e.g. Jokipii *et al.*, 2007), is a heliophysical cell. Coronal mass ejections, spewing solar cosmic rays from their leading shock front as they invade the interplanetary medium (e.g. Gopalswamy, 2007), are helio-spheric cells. Cometary magnetospheres, which, besides the usual byproducts of solar-wind–magnetosphere battles already mentioned, manifest a zoo of ultralow-frequency (ULF) waves generated by the pickup of newly ionized cometary gas (e.g. Cravens and Gombosi, 2004; Tsurutani, 1991), are heliophysical cells. Even corotating interaction regions (CIRs) in the solar wind – two solar wind streams pushing against each other hard enough to produce a pair of shock fronts, which accelerate particles into the MeV range (Richardson, 2004) – are heliophysical cells.

6.8 Summary

We have noted that the structures that naturally emerge from the magnetic orga-nization of matter, at least within the heliosphere, can be grouped into current sheets, flux tubes, and cells. For each class, the linear dimensions of the objects thus categorized range over many orders of magnitude. In fact, depending on how close one is to an instance of a magnetically organized structure, it can be called a current sheet (e.g. the cross-tail current sheet in a magnetospheric tail), a flux tube (the lobes of a magnetospheric tail), or a cell (the magnetosphere itself). How one classifies an example under study depends, therefore, on whether the purpose of the study is: to define boundaries, for which current sheets are useful; to define topologies or spatial connections, for which flux tubes are appropriate; or to investigate the origin of energetic processes in the heliosphere, for which cells are the breeding factories.

7

Turbulence in space plasmas

CHARLES W. SMITH

7.1 Preamble

In this chapter, we examine the concepts of hydrodynamic turbulence as they apply to space plasmas. We focus primarily on the solar wind at 1 AU in order to illustrate the basic ideas of magnetohydrodynamic fluid turbulence in a unified and coherent fashion, but we extend these lessons to other space plasma systems. Of particular interest is the energy cascade that leads to heating of the background plasma. There is growing evidence that a turbulent dynamic does account for the in situ heating of the solar wind from 0.3 to 100 AU. Turbulence may also provide an explanation for the coronal heating that accelerates the solar wind at the source. The discussion offered here is an attempt to explain how the fundamental ideas of turbulence apply to plasma dynamics. These thoughts are only a beginning, an introduction to the subject, but will lead the reader to the most recent and sometimes controversial ideas being discussed by the space physics and astrophysics communities.

Turbulence, by its rightful definition, is a branch of fluid dynamics. The term "plasma turbulence" has been used to denote the nonlinear evolution of dynamically coupled wave and particle populations, but this is something different. However, this raises an important point: the magnetohydrodynamic (MHD) systems of space physics are not true fluids but are plasmas that mimic magneto-fluid behavior at low frequencies. They retain their ability to exhibit purely plasma kinetic behavior such as the resonance of plasma waves with charged particle populations. Also from the plasma lexicon there is the term "weak turbulence", which is taken to be the nonlinear interaction of linear plasma modes derived from second- and higher-order expansions of the plasma equations (wave–wave interactions as opposed to wave–particle interactions). Again, this is not the focus of the present chapter. It is vitally important when studying any space plasma that we distinguish between the diverse kinetic plasma processes and fluid-like turbulence. With that in mind, the observation I tell my students is "Whatever can happen, will happen!" and

Fig. 7.1. (Left) Schematic illustration of laminar fluid flow. The behavior is orderly and predictable. (Right) Schematic of turbulent flow downstream of a generating grid (off the image, to the left). The flow is disordered and composed of many circulation patterns of varying scale.

this has proven itself true time and time again. We should note that kinetic plasma processes tend to evolve faster than MHD turbulence, but that turbulence often wins the day in the end. The ultimate fate of the coherent energy we recognize in plasma fluctuations is nonlinear decay into heat, and in many cases this energy conversion is accomplished through turbulent dynamics. This chapter is meant to be more a road map than a tutorial designed to convey deep understanding. It is my hope that you will find here something useful that will prompt you to read further on the subject. To that end, I can recommend several particularly good sources: Davidson (2004) for the basics of Navier–Stokes turbulence; Batchelor (1953) and Frisch (1995) for the more advanced topics; and Biskamp (2003) for magneto-hydrodynamic turbulence theory. I have drawn on each of these texts in preparing this chapter. Since space plasmas are complicated systems, I recommend Kallenrode (2001) for a grounding in the fundamentals of space plasma physics.

7.2 Introduction

Turbulence is the nonlinear evolution of a fluid in a way that lies outside perturbative dynamics. Simply put, the system has no rest state and cannot be described as remaining in any sense close to the initial state in which it is created. Fortunately, we can make predictions that describe the system as it evolves toward an asymptotic statistical state. When energy is added to the turbulent flow, the injected energy is transported by the turbulence to different length scales via dynamics that are both causal and non-reproducible. This is the first characteristic of a turbulent flow: the energy of the fluctuating fields, i.e. the velocity, magnetic, density, etc., is transported to a range of spatial (length) scales in a manner that has less to do with how the energy is injected and more to do with the nonlinear evolution of the fluid equations. For this reason, the tools and predictions of turbulence studies are statistical in nature.

Figure 7.1 demonstrates the basic nature of turbulence. The laminar flow to the left is how we imagine a fluid to behave. If we see a stream from a distance we might imagine that the water is flowing smoothly around rocks in a uniform laminar manner that yields readily to predictive calculations. The turbulent flow to the right is what actually happens in most real-world circumstance. The water moving around the rocks is subject to viscous drag and boundary conditions that limit the flow on the surfaces. This creates gradients in the flow, called "shear", which lead to vortices: circulation patterns within the flow. These circulation patterns detach from the sources, become embedded in the flow, and interact. One can often observe this basic turbulent behavior by watching the circulation in a stirred cup of coffee: it is this basic circulation that occurs repeatedly in everyday fluids. Although turbulence is a ubiquitous fluid process, and one of the most famous turbulence studies was made by towing an instrument behind a boat in the ocean (Grant *et al.*, 1962), turbulence studies are most often performed in wind tunnels and other laboratory devices. Under these laboratory conditions an obstacle of prescribed shape may be placed in the flow (the model of a car, for instance) in order to study the turbulence associated with it, or a wire grid may be used to generate the localized instabilities that lead to homogeneous turbulence. Measurements close to the wire grid will be dominated by the grid, the grid spacing, its orientation, etc., but more distant measurements will take on characteristics of turbulence in which the memory of how the turbulence was generated is lost. Turbulence can be excited by a variety of sources. In the above example we considered shear flow around obstacles, but temperature gradients can also be effective in driving turbulent flows and this is a leading source of turbulence in solar dynamics.

Figure 7.1 also demonstrates an important advantage of turbulent flow: mixing is accomplished via convection rather than diffusion. Consider the refrigerator as one example of convective diffusion: the coolant in the coils of the refrigerator carries heat from the food chamber to a reverse heat engine, which extracts the heat and expels it outside the box. Heat must flow through the cross section of the cooling coils at two stages: first into the coolant as heat is removed from the box and then back to the walls of the coils to be expelled into the exterior for the cycle to be completed. It is the circulation pattern driven by gradients in the flow at the boundaries of the coils that allows for the efficient removal of thermal energy from the wall surface. The convection of heat via the motion of the fluid across the diameter of the tubes accelerates the transport of thermal energy and permits refrigeration units to be a practical size. Now, consider the problem of our hot Sun. Although nuclear fusion occurs within the deep interior, heat must pass out of the Sun through a relatively cool photosphere. This leads to convection cells, which overturn the fluid in what is often described as a boiling motion seen across the photosphere. The forced flow of coolant drives the instabilities that result in vortex motion in refrigerant coils whereas thermal instabilities drive solar convection cells, but the basic underlying principle is the same: vortices enhance diffusion. As an exercise, the reader is invited to list the number of physical systems where transport is accomplished via diffusion rather than convection.

In fast laboratory flows the time scale for convecting a given turbulent structure past the measurement device is short compared with the inherent time scale of the fluctuations. This leads to the Taylor frozen-in-flow assumption whereby the spatial scales of the fluctuations are converted into frequency measurements in the laboratory frame: $2\pi f = \mathbf{k} \cdot \mathbf{v} + \omega$, where f is the frequency in the laboratory frame, $\mathbf{k}$ is the wave vector associated with the spatial scale of variation, $\mathbf{v}$ is the flow velocity, and ω is the frequency of the turbulent oscillation in the frame of the fluid. Most theories of turbulent dynamics are based on interacting wave vectors and not on simple frequencies. Measurements, however, are most often frequency based. In high-speed flows $2\pi f \simeq \mathbf{k} \cdot \mathbf{v}$. The inability to make this association between wave vector and frequency, for instance, under many magnetospheric conditions is a major obstacle to applying the ideas of space-based turbulence to some space plasmas.

The most notable successes of turbulence theory to date include prediction of the power spectra of turbulent fluctuations. For instance, Kolmogorov (1941a) used the basic ideas of interacting hydrodynamic fluctuations to drive a cascade of energy from the largest to the smallest spatial scales. From this he successfully predicted the form of the turbulent hydrodynamic velocity spectrum. We discuss this point in some detail below.

The solar wind provides an ideal wind tunnel to extend hydrodynamic turbulence to include magnetic fields in the vastness of outer space. It also enables us to frame the controversy that has plagued such efforts for over 40 years. For example, Coleman (1966) observed that magnetic and velocity fluctuations are correlated in the solar wind and that they lie preferentially within the plane perpendicular to the mean magnetic field. From this he argued that the fluctuations are hydromagnetic plasma waves propagating away from the Sun. Two years later, in Coleman (1968), he modified his opinion, on the basis of the form of the energy spectrum and other attributes, to argue that the fluctuations must arise in situ in a manner analogous to hydrodynamic turbulence. He wrote *From the spectral properties of the plasma fluctuations, it is tentatively concluded that the solar wind flow is often turbulent in the region near 1 AU. A heuristic model of the turbulent flow that is consistent with the observations is one in which the energy for the turbulence is derived from the differential motion of the streams in the solar wind plasma. Instabilities associated with the differential streaming produce long wavelength Alfvén waves. The energy extracted from the differential motion cascades through a hierarchy of Alfvén waves until it reaches waves short enough for dissipation by proton cyclotron damping.* In other words, the large-scale motions of fast-moving plasma passing over slow-moving plasma drives a cascade of energy from the largest to the smallest spatial scales where, ultimately, dissipation converts the coherent plasma fluctuations into heat. Coleman cited the vehicle for this cascade to be the interaction of Alfvén waves, consistently with his earlier paper, but time and further study raise some important questions in this regard. It is true that the MHD equations support wave modes not seen in hydrodynamics. However, we shall show below that not all fluctuations with traits resembling known wave modes need be waves at all.

Having said all that, it is disquieting to realize how little we know about the subject of turbulence; this statement is especially true of the MHD turbulence as it applies to solar, solar wind, magnetospheric, and interstellar observations. So many questions remain unanswered and debated within the community that no one view can be represented as "standard" or "widely agreed", and I will not attempt to provide a unified view here. The level of disagreement is so profound that there remains an active debate on the merits of weakly interacting waves as a model for turbulent evolution, and we are only now beginning to assess the rates and time scales needed to parameterize that debate accurately. In this chapter we will seek to motivate a need for, and then examine, how MHD turbulence theory is descended from traditional hydrodynamic turbulence ideas. We will focus almost entirely on the solar wind in an effort to develop general ideas from a particular set of observations and to test the applicability of those ideas to one well-studied system. The application of the developed ideas to other space plasma systems will be outlined briefly at the end of the chapter.

7.3 What observations characterize the solar wind?

From the earliest observations of the near-Earth solar wind to the latest probes of the outer heliosphere, Eugene Parker's prediction for the basic properties of the solar wind are validated (see Chapter 9). From a few solar radii outward, the solar wind is an electrically neutral supersonic flow of coronal plasma that carries with it the magnetic field of the Sun (Biermann, 1951, 1957; Parker, 1958, 1963). The flow is radially outward from the Sun and essentially constant for heliocentric distances beyond ~ 1 AU (Richardson *et al.*, 1995a). As a result, the plasma density decreases with heliocentric distance as R^{-2}. The magnetic field of the Sun is convected outward by the electrically conducting plasma so that the radial component varies as R^{-2}. The azimuthal component is formed via solar rotation and behaves asymptotically as R^{-1}. The intensity of the resulting interplanetary magnetic field (IMF) decreases according to prediction (Parker, 1958), as R^{-2} inside ~ 2 AU where the radial component of the field dominates and as R^{-1} outside ~ 2 AU where the azimuthal field dominates. The interplanetary magnetic field (IMF) is convected by the solar wind and remains rooted on the Sun. This results in an Archimedean spiral: the radial IMF near the Sun is converted into a nearly azimuthal field in the outer heliosphere (see Section 9.2). The mean IMF direction at 1 AU in the ecliptic is approximately 45° away from the radial direction from the Sun.

What makes the study of the solar wind so interesting today is that it originates from multiple sources and possesses transient ejecta embedded within the flow. At a solar minimum there is a clear pattern of latitudinal structure with a high-speed wind over the solar poles and a low-speed wind originating near the solar equator (McComas *et al.*, 2000). At a solar maximum the structure is more complex, as multiple high-speed and low-speed sources lead to a highly variable interplanetary flow and an enhanced interaction of high-speed and low-speed streams. Add to this coronal mass ejections (CMEs) that occur during solar active times and the corotating interaction regions (CIRs) present during quiet times and their resulting interaction with the ambient solar wind structure, and the result is a highly dynamic flow with large-scale wind shears, shocks, and a variety of in situ transient dynamics, all of which are sources of turbulent energy (Matthaeus *et al.*, 1995).

Initial characterizations of the interplanetary magnetic, velocity, and density fluctuations have likewise formed a strongly influential basis for later studies. Magnetic field and velocity fluctuations are largely transverse to the mean IMF and strongly correlated in a manner consistent with the outward propagation of Alfvén waves (Unti and Neugebauer, 1968; Belcher and Davis, 1971). Density fluctuations are finite but generally small (Barnes, 1979). For these reasons solar wind fluctuations have often been characterized as being dominated by outward propagating Alfvén waves but with a minority component of fast-mode waves to account for the observed density fluctuations. However, the nonlinearities of the MHD equations can couple wave modes to produce fluctuations with properties that are distinct from the individual modes. Further, there is observational evidence of a nonlinear cascade leading to asymptotic states; this challenges the wave view at a fundamental level.

7.3.1 Which basic questions remain unanswered?

Of course, any author when asked which basic questions remain will provide different answers. Other authors in this text would say "What is the nature of magnetic reconnection?", "How are CMEs ejected?", "What is the dynamics of magnetospheric storms?", or "How are the most energetic cosmic rays accelerated?" For me and for many others, the most prominent question regarding the solar wind is "What are its source and dynamics?" Restating this, we can ask "What heats the solar atmosphere and accelerates the solar wind?" The general formalism for accelerating the wind is widely recognized (Hundhausen, 1972) and the accelerating solar corona can be imaged (Sheeley *et al.*, 1997). However, we need to

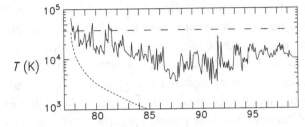

Velocity range (km/s)	γ (0.3–1 AU)
< 300	1.331 ± 0.129
300 to 400	1.223 ± 0.087
400 to 500	1.033 ± 0.095
500 to 600	0.826 ± 0.099
600 to 700	0.762 ± 0.092
700 to 800	0.808 ± 0.169

Fig. 7.2. (Left) Radial dependence of thermal proton temperatures $T_p \sim R^{-\gamma}$ from 0.3 to 1 AU and corresponding wind speed ranges. Variations shallower than $R^{-4/3}$ imply in situ heating. Reproduced from Schwenn and Marsch (1991). More recent analyses indicate that in situ acceleration of the slow wind implies that the radial dependence of slow-wind temperatures is shallower than that shown. (Right) Proton temperatures observed by Voyager 2 (solid line) from 1 to 70 AU. The average 1 AU temperature is represented by the horizontal broken line. The adiabatic expansion from 1 AU is shown by the descending short-broken line. Reproduced from Smith *et al.* (2001a).

discover the heat source that elevates the temperature of the solar atmosphere above the chromospheric level and how that heat energy is applied, since this determines the characteristics of the resulting wind. We know that there exists some source for heating the corona that involves energy transport from lower altitudes (Gabriel, 1976; Schrijver *et al.*, 1999). Electron heat conduction was the original suggestion for the mechanism accelerating the wind (Parker, 1958). Ion cyclotron damping may account for the fast wind (Hollweg and Isenberg, 2002). Magnetic reconnection is a popular suggested energy source (Parker, 1994). The source dynamics for the slow wind is unresolved and it would be an exaggeration to claim that much is known about the acceleration mechanism for the fast wind. McComas *et al.* (2007b) provided a good and up-to-date review of the solar wind problem within the context of a proposed solar probe.

While some heating mechanism is responsible for the initial acceleration of the wind, heating continues in the region of supersonic flow, beyond 0.3 AU, and into the outer heliosphere. If a wind expands adiabatically, the proton temperature would vary with heliocentric distance as $R^{-4/3}$. Inside 1 AU the temperature varies less steeply, as $R^{-0.9}$, and the slow wind undergoes continued, albeit minimal, acceleration (Arya and Freeman, 1991; Freeman *et al.*, 1992; Totten *et al.*, 1995; Verma *et al.*, 1995; Vasquez *et al.*, 2007b). Unraveling the mechanism of the continued acceleration leads to the conclusion that both fast and slow winds undergo in situ heating inside 1 AU. By ~ 10 AU the radial variation of the temperature is flattening to $R^{-0.7}$ (Gazis *et al.*, 1994) and over the range from 1 to 43 AU the best-fit proton temperatures vary as $R^{-0.49}$ (Richardson *et al.*, 1995b; Richardson and Smith, 2003). See Figure 7.2. Beyond this, the radial dependence of the proton temperature is often deceptively flat. By 30 AU the solar wind is ten times hotter than adiabatic expansion from 1 AU would predict; by 70 AU the wind is a hundred times hotter. Folded into these behaviors is a measure of solar cycle variability (Smith *et al.*, 2006b) that leads to the long-term variability of local heating as a function of the solar cycle. If we can understand the means by which energy is provided to the background plasma in the solar wind then perhaps we can find insights into the dynamics that accelerates the solar wind closer to the Sun.

Although some bulk plasma properties (IMF intensity and direction, flow velocity, density, etc.) but not the temperature seem well described by early theories, the fluctuation spectra (for the magnetic field, velocity, and density, for instance) have all been the subject of contradictory interpretations and assertions. The spectrum of the magnetic fluctuations can be divided into three subranges, and we discuss these in detail in Section 7.6. For now, we focus on frequencies from $\sim 5 \times 10^{-5}$ Hz to ~ 0.3 Hz at 1 AU in the ecliptic plane. We will see later that these frequencies are often called the "inertial range", in analogy with hydrodynamic turbulence. The solar wind is supersonic, so that wave vectors $\mathbf{k}$ are convected by the solar wind velocity $\mathbf{v}_{SW}$. This yields frequencies in the spacecraft frame f_{sc} that are readily described by the aforementioned Taylor frozen-in-flow assumption (see Fig. 7.5 for an example): $2\pi f_{sc} = \mathbf{k} \cdot \mathbf{v}_{SW} + \omega_r$, where ω_r is the wave frequency in the rest frame of the plasma (generally seen as small compared with the convection term). We can invert the Taylor frozen-in-flow equation and, for a typical 450 km/s wind with an angle Θ_{BR} between the mean field and the radial direction (that of the wind velocity) equal to $45°$, the above frequency range implies that the projection k_r of the wave vector onto the radial direction, generally called the *reduced* component of the wave vector, satisfies $7 \times 10^{-7} < k_r < 4 \times 10^{-3}$ km^{-1} (with equivalent length scales from 1500 km to 9×10^6 km). This is important: the actual wave vectors contributing to these measurements could have any component perpendicular to the flow and produce the same measured frequency subject to minor variations due to ω_r. For this reason a fundamental question in solar wind studies has been "What is the orientation of the wave vectors that we measure?" To this we can add "... and how do they evolve?"

As stated in the previous section, the magnetic and velocity fluctuations at these scales are strongly correlated and consistent with the outward propagation of Alfvén waves (Unti and Neugebauer, 1968; Belcher and Davis, 1971). However, Roberts *et al.* (1987) showed that this correlation degrades with heliocentric distance and by ~ 10 AU the correlation $\langle \delta \mathbf{v} \cdot \delta \mathbf{B} \rangle \to 0$ (the δ symbol is used here, and below, to indicate fluctuations relative to a slowly varying background). Since the correlation between the magnetic and velocity fluctuations is a measure of the fractions of the waves propagating outward and inward (anti-sunward and sunward), it appears that the ratio of these two populations is changing as the plasma evolves outward from the Sun. It is not clear which mechanism has sufficient energy to effect this change if the fluctuations are non-interacting waves. The fluctuations are largely transverse to the mean magnetic field and they remain perpendicular to the mean field as the field rotates in the outer heliosphere. Alfvén waves propagate in a straight line and are insensitive to rotations of the background field. Therefore, non-interacting waves in the outer heliosphere should have fluctuations that are perpendicular to the radial direction and not the mean-field direction, which is azimuthal. Perhaps most central to this discussion, the spectrum of fluctuations is seen to display a power law with indices varying over only a narrow range of values. The indices of the magnetic spectra are centered on $-5/3$ (Smith *et al.*, 2006a) while the indices of the velocity spectrum are centered on $-3/2$ (Podesta *et al.*, 2006). Both forms are predictions of different magnetohydrodynamic (MHD) turbulence theories discussed below, but neither theory, or apparently any other theory at present, predicts the coexistence of both.

7.3.2 Why turbulence?

If the fluctuations are composed of Alfvén waves then their evolution may be understood by a two-step process often referred to as "weak turbulence theory". First, Alfvén waves propagate according to linearized first-order expansions of the plasma equations, with dispersion relation $\omega_r = \mathbf{k} \cdot \mathbf{v}_A$, where $\mathbf{v}_A \equiv \mathbf{B}_0/\sqrt{(4\pi\rho)}$ is the Alfvén velocity. The velocity and magnetic fluctuations $\delta\mathbf{v}$ and $\delta\mathbf{B}$ are largely transverse to the mean-field direction, and $\delta\mathbf{v} = \pm\delta\mathbf{B}/\sqrt{(4\pi\rho)}$, the sign depending on the wave propagation direction and the direction of the mean magnetic field. Neither of these characteristics changes as the wave propagates. Second, higher-order expansions of the plasma equations lead to coupling terms, so that energy can be exchanged between the waves. To remain consistent with the derivation and the concept, the time scale for energy transfer between modes must be long compared with the period of the waves. In other words, the wave identity must be the leading dynamic that controls the evolution of the fluctuation, and so energy exchange is secondary. This means that the nonlinearity is comparatively weak in weak turbulence theories (hence the name) and energy transfer rates are relatively slow. Adding some fraction of compressive modes will not change this basic relationship of the time scales. Still, weak turbulence theory provides a potential means of understanding why $\langle\delta\mathbf{v} \cdot \delta\mathbf{B}\rangle \to 0$ in the outer heliosphere, why fluctuations remain transverse to the local mean magnetic field, and how energy is moved from large to small scales for dissipation and heating of the background plasma. The question we must ask is "Does weak turbulence theory provide the correct rates and spectra?" Then one must wonder what happens when there are more modes interacting with one another, so that energy is moved at a higher rate and the overall time scale for energy transfer through a given mode becomes small. Can we still describe the problem as a system of interacting waves if a given wave no longer lives long enough to accomplish a complete oscillation? Traditional hydrodynamic turbulence attempts to explain exactly this situation: fluctuations are short-lived and energy transport between the modes is the dominant dynamic. So we will now examine traditional hydrodynamic turbulence theory for insights into the MHD turbulence problem.

7.4 The Navier–Stokes equation and hydrodynamic turbulence

Hydrodynamic turbulence is described by the Navier–Stokes (N–S) equation,

$$\frac{\partial \mathbf{v}(\mathbf{x}, t)}{\partial t} + (\mathbf{v} \cdot \nabla)\mathbf{v} = -\frac{1}{\rho}\nabla p + \nu\nabla^2\mathbf{v} \tag{7.1}$$

where $\mathbf{v}$ is the velocity field of the fluid, ρ is the mass density, p is the fluid pressure, and ν is the kinematic viscosity. To this we add incompressibility:

$$\nabla \cdot \mathbf{v} = 0. \tag{7.2}$$

Simply put, even incompressible turbulence is difficult, but it is the simplest case and this is where the most progress has been made. So long as density variations remain a 10% effect, as is the case in many space plasma systems, incompressible turbulence ideas should provide a good starting point from which to grow more advanced theories.

The basic "mode" of the N–S equation is one that creates vortices (often called "eddies"). We extract the dynamics of vortices by computing the vorticity $\omega = \nabla \times \mathbf{v}$. Equation (7.1) can be expressed as

$$\frac{\partial \omega(\mathbf{x}, t)}{\partial t} = \nabla \times (\mathbf{v} \times \omega) + \nu\nabla^2\omega. \tag{7.3}$$

We can rewrite $\nabla \times (\mathbf{v} \times \omega)$ as $(\omega \cdot \nabla)\mathbf{v} - (\mathbf{v} \cdot \nabla\omega)$ and find that a leading dynamic of the nonlinear term in the N–S equation is the advection of vortices (as is evident from $\partial\omega/\partial t + \mathbf{v} \cdot \nabla\omega = \cdots$). If one vortex places a velocity field across another in such a way that the applied field has small spatial gradients, the subject vortex will simply advect in the field of the first. However, when one vortex applies a velocity field with strong gradients across a second vortex ($\partial\omega/\partial t \sim (\omega \cdot \nabla)\mathbf{v}$) then the former can distort the latter if the local gradients are sufficiently strong. We will examine that interaction and distortion in the next section, but it was first proposed by Kolmogorov (1941a) that the local gradient of one vortex acts on another so as to shed smaller vortices, which in turn interact to shed still smaller vortices until a broad spectrum of vortex energy is created.

Two important processes are accomplished by these dynamics. First, the fluid is "stirred", so that mixing is accomplished by convection rather than diffusion.

We alluded to this in Section 7.2. The same behavior can be observed in movies of the solar photosphere and simulations of the solar convection zone. Numerous simulations of magnetospheric transients reveal similar convected circulation patterns. These and many other systems experience enhanced mixing via convection.

Second, a process is put in place whereby kinetic energy in the form of the organized motion of the fluid is moved from one scale to another *en route* to its dissipation into heat. The leading dynamic behind the movement of energy to smaller scales is vortex-line stretching, or the distortion of one vortex by another, which is well described in Tennekes and Lumley (1972).

One last idea remains to be explored before proceeding further: we must define the Reynolds number. The N–S equation (7.1) can be rewritten if we adopt characteristic (large) scales for the flow v_0 and L_0, so that time is also rescaled in units $t_0 = L_0/v_0$. We can then define the Reynolds number $R \equiv L_0 v_0/\nu$ and the entire N–S equation can be rescaled in terms of L_0 and v_0 units ($v \rightarrow v'/v_0$, $\partial v/\partial t \rightarrow (L_0/v_0^2)\partial v'/\partial t'$, $\partial/\partial x \rightarrow L_0 \partial/\partial x'$, etc.). Then the rewritten N–S equation has the same form as Eq. (7.1) with the substitution $\nu \rightarrow R^{-1}$. Large Reynolds numbers imply large characteristic scales and small viscosities. It is risky and subjective (or at the least highly dependent on context) to state how large R must be to yield turbulent behavior, but $R > 10^2$ seems to be a minimum for most discussions; we will see in the next section that this is the condition needed to establish the most famous prediction in turbulence theory.

7.4.1 Basic ideas of Navier–Stokes turbulence theory

We can compute the rate of change of the energy E_v in a turbulent incompressible flow from the N–S equation as follows:

$$\frac{\partial E_v(\mathbf{x}, t)}{\partial t} = \mathbf{v} \cdot \frac{\partial \mathbf{v}}{\partial t} = -\mathbf{v} \cdot [(\mathbf{v} \cdot \nabla)\mathbf{v}] - \mathbf{v} \cdot \left(\frac{1}{\rho}\nabla p\right) + \nu \mathbf{v} \cdot \nabla^2 \mathbf{v}. \qquad (7.4)$$

If we compute an ensemble average or perform a volume integral and apply reasonable boundary conditions that isolate the flow from sources and sinks, we find that the nonlinear terms of the N–S equation (7.4), $-\mathbf{v} \cdot [(\mathbf{v} \cdot \nabla)\mathbf{v} + \rho^{-1}\nabla p]$, conserve energy. The diffusive term $\nu \mathbf{v} \cdot \nabla^2 \mathbf{v}$ does not conserve energy. This means that we have a dynamic for moving energy through the system to different spatial scales and a separate means of dissipating energy to form heat.

We can examine better the separate processes of transporting and dissipating energy if we transform the equations to wave number (Fourier) space. We make use of the simpler discrete transform with $\mathbf{v}(\mathbf{x}) = \sum_{\mathbf{k}} \mathbf{v}_{\mathbf{k}} e^{i\mathbf{k}\cdot\mathbf{x}}$ with Fourier components

$\mathbf{v_k}$. The N–S equation (7.1) then becomes

$$\frac{\partial \mathbf{v_k}}{\partial t} = -\sum_{k=k_1+k_2} \left[\left(\mathbf{v_{k_1}} \cdot i\mathbf{k_2} \right) \mathbf{v_{k_2}} - ik p_k - \nu k^2 \mathbf{v_k} \right] \tag{7.5}$$

and the incompressibility equation (7.2) is now

$$\mathbf{k} \cdot \mathbf{v_k} = 0. \tag{7.6}$$

It is then a straightforward, if laborious, exercise to show that on the one hand suitably closed triads $\mathbf{k}$, $\mathbf{k_1}$, $\mathbf{k_2}$ for which $\mathbf{k} = \mathbf{k_1} + \mathbf{k_2}$ conserve energy. On the other hand $-\nu k^2 \mathbf{v_k}$, which becomes $\partial E_k / \partial t \sim -\nu k^2 E_k$, is increasingly dissipative as $|\mathbf{k}|$ becomes large. Energy dissipation is a small-scale process.

Kolmogorov (1941a) understood the implication of these arguments and the balance between advection and distortion in vortex interactions. He asserted that the large vortices will primarily convect smaller vortices, but that two vortices of roughly equal size will interact through their velocity gradients and distort one another. Furthermore, he offered a time scale for the distortion to take place. Kolmogorov first notes that a vortex of characteristic speed v and size λ will circulate, or "turnover", in the time scale $\tau = \lambda/v$. Next, he postulates that the time scale for two vortices of equal size to distort one another is that same turnover time τ if the distortion of one vortex is tied to the circulation pattern of the other. Since the energy contained within either vortex is v^2, the resulting rate at which energy moves through the spatial scale λ becomes $v^2/\tau = v^3/\lambda$.

Kolmogorov (1941a) also proposed his "first hypothesis of similarity" (Frisch, 1995): *At very large, but not infinite, Reynolds numbers, all the small-scale statistical properties are uniquely and universally determined by the scale λ, the mean energy dissipation rate ϵ, and the viscosity ν.* Then the omnidirectional energy spectrum can be described from simple dimensional analysis, as it is dependent only on these three quantities:

$$E(k) = \epsilon^{2/3} k^{-5/3} F(\eta k) \tag{7.7}$$

where $\eta \equiv (\nu^3/\epsilon)^{1/4}$. Here $F(\eta k)$ is a universal dimensionless function that is a constant C_K at small k and forms the dissipation range at $\eta k \geq 1$ and $E(k)$ is the integral over angles of the fully three-dimensional (3D) spectrum, which represents the energy in concentric layers dk in thickness, the total energy being given by $E = \int_0^\infty E(k) dk$. Kolmogorov assumed homogeneity (the statistics of all points in space are the same) and isotropy (there is no preferred direction). He also assumed a stationary spectrum, so the energy cascade rate equals the energy dissipation rate; this also equals the rate at which the fluid is heated via the turbulent cascade. Experiment has shown that $C_K \simeq 1.62 \pm 0.17$ (Sreenivasan, 1995). In Fig. 7.3 the

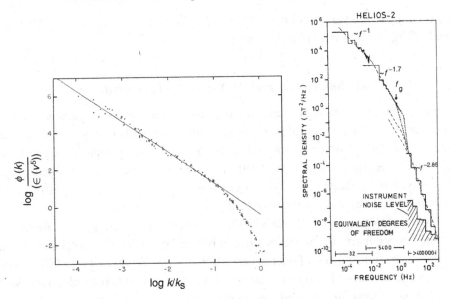

Fig. 7.3. (Left) Inertial and dissipation-range spectrum for hydrodynamic turbulence (From Pond *et al.* (1963), which gives the figure as an augmentation of the Grant *et al.* (1962) result.) (Right) Spectrum of the IMF measured at 0.3 AU on 14 April 1975 by the Helios spacecraft, showing inertial and dissipation range spectra. The latter extends to 200 Hz, which is the limit of the measurement. (From Denskat *et al.* (1983).)

spectra of N–S and interplanetary turbulence may be compared. Combining the spectral prediction with the above assumption of evolution on the eddy turnover time implies that the largest scales in the spectrum live the longest time while the smallest scales evolve most quickly.

Kolmogorov's prediction implies there should be a subset of the turbulent spectrum in which energy is conserved as it moves to smaller scales and that only at those small scales will dissipation become important. This energy-conserving region of the spectrum is termed the "inertial range". This same term is used to describe the spectrum of interplanetary fluctuations, described in Section 7.3.1, for which magnetic and velocity fluctuations are correlated and a similar power law form is observed; it is also used in scintillation measurements of interstellar spectra. The first validation of the predicted hydrodynamic spectral form came from Grant *et al.* (1962), when they towed a flow meter behind a boat in a tidal channel to record the fluctuations in the fluid velocity. The left-hand panel in Fig. 7.3 shows the results of that experiment. The prediction holds for large Reynolds numbers, which means that a significant separation in scale size between injection and dissipation is required to observe the inertial range. The $-5/3$ form of the spectrum is confirmed, with the steepening of the spectrum at small scales associated with dissipation. We see very similar behavior in MHD space plasmas in the right-hand panel of Fig. 7.3 and later in this chapter.

7.4.2 Spectral cascade in Navier–Stokes turbulence

At this point the motivated reader should investigate topics including "rugged invariants" and "absolute equilibrium ensembles", which can yield surprising insights into the evolution of the turbulent spectra of two-dimensional (2D) fluids (Kraichnan and Montgomery, 1980) and their associated weather phenomena (Pedlosky, 1987). These topics have a growing relevance to space applications of magnetohydrodynamics but are outside the scope of this chapter.

While Kolmogorov (1941a) provided an estimate for the energy cascade rate in well-developed inertial-range turbulence, it is based on an heuristic argument. That argument postulates a rate of energy cascade based on the interaction of like-sized vortices and presumes a fast time scale for the interaction. It is possible to measure the nonlinear terms directly and also to measure the rate of energy cascade along with the associated time scale in any geometry even when the inertial range is not fully developed. Kolmogorov (1941b) derived an expression based on third-order structure functions that in 3D can be written:

$$S_3(\mathbf{L}) \equiv \left\langle \left[v_{||}(\mathbf{x}) - v_{||}(\mathbf{x} + \mathbf{L}) \right]^3 \right\rangle = -\tfrac{4}{5}\epsilon |\mathbf{L}| \qquad (7.8)$$

where $\mathbf{L}$ is the separation vector between the two points and $v_{||}$ is the component of $\mathbf{v}$ parallel to $\mathbf{L}$. See Frisch (1995) for a more detailed discussion and a derivation of this result based on the Kármán–Howarth–Monin relation. See also Hill (1997) and Antonia *et al.* (1997) for modern extensions and criticisms of the formalism. The relation (7.8) has been generalized to anisotropic turbulence (Monin, 1959; Monin and Yaglom, 1975). Although slow to converge and requiring many samples to yield statistically significant results, this expression provides a reliable estimate for the energy cascade rate ϵ that is based on a rigorous derivation and verifiable assumptions. We will find below that there exists an extension of this idea for MHD turbulence, and its application to solar wind measurements yields excellent agreement with the local heating rate.

7.5 Magnetohydrodynamic fluid turbulence

As we noted at the beginning, interplanetary space is not filled with a collisional fluid; rather, space contains a mixture of ionized gas(es), neutral atoms, molecules, lightly charged dust, and photon radiation. To bring the subject of fluid turbulence to the space environment, we begin with the simplest possible magnetized-fluid equations and so write down the incompressible form of the single-fluid MHD equations (compare with the fully compressible versions in Section 3.2.1):

$$\frac{\partial \mathbf{v}(\mathbf{x})}{\partial t} + (\mathbf{v} \cdot \nabla)\mathbf{v} = -\frac{1}{\rho}\nabla p + \frac{1}{4\pi}(\nabla \times \mathbf{B}) \times \mathbf{B} + \nu\nabla^2\mathbf{v}, \tag{7.9}$$

$$\frac{\partial \mathbf{B}(\mathbf{x})}{\partial t} = \nabla \times (\mathbf{v} \times \mathbf{B}) + \frac{c^2}{4\pi\sigma}\nabla^2\mathbf{B}, \tag{7.10}$$

$$\nabla \times \mathbf{B} = \frac{4\pi}{c}\mathbf{j}, \tag{7.11}$$

$$\nabla \cdot \mathbf{v} = 0, \tag{7.12}$$

$$\nabla \cdot \mathbf{B} = 0, \tag{7.13}$$

where $\mathbf{j}$ is the current density and σ is the electrical conductivity. The similarity to the N–S equation (7.1) is unmistakable, as the momentum equation (7.9) contains the N–S equation! Our goal is to generalize the concepts of the previous section to address the above equations. In the future, a viable description of compressible MHD turbulence may be found that permits the same degree of detailed prediction and discussion as that given below, but such a description is beyond the scope of this chapter and present research. Others have investigated interactions with the neutral gas or the possibility of an electron MHD regime at smaller scales (Kingsep *et al.*, 1990), but that, too, is beyond what we are considering here.

It is possible to define characteristic scales as we did in defining the Reynolds number. This previously defined Reynolds number becomes the kinetic Reynolds number and the magnetic Reynolds number is

$$R_{\mathrm{m}} \equiv \frac{v_0 L_0}{c^2/(4\pi\sigma)}.$$

As before, these two numbers define the scale separation for MHD turbulence, and large values favor the development of spectra dominated by nonlinear dynamics. We can also define Elsässer variables $z^{\pm} \equiv v \pm b/\sqrt{(4\pi\rho)}$ and the MHD equations can be rewritten in terms of the interaction of the $z^{\pm}$ fields (Elsässer, 1950). This naturally speaks to the role of Alfvén waves (or more correctly "Alfvénic fluctuations") as one z-field distorts the other in a manner reminiscent of vortices in hydrodynamics.

7.5.1 *Spectra in single-fluid MHD*

Examination of the MHD equations (7.9)–(7.13) reveals that, as with the N–S equation, the nonlinear terms conserve energy while the viscosity and resistivity dissipate energy. If we assume first that there is no mean magnetic field, the arguments of the Section 7.4 transport readily to the MHD equations. All the arguments put forward by Kolmogorov that led to Eq. (7.7) hold if one is now discussing the spectrum of the total energy (magnetic plus kinetic) and it is therefore reasonable to postulate that the spectra of MHD turbulence should follow the $k^{-5/3}$ form (Fyfe *et al.*, 1977; Mininni *et al.*, 2005; Biskamp and Müller, 2000).

If we add a mean magnetic field, we must then consider a major complication: there are now two time scales in the problem. The first is the eddy turnover time and the second is the time for two Alfvén waves to propagate past one another. Iroshnikov–Kraichnan theory (Iroshnikov, 1964; Kraichnan, 1965) postulates that the passage time of two Alfvén waves propagating at the Alfvén speed, rather than the eddy turnover time of HD turbulence, yields the critical time scale for MHD turbulence and so leads to a different prediction for the inertial-range spectrum in MHD turbulence:

$$E_{\mathrm{T}}(k) = A_{\mathrm{IK}}(\epsilon v_{\mathrm{A}})^{1/2} k^{-3/2} \tag{7.14}$$

where A_{IK} is a constant which can be shown to be given by $A_{\mathrm{IK}} = C_K^{3/4} = 1.42$ (see the discussion after Eq. (7.7)) (Matthaeus and Zhou, 1989). Iroshnikov–Kraichnan theory is based on Alfvén wave dynamics and so predicts the equipartition of energy between the magnetic and kinetic fluctuations. Although IK theory was presented as a prediction for isotropic MHD turbulence, Alfvén wave propagation is strongly anisotropic: the Poynting flux vector is always directed along the mean magnetic field. It therefore may be wondered whether this prediction is more appropriate for field-aligned wave vectors than for perpendicular wave vectors.

The above predictions apply to isotropic MHD turbulence. Experiments with liquid metals and an applied DC magnetic field show that the turbulence evolves to nearly 2D flow confined to the plane perpendicular to the applied DC field (Lielausis, 1975; Tsinober, 1975; Volish and Koliesnikov, 1976). Magnetohydro-dynamic simulations by Matthaeus *et al.* (1983), Ghosh *et al.* (1998), Oughton *et al.* (1994), and Matthaeus *et al.* (1996a), showed that the energy in inertial-range fluctuations moves to perpendicular wave vectors in the presence of a mean magnetic field. This behavior was also shown to exist in Hall-MHD (Ghosh and Goldstein, 1997), where electrons and ions are treated as separate species in the fluid limit. In this way, the initial distribution of energy is overwritten by the turbulent cascade to produce a 2D MHD system in the plane perpendicular to the mean field. The above generalization of the Kolmogorov (1941a) theory (Eq. 7.7) would seem to be equally valid for 2D MHD turbulence, where energy is described in the above manner. This led Montgomery and Turner (1981) to suggest that perpendicular wave vectors should evolve according to turbulence dynamics while parallel wave vectors may be more wave-like.

With a 2D geometry evolving from the 3D MHD turbulence, owing to the presence of an externally applied mean magnetic field (such as is imposed by the Sun on its atmosphere and solar wind or by the Earth on the magnetosphere), it becomes necessary to understand better the coupling of the 2D state to the field-aligned wave vectors. Although we have argued that wave propagation from the Sun to the interstellar medium seems to violate observations in the solar wind, it is difficult to deny that wave generation is probably involved in any process that disturbs magnetic field lines. Therefore, whether we adopt a weak turbulence or strong turbulence viewpoint, we must find ways to couple wave activity to the nonlinear processes that evolve the fluid (otherwise we have to accept that the wave activity represents a decoupled component that greatly complicates the interpretation of observations).

One approach to this problem has been through the reduced MHD (RMHD) equations, sometimes called the Strauss equations (Rosenbluth *et al.*, 1976; Strauss, 1976, 1977; Montgomery, 1982). This approach attempts to separate the turbulent 2D time scales from the more slowly evolving wave spectrum; see the review by Zank and Matthaeus (1992). Higdon (1984) used the ideas of RMHD (Strauss, 1976) and the prediction of a $-5/3$ spectrum for the 2D component provided by Fyfe *et al.* (1977) to propose that the spectrum for parallel wave vectors should vary as k^{-3}. Higdon argued that the separation between wave vectors represented by the differing parallel and perpendicular dynamics is given by the relation $k_z = k_\perp^{2/3}$. In this way the 2D component becomes increasingly significant as k becomes larger. Goldreich and Sridhar (1995) adopted the same idea, calling it the "critical balance assumption", but derived $k^{-5/3}$ for the perpendicular spectrum and k^{-2} for the measured parallel spectrum. (Biskamp (2003) mistakenly stated that the predicted spectrum for parallel wave vectors was $-5/2$.) Boldyrev (2005, 2006) predicted that the perpendicular wave vectors alone assume a $-3/2$ spectral index. With each distinct prediction for a spectral form comes a different prediction for the rate of energy cascade and a different resulting rate of heating. As we will see in the next section, observations do not yet conclusively resolve the question of what is the spectral form for inertial range fluctuations in MHD turbulence.

Most simulations of MHD turbulence provide dissipation within the fluid approximation in the spirit of HD turbulence. As such, viscosity and resistivity provide the dissipation through the formation of velocity gradients and current sheets. In most space and astrophysical plasmas dissipation occurs outside the fluid approximation at kinetic scales that mark the breakdown of the single-fluid theory, and we will explore this idea in the next section.

The structure function analysis of Kolmogorov (1941b) has been extended to include isotropic MHD (Politano and Pouquet, 1998b, 1998a; Podesta, 2007) and anisotropic MHD (MacBride *et al.*, 2008; Podesta *et al.*, 2008), so that formal calculations of the energy cascade without regard for the spectral form can be performed. The isotropic 3D form is

$$D_3^{\pm} \equiv \left\langle \delta z_{\parallel}^{\mp}(\mathbf{L}) \left\{ \sum_{i=1,3} \left[\delta z_i^{\pm}(\mathbf{L}) \right]^2 \right\} \right\rangle = -\frac{4}{3} \epsilon^{\pm} |\mathbf{L}| \tag{7.15}$$

where we again define the Elsässer variables $\mathbf{z}^{\pm} \equiv \mathbf{V} \pm \mathbf{B}/\sqrt{4\pi\rho}$ (Elsässer, 1950) and $\delta \mathbf{z} \equiv z(\mathbf{x}) - z(\mathbf{x} + \mathbf{L})$; we adopt the notation D_3 because the structure function is no longer the familiar symmetric form S_3 used in hydrodynamics.

7.6 The spectrum of interplanetary turbulence

We now turn to a more detailed analysis of solar wind turbulence. We choose the solar wind for a focused examination for reasons that are by now obvious: it is a supersonic and super-Alfvénic flow very similar to the flows in traditional wind tunnels where single probes can produce space-based analyses derived from inherently frequency-based measurements. Simply put, the solar wind is the ideal MHD wind tunnel. The ideas we develop here should have natural application to magnetospheres, to the Sun and the solar atmosphere, and to interstellar space, but we need a theory and observations that can inform our interpretation of subsonic or remote measurements.

7.6.1 *The energy-containing range of interplanetary turbulence*

In the introduction to this chapter we referred to laboratory experiments on N–S turbulence in wind tunnels where the turbulence is generated by flow past a simple wire screen. We also discussed how the initial turbulence evolves downstream to lose memory of the grid structure and exhibit uniform isotropic turbulence that is independent of the source, which is the foundation of the studies discussed above. Measurements recorded too close to the grid retain a memory of the source mechanism and fail to exhibit isotropy, homogeneity, or the spectrum predicted by Kolmogorov (1941a) (Eq. 7.7). If the solar wind is a natural wind tunnel for the study of MHD turbulence then the Sun is the source for the long-lived large-scale disturbances that drive the turbulence. These large-scale disturbances are the wind shear regions, where fast winds collide with slow winds, the coronal mass ejections (CMEs) that push their way through the preceding plasma, shocks, all forms of varying solar sources, and whatever turbulence might exist within the acceleration region of the solar wind. The latter is particularly important for observations at 1 AU since there continues to be evidence that the solar turbulence imparts to the solar wind a signature related to the turbulence of the acceleration region. We will address that issue below.

Thus these large-scale transients form the sources for the solar wind turbulence (Coleman, 1968; Goldstein *et al.*, 1984). Their interaction represents large energy reservoirs that can continue to drive fluctuations within the solar wind for a distance of many astronomical units (AU), and it is these large-scale fluctuations which, in turn, interact to produce the turbulent cascade. To estimate the rate of the energy cascade driven by the interaction of these large-scale transients, we can return to the hypothesis of Kolmogorov (1941a) as applied to MHD (Eq. 7.7) and determine the energy in the Elsässer fields $\delta v^2 + \delta B^2/(4\pi\rho) = \delta z^2$ and the scale size λ of the characteristic energy-containing transients. The resulting rate of energy driving is $\delta z^3/\lambda$. Matthaeus *et al.* (1996a, 2004) and Breech *et al.* (2007) have developed this idea in detail for a turbulent transport model based on the rate at which the turbulence is driven by the energy-containing scales, and Smith *et al.* (2001a) tested the theory against observations by Pioneer 11 and by Voyager 2.

Inside ~ 30 AU the dominant heating source is the interaction of the large-scale transients described above, which drives the cascade of energy to smaller scales. Beyond this point, entrainment (the overtaking of slow streams by fast streams, producing the merging of the transient flows) is sufficiently complete that local gradients are minimized and transient behavior is largely lost as a source for driving interplanetary turbulence. However, wave excitation and energy injection by newborn pickup protons (described below) starts at ~ 6, AU, and by ~ 30 AU it becomes the dominant source of energy for heating the background plasma.

Pickup ions (the result of charge-exchange collisions of the solar wind ions with neutral particles from the interstellar medium that have penetrated deeply into the heliosphere) form the dominant pressure term in the outer heliosphere and thereby possess considerable energy for possible energization of the thermal component. There is no evidence that all the energy of the pickup ions enters into the thermal population; estimates place that energy transfer at only ~ 1 % of the total pickup ion energy. Newborn pickup ions are given speeds twice the solar wind speed by virtue of the pickup process and instabilities arising from particle scattering lead to wave generation and enhanced power in the fluctuation spectra (Lee and Ip, 1987). The resulting wave energy appears at frequencies comparable with the cyclotron frequency, normally considered to be part of the inertial range, but in fact they represent new energy sources and as such function much like a second energy-containing range at smaller scales. Early theories for wave generation by interstellar pickup ions (Lee and Ip, 1987) were based on lessons learned from cometary observations (Goldstein and Wong, 1987; Tsurutani, 1991) and predicted large buildups of wave energy within a narrow range of spacecraft frequencies.

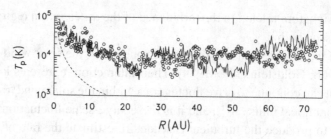

Fig. 7.4. Solar rotation averages of Voyager 2 proton temperatures (circles) and turbulence theory prediction (solid line) derived from the Omnitape 1 AU boundary condition. The adiabatic expansion from 1 AU is also shown (broken line).

This buildup of wave energy should lead to the scattering of newborn ions into a bisphere, but the waves are not seen in this form (Murphy *et al.*, 1995).

The expected dramatic enhancement in the power spectra due to the accumulation of pickup ion wave energy is never seen. Instead, the wave energy is broadly distributed over a larger range of frequencies and the enhanced power is small. Isenberg *et al.* (2003) showed that by assuming the wave energy enters into the turbulence cascade, the ion-scattering dynamics and the rate of energy obtained from the ion populations are modified, reducing the energy injection by the pickup ions. This point is important as the waves excited by the newborn pickup ions do not themselves interact directly with the thermal ions, so that a cascade of the wave energy to smaller scales is required for the wave energy to become heat. Incorporating this into the heating theory (Zank *et al.*, 1996; Smith *et al.*, 2001a) leads to predicted energies within a factor ~ 2 of observations (Smith *et al.*, 2006b). In Fig. 7.4 the turbulence prediction may be compared with the observed temperature of solar wind protons as measured by Voyager 2. The success of this turbulent transport theory, driven first by the large-scale energy-containing fluctuations of the wind and then by wave generation by newborn pickup ions, points to the value of a turbulent cascade in two regards. First, it can account for the heating of the interplanetary plasma from 0.3 to 100 AU; second, it shows that turbulent dynamics can be competitive with kinetic physics in some cases.

If we look for analogies between the energy-containing range of interplanetary fluctuations and other space plasma observations, we are immediately drawn to images of the Sun's photosphere. Driven by thermal energy below the surface, the convection cells visible on the surface overturn in a vast array of "cells abutting cells" (see e.g. Fig. 8.3). Although they are driven by a different energy source, can we imagine that there are no shears along the surfaces of these convection cells? Can we postulate that the large vortices evident on the Sun's photosphere are non-interacting? Can we really claim that these large-scale fluctuations do not drive a form of fluid turbulence leading to a predictable spectrum of fluctuations and some

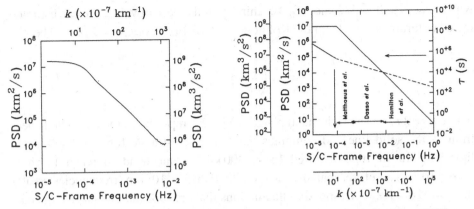

Fig. 7.5. (Left) Computed spectrum of total power for ACE observations from 1998 through 2004 (from MacBride *et al.*, 2008). (Right) Representation of the total power spectrum (solid curve) and computed eddy turnover time (i.e. the lifetime) as a function of scale. The horizontal arrow marks the average convection time from the Sun to 1 AU. Frequency intervals representing the Matthaeus *et al.* (1990) and Dasso *et al.* (2005) analyses are shown (from Hamilton *et al.*, 2008.)

form of local energy cascade? It seems unthinkable that turbulence does not play a significant role in these dynamics.

7.6.2 *The inertial range of interplanetary turbulence*

The inertial range is the most extensively studied part of the interplanetary spectrum. It is this part of the magnetic spectrum that Belcher and Davis (1971) observed to be characteristic of outward-propagating Alfvén waves with largely transverse magnetic and velocity fluctuations and that Coleman (1968) observed to have an $f^{-5/3}$ form. It was once popular to argue that the source region for the solar wind is turbulent but the interplanetary fluctuations are not. In this view the interplanetary waves were imprinted with the spectrum of the turbulent acceleration region. If we instead adopt the view that the interplanetary spectrum is that of an evolving magneto-fluid, we must still recognize that it is possible to make measurements at a distance sufficiently close to the source that the observations are still strongly imprinted by the source dynamics and have not yet evolved as a result of in situ solar wind dynamics. One measure of this distance is that it is the distance where the transit time from the source exceeds the lifetime of a turbulent fluctuation. Then fluctuations at this and smaller scales are likely to be the result of in situ dynamics, while larger scales may still retain a memory of the source region.

The left-hand panel of Fig. 7.5 shows the spectrum of the total energy $E_T(k)$ (magnetic plus kinetic) as derived from ACE observations over the period 1998–2004. The spectrum can be fitted to a power law with resulting index -1.6 ± 0.15, which places both $-3/2$ and $-5/3$ within 1σ of the best-fit value. We can estimate the eddy turnover time from the Kolmogorov (1941a) theory, $\tau = \lambda/\delta v$, to be

$$\tau = \left(\frac{2\pi}{k}\right)\sqrt{\frac{3}{kE_k}}$$

where $k = 2\pi/\lambda$. The right-hand panel of Fig. 7.5 reproduces the computed spectrum as the solid line and computes the resulting estimate for τ as the dashed line. The transit time required for a 400 km/s solar wind to reach Earth is $\sim 4 \times 10^5$ s. Therefore, spatial scales smaller than 5×10^6 km (frequencies greater than 8×10^{-5} Hz) correspond to fluctuations that are too small to survive the cascade dynamics of the inertial range during the transit from the source region to Earth (their turnover time is small compared with the lifetime of the flow); therefore they must arise via the in situ dynamics. We might call this scale, whose turnover time matches the lifetime of the flow, the transit time scale but in fact it agrees very well with the computed correlation scale. In other words, fluctuations at scales smaller than this exhibit a loss and replenishment of energy as they participate in the turbulent energy cascade while fluctuations at larger scales remain long-lived on the time scale of the lifetime of the solar wind at 1 AU. Therefore, as mentioned above, large scales retain a memory of the source while smaller scales are no longer dependent on how the flow was created. Fluctuations at scales smaller than the correlation scale may manifest characteristics different from those at larger scales because they arise via in situ dynamics.

The left-hand panels of Fig. 7.6 show the trace of the magnetic fluctuation spectrum for the period of time observed by the Advanced Composition Explorer (ACE) spacecraft. This spectrum focuses on the high-frequency end of the inertial range and extends into the dissipation range to be discussed in the next subsection. There is nothing especially unique about the former interval, and it is well fitted by a $-5/3$ power law. The right-hand panels of Fig. 7.6 show the distributions of power law indices derived from the same data base of events. Note that the distribution is centered on $-5/3$ but includes $-3/2$ at the 1σ level.

To complicate matters, Podesta *et al.* (2006, 2007) showed that the power spectrum of the Helios and Wind magnetic field fluctuations at 0.3 and 1 AU, respectively, in the range 10^{-4} to 10^{-1} Hz display a $-5/3$ index while the velocity spectra possess a $-3/2$ spectral form; see Fig. 7.7. To further complicate the discussion, there appears to be a dependence of the power law index on the orientation of the wave vector. Ulysses observations at high latitude during solar minimum yield a $-5/3$ spectral index for magnetic fluctuations between 0.01 and 0.1 Hz except for the most field-aligned wave vectors, which have an index -2 (T. S. Horbury, private communication, 2007). The latter is revealed when the angle Θ_{bv} between

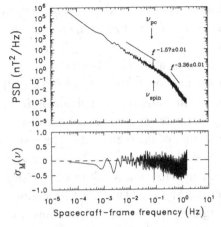
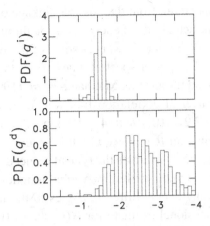

Fig. 7.6. (Left-hand panels) Spectrum of ACE magnetic field interval for day 213 in 2000, 0–6 UT. The inertial range spectral slope is −1.57, midway between −5/3 and −3/2 expected values. (Right-hand panels) Distribution of inertial-range spectral index q^i (upper right) and dissipation range spectral index q^d (lower right) from ACE study. Both the normalized magnetic helicity (lower left) and dissipation range spectral indices (lower right) are discussed in Section 7.6.3. Both figures are from Hamilton *et al.* (2008).

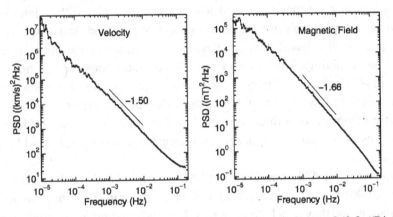

Fig. 7.7. (Left) Trace spectrum of velocity fluctuations showing a −3/2 fit. (Right) Trace spectrum of magnetic fluctuations showing a −5/3 fit. Both figures are from Podesta *et al.* (2007).

the mean field and solar wind velocity tends to 0. This would seem to agree with the prediction of Goldreich and Sridhar (1995). However, an analysis at low latitudes using ACE data focusing on spacecraft frequencies from 0.001 to 0.01 Hz (J. A. Tessein, private communication, 2007) does not produce the same result but shows a general lack of dependence of the power law index on Θ_{bv}. The same study shows that the velocity spectrum likewise lacks any dependence on Θ_{bv}.

So, it is difficult at present to determine which approximation of MHD turbulence best describes the solar wind inertial-range spectra. Thus everything is uncertain, from the possible linkage of parallel and perpendicular wave vectors to the actual rate of energy cascade through the spectrum. Fortunately, there are ways to determine answers to some of these questions.

The power spectral index is not the only form of anisotropy observed in solar wind turbulence. Matthaeus *et al.* (1990) used a method of building an (assumed) universal correlation function for magnetic fluctuations for spatial scales from $\sim 7.2 \times 10^5$ km to 4×10^6 km by comparing estimates of the reduced correlation function $R(r = v_{SW}\tau)$ while retaining the direction of the mean field. The resulting analysis treats the typical reduced correlation function analysis as a sample cut through the multidimensional function and attempts to rebuild that function using many cuts at different angles to the mean magnetic field. The resulting multidimensional estimate for $R(\mathbf{r})$ shows two energized wave vector populations, one field-aligned and one perpendicular to the mean magnetic field. This has come to be called the Maltese Cross result. However, Dasso *et al.* (2005) repeated the analysis for spatial scales from 5×10^4 to 10^6 km and showed that the two populations exist under different conditions: the field-aligned wave vectors dominate $R(\mathbf{r})$ in high-speed winds $v_{SW} > 500$ km/s and the perpendicular wave vectors dominate in low-speed winds $v_{SW} < 400$ km/s. Are these remnant signatures of the different solar sources that produce high- and low-speed winds, or are they intermediate states in the interplanetary turbulence where high-speed winds arrive at 1 AU more quickly and thereby reveal earlier states in the evolution of the turbulence away from its initial condition? We do not know. However, the computed lifetimes in the right-hand panel of Fig. 7.5 suggest that these large-scale fluctuations have experienced relatively few overturnings via the turbulent cascade since their original acceleration from the Sun three days earlier. Therefore, if any part of the inertial range retains a knowledge of its acceleration dynamics and origin, it would be these scales.

Leamon *et al.* (1998b) and Hamilton *et al.* (2008) examined fluctuations from 4000 to 50 000 km at the smallest scales of the inertial range just prior to the onset of dissipation, and they concluded that all solar wind conditions are an admixture of field-aligned and 2D wave vectors, the 2D wave vectors dominating the magnetic energy content at these scales. Hamilton *et al.* (2008) concluded that the admixture is less dependent on wind speed than is suggested by the large-scale analysis of Dasso *et al.* (2005). In particular, the dominant field-aligned geometry of fast winds seen by Dasso *et al.* (2005) is no longer evident at the smaller scales (Hamilton *et al.*, 2008). Again, since these fluctuations are much smaller than the correlation scale, it follows that they have undergone the greatest amount of regeneration via the turbulent cascade; to say it another way, they are more evolved than the larger scales and offer a better representation of the state to which the large-scale fluctuations are evolving.

We will find an explanation for this contrast between the wave vector orientations of the large- and small-scale inertial-range fluctuations below. First, though, we must obtain an independent estimate for the rate at which the thermal ions are heated in the solar wind.

Vasquez *et al.* (2007a) examined the radial dependence of solar wind proton temperatures and concluded that the in situ heating rate can be obtained from a simple formula,

$$\epsilon_{\text{heat}} = 3.6 \times 10^{-5} T_p v_{\text{SW}} \qquad \text{(J/(kg s))} \qquad (7.16)$$

where ϵ_{heat} is the rate of heating, T_p is the proton temperature, and v_{SW} is the solar wind speed. If one assumes that all the cascading energy goes into protons (probably an oversimplification) and that all the heating is achieved by the turbulent cascade (definitely not true, but probably a close approximation for plasma that has not experienced shock heating), then Eq. (7.16) becomes an estimate for the expected rate of energy cascade through the turbulent spectrum. Vasquez *et al.* (2007a) showed that the computed energy cascade rate as predicted for the $-5/3$ spectral form by the Kolmogorov (1941a) theory exceeds the observed heating rates and, if applicable, would mean that the solar wind temperature would rise with distance from the Sun even as far out as 1 AU. They found that the predicted cascade of the $-3/2$ spectrum (Kraichnan, 1965) yields appropriate values for the local heating rate, yet one wonders why either prediction should work given the dichotomy of velocity and magnetic field spectral indices reported recently (Podesta *et al.*, 2006, 2007).

MacBride *et al.* (2008) take a more rigorous approach than that of the power spectrum predictions; they applied the structure function formalism of Kolmogorov (1941b), and Politano and Pouquet (1998a, b) to assess the rate of energy cascade, since this formalism is independent of power spectral form and valid in the most general sense. See Fig. 7.8. MacBride *et al.* found cascade rates consistent with both the inferred rates derived from the local gradient of proton temperatures and the rate derived from the turbulent transport theory of large energy-containing scales. Their analysis agrees with observations for both high- and low-speed wind conditions that were observed to have heating rates differing by an order of magnitude. See Table 7.1 and the right-hand panel of Fig. 7.8. MacBride *et al.* (2008) demonstrated that fluctuations with values of the Elsässer variables $z^{\pm}$ consistent with outward propagation will cascade more aggressively (and presumably damp more rapidly) than fluctuations consistent with inward propagation, in agreement with the observation that $\langle \mathbf{v} \cdot \mathbf{b} \rangle \to 0$ by ~ 5 AU (Roberts *et al.*, 1987). They also found that the energy cascade rate perpendicular to the mean field is consistently larger than the rate of cascade along the mean field (Fig. 7.8, right-hand panel). The latter was especially true in high-speed wind conditions where the inertial-range

Table 7.1. *Computed energy cascade rates for* D_3

	$\epsilon \pm \sigma_\epsilon^a$	$\times 10^3$ J/(kg s)
Subset	MHD	Hydrodynamic
All data 1 AU	$6.50 \pm 0.14(1.45)$	$4.96 \pm 0.07(0.76)$
Low-speed 1 AU	$3.28 \pm 0.04(0.47)$	$1.44 \pm 0.02(0.24)$
High-speed 1 AU	$12.01 \pm 0.35(3.74)$	$10.70 \pm 0.15(1.61)$
Ulysses solar min.	0.15	

[a] The errors σ_ϵ in the mean values ϵ are derived from the standard deviations in the computed distributions (these are given, for comparison, within parentheses).

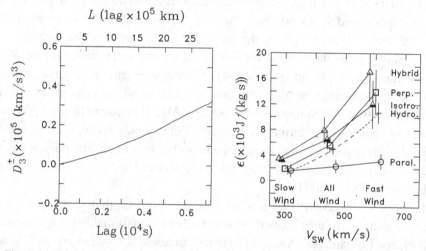

Fig. 7.8. Third-order structure function analysis for ACE data. (Left) Plot of $D_3^\pm$ (see Eq. 7.15) in the isotropic formalism using ACE data for the seven years surrounding solar maximum. The curve is for an energy cascade of the total power. (Right) Summary plot of hybrid D_3 analysis at 1 AU using ACE observations, showing that the strongest cascade is in the perpendicular direction (from MacBride *et al.*, 2008).

cascade is moving energy away from the field-aligned wave vectors and into the 2D turbulence. Therefore, it is true that the turbulence is evolving in all cases toward a 2D state. A turbulent cascade through the inertial range provides the correct energy input for heating regardless of the spectrum of the turbulence. The task then becomes to obtain a better understanding of the physics that yields anisotropic spectral indices.

Sorriso-Valvo *et al.* (2007) applied the isotropic form of the third-order structure function method to Ulysses observations at high latitude during solar minimum and found a cascade rate ~ 150 J/(kg s). The significantly different heating rates at high and low latitudes suggest numerous interesting studies of the turbulent cascade in the future, starting with an examination of turbulent sources at high latitudes (stream shears, etc.) and including the effects of large-amplitude waves and $\langle \mathbf{v} \cdot \mathbf{B} \rangle$ correlations in reducing the rate of turbulent cascading.

So, we find that we can measure the active cascade without resorting to spectral predictions that, at this point, possess limited ability to predict the observations. The next question is "What happens when dissipation sets in?"

7.6.3 The dissipation range of interplanetary turbulence

We began this chapter by alerting the reader to the possibility that plasma kinetic processes can interfere with the fluid processes described here. This was a warning against ignoring processes such as ion beam instabilities upstream of shocks wherein segments of the background energy and polarization spectra are artificially enhanced by a process outside the turbulence dynamics. However, it is just as true that a description of the ultimate fate of the energy transported through the turbulent spectrum may lie outside the fluid equations. In the traditional hydrodynamics described in section 7.4.1 the dissipation dynamics lies entirely within the fluid equations: the turbulent cascade and the conversion of energy into heat is fully described by the N–S equation. Our description of the MHD equations in Section 7.5 was no different: dissipation is accomplished by viscosity and resistivity. In naturally occurring plasmas such as the solar wind the single-fluid MHD equations are only valid at length scales greater than the ion inertial scale $\lambda_{ii} = v_A / \Omega_{ic}$ where v_A is the Alfvén speed and Ω_{ic} is the ion cyclotron frequency (Formisano and Kennel, 1969). At 1 AU the convection of the spatial scale λ_{ii} appears at ~ 0.5 Hz. The power spectrum steepens at this frequency to form a new power law spectrum of variable index from ~ -2 to ~ -5 (see the lower right-hand panel of Fig. 7.6), which is seen to extend at least to 200 Hz (Denskat *et al.*, 1983) (see Figure 7.3, (right-hand panel). If the breakdown of single-fluid MHD marks the end of the inertial range spectrum then dissipation must be accomplished either by kinetic proton dynamics or by electron dynamics.

Goldstein *et al.* (1994) noted that the magnetic helicity of the dissipation-range spectrum is biased in a sense consistent with ion cyclotron damping (see Figure 7.6, lower left-hand panel). Leamon *et al.* (1998a) and Hamilton *et al.* (2008) refined this observation to conclude that cyclotron damping provides about half the total dissipation process while the remaining dynamics (Landau or transit-time damping, current sheet formation, etc.) is not sensitive to polarization. Stawicki *et al.* (2001) argued that cyclotron damping removes fluctuations from the Alfvén branch, and the remaining fast-mode waves transition to whistler-mode waves. They also argued that dispersive wave propagation can account for the average -3 spectral index of the dissipation range. Smith *et al.* (2006a) stated that the dissipation-range spectral index depends on the rate of energy cascade, as computed from the power spectrum via Eq. (7.7). Howes *et al.* (2007) pointed out that the dissipation-range spectrum measured by flux-gate magnetometers is subject to errors deriving from instrument noise so that the lower the amplitude of the true spectrum, the flatter the measured dissipation range. They argued that the true dissipation range is exponential and represents a second inertial range described by electron MHD (eMHD) (Kingsep *et al.*, 1990). Gary *et al.* (2008) considered a whistler-mode inertial range at frequencies greater than the Doppler-shifted ion inertial scale and showed via numerical simulation a cascade toward a new 2D spectrum in this range.

Leamon *et al.* (1998b) and Hamilton *et al.* (2008) used the technique of Bieber *et al.* (1996) to compare the fraction of energy associated with 2D wave vectors with the energy content of the field-aligned wave vectors. Although they find that the 2D wave vectors contain most of the energy in the highest frequencies of the inertial range, the ratio is closer to 50 : 50 in the dissipation range; the field-aligned wave vectors contain a larger fraction of the total energy in the dissipation range. This means that either dissipation acts preferentially on the 2D component or there exists a dynamic that moves energy associated with 2D wave vectors into the field-aligned wave vectors for dissipation. This only applies to the lowest frequencies in the dissipation range and does not preclude subsequent evolution of the spectrum over frequencies leading to the electron gyrofrequency.

7.7 Non-Gaussianity in turbulent space plasmas

Statistical-mechanics states with Gaussian distributions are the end states of all thermal processes and yet non-Gaussianity abounds in plasma physics and the solar wind. We see significant and long-lasting departures from Gaussianity in thermal particle distributions (Tu and Marsch, 2002; Marsch *et al.*, 2004; Heuer and Marsch, 2007) that have been interpreted as resulting from cyclotron damping and there are long-lived beams in the thermal particle distributions associated with acceleration sources (Lee, 1984). The evolution of these particle populations can be studied with kinetic Vlasov theory (Stix, 1992) and often results in the growth of coherent wave energy where the "free energy" of the particles is sufficient to drive instabilities that repopulate the wave spectrum (Gary *et al.*, 2001; Kasper *et al.*, 2002; Hellinger *et al.*, 2006). The resulting waves can strongly affect the propagation and scattering of the energetic particles, producing hotter energetic particle populations than the original source. In many instances it is the free energy associated with anisotropies of the thermal ion population that drives the wave growth (Gary *et al.*, 2001). This two-way exchange of energy between the thermal characteristics of the background particles and the coherent behavior of the plasma greatly complicates discussions of the dissipation-range dynamics. If dissipation is provided by the removal of coherent plasma fluctuations via a single kinetic process then it is most often the case that anisotropies (sources of free energy) of the thermal ion population can grow. The result may be a return of energy from incoherent to coherent processes, complicating the fluctuation spectrum and leading to distinctly non-fluid observations in association with dissipation.

There is another form of non-Gaussianity in turbulence: when the fluctuations are distributed according to a Gaussian distribution function the nonlinear terms in the incompressible N–S equation (7.1) can become zero (Monin and Yaglom, 1975; Lesieur, 1990). The same is true of the incompressible MHD equations in Eq. (7.13) (Grappin *et al.*, 1983) and of nearly incompressible MHD (Zank and Matthaeus, 1993). The Gaussian distribution of fluctuations is inconsistent with active turbulence; however, it is not essential that the departure from Gaussianity be large. The distribution functions of solar wind fluctuations have been described as nearly Gaussian (Whang, 1977; Feynman and Ruzmaikin, 1994), highly non-Gaussian (Ruzmaikin *et al.*, 1995), and everything in between (Burlaga, 1991; Kabin and Papitashvili, 1998; Sorisso-Valvo *et al.*, 1999; Bruno *et al.*, 1999). There is evidence that the distribution functions become more Gaussian at larger lags (Marsch and Tu, 1994; Smith *et al.*, 2001a). Non-Gaussianity of this type is called "intermittency" and is the self-consistently generated concentration of the dynamics into small regions (Bruno and Carbone, 2005). This is distinct from "homogeneity", which is the large-scale uniformity of the turbulence and most often arises from source and boundary effects. A common measure of intermittency derives from the comparison of high-order structure functions where the nth-order structure function for the time series of component $b_i(t)$ is $S_n(\tau) \equiv \langle [b_i(t) - b_i(t + \tau)]^n \rangle$ (Monin and Yaglom, 1975). One key feature of intermittent turbulence among many is that it leads to spatially non-uniform dissipation processes.

7.8 Turbulence in the solar corona and solar wind acceleration

The concepts we have discussed above are quite general and stem from considerations of the N–S and MHD equations alone. Application of these ideas to the solar wind is immediate and straightforward for at least two reasons: the flow is super-Alfvénic and supersonic, so that measured frequencies can be converted into spatial scales in the same manner as is customary in traditional wind tunnel experiments. The latter is where turbulence ideas are first developed since the flow is relatively free of boundary effects, which can greatly complicate the discussion. Application of these same cascade and heating concepts to the acceleration region of the Sun is more complicated. The energy source exists largely below and outside the acceleration region, although wave propagation from these sources into the acceleration region is often cited as the source of solar wind acceleration energy, and reflection effects act to contain some of the energy.

A major problem in some theories of solar wind acceleration derives from the energy budget. It is necessary for there to be sufficient energy, over a narrow range of spatial and temporal scales, for processes such as cyclotron damping to convert wave energy into thermal energy, thereby accelerating the plasma. A static wave spectrum lacks sufficient energy to provide the necessary heating unless the energy taken up by the damping process is somehow replenished. The application of turbulent dynamics avoids this problem by replenishing the small scales through the action of the turbulent cascade. Therefore, the acceleration region need only provide a turbulent cascade of sufficient strength to heat the plasma at the prescribed rate to achieve acceleration. It does not need to provide an excess reservoir of energy at the smallest scales.

Matthaeus *et al.* (1999) presented one application of the turbulence-driven wind idea that demonstrates the opportunities afforded by the new dynamic and how traditional views and problems can be addressed in a new way. They assumed, as is often the case, that the photosphere or chromosphere are sources of outward propagating Alfvén waves that propagate into the corona. In the traditional view these waves would cyclotron-damp (in part) and the remaining wave energy would propagate into the solar wind formed by the resonant damping of the wave energy and resultant heating of the plasma. However, in the model of Matthaeus *et al.* low-frequency waves outside the normal resonant scales are reflected back into the sunward-propagating direction, as happens for waves in regions of changing refractive index. This produces the combination of $z^{\pm}$ waves necessary to drive MHD turbulence. The resulting cascade (assumed to be primarily in the two-dimensional plane) leads to dissipation by means other than cyclotron damping at a rate that is dictated by the nonlinear cascade dynamics driven by large-scale fluctuations. Dissipation may be accomplished by many means (cyclotron, Landau, and transit-time damping, current sheet formation, etc.) and the resulting "resonance" is broadly distributed over many spatial scales, which form the dissipation range for the turbulence in this region. That dissipation provides the heat necessary to accelerate the wind, and interplanetary fluctuations are more the result of the in situ interactions in large-scale transient behavior than the escape of wave energy from the solar chromosphere. This is only one application of the ideas presented here, but it does demonstrate how turbulence can accelerate the solar wind.

The detailed in situ measurements described in this chapter in relation to solar wind measurements are obviously lacking in the solar context. However, many solar imaging techniques provide multidimensional movies of the flow and magnetic field. As measurement techniques advance and resolution improves the Sun may prove to be a key laboratory for the study of MHD turbulence.

7.9 Interstellar turbulence

Measurements of the solar atmosphere using scintillation observations (Cole and Slee, 1980) have provided information about the solar atmosphere above the photosphere. Much the same techniques are used to study the interstellar region. More for the latter than the former, a leading problem is that such measurements are line-of-sight integrations of all regions between the source and observer. This places some limitations on the applicability of the method.

Scintillation measurements provide an inferred spectrum of the interstellar density fluctuations; this spectrum displays a $-5/3$ power law (Armstrong *et al.*, 1995; Spangler, 2007). Scintillation measurements have been used to infer a 2D geometry and intermittency for interstellar MHD turbulence (Spangler, 1999). It is routine to refer to these observations as "Kolmogorov spectra", although by now the reader will understand that Kolmogorov never predicted a spectrum for density variations or advocated its use as a proxy for the spectrum of other turbulent fields. The assumptions of a weakly compressible MHD density spectrum (Montgomery *et al.*, 1987) and other related and unrelated theories include the use of passive convective scalars to express the relationship between the fluctuations in energy and density in both solar wind and interstellar investigations (Goldstein and Siscoe, 1972; Armstrong *et al.*, 1981), but the reader is cautioned to appreciate that this remains the subject of ongoing research with many unanswered questions; he or she might examine Trang and Rickett (2007) for other viewpoints.

Measurement of the interstellar magnetic field relies upon Faraday rotation fluctuations and to date has not been extended to spatial scales into the inertial range. Therefore, these measurements provide a diagnostic of the energy-containing range of the spectrum and do not yield $-5/3$ spectra (Minter and Spangler, 1996; Haverkorn *et al.*, 2004). Keeping this in mind, the interstellar turbulence spectrum spans the broadest range of spatial scales of any measured turbulent spectrum, from a few parsecs (Minter and Spangler, 1996) to scales on the order of the proton gyroradius (200 km or less; Armstrong *et al.*, 1995). There is a potential problem with this: how do fluctuations separated by parsecs interact in a causal dynamic? This can only happen over the longest time scales, allowing communication between the most distant regions, if the turbulence is very old and slowly evolving, the fluctuations are remnant from an earlier time, or the largest scales are broken into independent and isolated regions exhibiting a common dynamic from a similar source. If the turbulence is evolving and representative of an active energy cascade at these largest scales, the large-scale energy-containing flows must be very long-lived and represent an enormous energy reservoir. The implications for the in situ heating of the interstellar medium by turbulent cascade are compelling (Spangler, 1991; Minter and Spangler, 1997; Minter and Balser, 1997).

7.10 Conclusion

We began this chapter by citing several excellent texts on the subject of turbulence and we will end it in the same way. On the subject of hydrodynamic turbulence there are numerous very good texts and, in approximate order of increasing degree of difficulty and completeness, they are: Tennekes and Lumley (1972), Pope (2000), Hinze (1975), and Monin and Yaglom (1971, 1975). Herring and Tennekes (1989) provides a broad overview of turbulence and includes an article by D. C. Montgomery on the subject of MHD turbulence. Regretably, there are very few texts on MHD turbulence beyond Biskamp (2003), but good reviews are Matthaeus *et al.* (1995) and Bruno and Carbone (2005). For interstellar turbulence Elmegreen and Scalo (2004) and Scalo and Elmegreen (2004) are recommended.

In this chapter we have focused on questions surrounding the formation of predictable fluctuation spectra that derive from a cascade of energy, for the purpose of presenting a view of the turbulent heating of space plasmas. This is only one view of MHD turbulence, although it is central to most of what follows in the subject and provides a useful motivation for the application of turbulence ideas to these systems. We have on the whole avoided discussion of weak turbulence, compressible turbulence, or eMHD theories. All are rich subjects of their own at various stages of development. We have only briefly examined the application of these ideas to the acceleration of the solar wind, to interstellar measurements, and magnetospheres due to limitations of space and the simple fact that these areas are less well developed owing to the difficulty of making the relevant measurements. The reader is encouraged to read further, starting with the texts listed above, with the expectation that ideas useful to the problems at hand will present themselves. Turbulence is a universal process, which is likely to present itself whenever a fluid system is given adequate time to evolve. It may not always be manifested in the same form, and it is true that the predictions of turbulence theory can depend on symmetry or, in some instances, on the manner in which the flow is driven. In the end very few observable processes in a naturally occurring plasma will remain linear and so fail to evolve into a broad spectrum of fluctuations that we can describe as turbulence.

8

The solar atmosphere

VIGGO H. HANSTEEN

8.1 Introduction

The Sun is an average main-sequence star in middle age and of spectral class G2 v.
The Sun has mass $M_\odot = 2 \times 10^{30}$ kg, radius $R_\odot = 700$ Mm, and effective temperature $T_{\text{eff}} = 5780$ K. The spectral class G2 v implies that the energy generated by
hydrogen fusing into helium in the solar core is carried by convection in the outer
one-third of the Sun's envelope. Lighter main-sequence stars, of spectral classes K
and M, have convection zones that extend steadily more deeply as the stars become
less massive. Conversely, main-sequence stars heavier than the Sun, of spectral
classes F and A, are convective only in a thin layer near their surface. The heaviest
main-sequence stars, of spectral classes B and O, have convective cores but no
convection near the surface. It is the convective motions in a stratified environment, coupled with differential rotation, that are ultimately responsible for solar
and stellar dynamos and thus solar magnetic fields and solar and stellar activity.
As the Sun is a middle-aged star, the solar wind has had time to carry away a large
fraction of the initial angular momentum of the proto-solar cloud (see Vol. III). At
present the Sun's rotation period ranges from 35 days at the poles to 25 days at the
equator, but it was presumably much shorter early in the Sun's life, 4.5 Gyr ago.
Thus, solar activity now is less than that found in the younger main-sequence stars
or in the young Sun itself.

The structure on the visible surface of the Sun, the photosphere, is most obvious
in the form of sunspots, dark regions of lower temperature, roughly 4000–5000 K,
where the strong magnetic field, with field strengths up to several kilogauss, dominates the convective motions. Sunspots often appear in pairs or paired clusters
with spots of opposite polarity arranged west and east ("west" and "east" being
counterintuitively defined by their geocentric directions on the Earth's sky). The
spots are roughly circular, with radii on scales ranging from tens of megameters

down to a few megameters. On some young rapidly rotating stars, starspots may cover up to half a hemisphere in the most active cases, while polar or high-latitude starspots – which have never been observed on the Sun – are common. See e.g. Schrijver (2002) for a review comparing sunspots and starspots.

Sunspots are distributed on the solar surface according to the sunspot cycle, appearing at high latitudes early in the cycle then gradually appearing closer and closer to the equator as the cycle progresses; the leading spots in a given hemisphere usually have the same polarity, the opposite polarity leading in the other hemisphere. The sunspot cycle has a period of some 11 years (with a magnetic cycle of 22 years if we also consider the polarity of leading sunspots and the general background magnetic field). Convective motions continually perturb sunspots, limiting their lifetimes to some few solar rotations and eventually shredding and transporting the spots' magnetic flux into the background photospheric magnetic field.

Sunspots are surrounded by much smaller concentrations of somewhat weaker field (1–2 kG) (Fig. 8.1). The collection of flux concentrations that emerge together onto the solar surface, including within them the spots and the neighboring opposite-polarity field, are referred to as active or bipolar regions. There are many more small active regions than there are large ones; the spectrum ranges from below 10^{18} Mx to 10^{21} Mx and is distributed more or less like a power law, with index -2 for the active-region frequency spectrum. The largest of these active regions emerge close to the latitude of peak activity, which migrates equatorward as the cycle progresses (thus forming a "butterfly" diagram), with a clear preference for an east–west orientation. Smaller regions have a larger spread in latitude (up to the poles for the smallest, or ephemeral, regions) and in orientation (the smallest are almost, but not quite, randomly oriented relative to the equator). These and other properties of active regions are summarized in Schrijver and Zwaan (2000).

Looking at the solar photosphere, we see the top of the convection zone in a granulated form. Hot gas rising from the solar interior as part of the energy-transport process reaches a position where the opacity is no longer sufficient to prevent the escape of radiation. The gas radiates and cools, and in doing so loses its buoyancy and descends. At the surface the gas density is of order 10^{-4} kg/m^3 and the pressure is of order 10^4 Pa, but they decrease exponentially with height with a scale height of some 100 km. (The smallness of the scale height is the reason why the solar limb appears sharp when viewed with the naked eye.) The granular cells have dimensions of order 1 Mm, but numerical simulations indicate that the convective length scales rapidly become larger as one proceeds below the solar surface. These motions, ultimately driven by the requirement that the energy generated by nuclear fusion in the Sun's core be transported in the most efficient manner, represent a vast reservoir of "mechanical" energy.

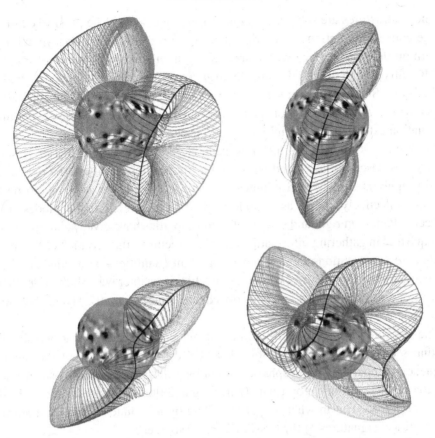

Fig. 8.1. The Sun's surface magnetic field comprises a multitude of dipolar regions of widely different fluxes, whose frequencies wax and wane with the solar cycle. The large-scale coronal magnetic field, the foundation of the heliospheric field, expands from regions of partly open magnetic field that enclose the closed-field corona. The figure shows the global topology of the Sun's field in a so-called potential-field source-surface approximation (see also Section 4.2.1 and Fig. 4.7). In particular, it shows four realizations of the "streamer belt" for a solar magnetic model. Shown are four phases of the simulated magnetic cycle: clockwise from the top left, $t = 3.1$, 3.6, 4.5, 6.0 y in a sunspot cycle of 11 y. Each panel shows a magnetogram of the solar surface, the neutral line(s) (bold) at the source surface, and the highest closed field lines that reach up to the neutral line(s); the darkest lines are nearest to the "observer". The panels show, clockwise from the upper left, a near-quadrupolar situation; a strongly tilted dipolar case; a strongly warped current sheet; and another nearly dipolar case with less tilt relative to the solar equator. From Schrijver (2005).

Looking more closely, we see that granulation is not the only phenomenon visible at the solar surface. The quiet and semi-quiet photosphere is also threaded by magnetic fields that appear as bright points, as well as darker micro-pores and pores. While they are able to modify photospheric emission, these small-scale magnetic structures are subject to granular flows and seem to be passively carried by the convective motions. The bright points are organized in a honeycomb-like pattern on the solar surface with a size scale, larger than granulation, of roughly 20 Mm; this pattern defines the so-called *supergranular network* (often referred to as "the quiet-Sun network" or simply "the network"; the interiors of the convection cells are the *internetwork* regions) and is suggestive of convective cells larger than granulation extending deeper into the solar interior.

Convective flows are also known to generate the perturbations that drive solar oscillations. Oscillations, sound waves, with frequencies mainly in the band centered roughly at 3 mHz or 5 minutes, are omnipresent in the solar photosphere and are collectively known as p-modes ("p" for pressure). These p-modes are a subject in their own right and studies of their properties have given solar physicists a unique tool in gathering information on solar structure – the variation of the speed of sound c_s, the rotation rate, and other important quantities – at depths far below those accessible through direct observations. In this chapter we will consider them only insofar as they interact and possibly channel energy into the layers above the photosphere.

The shuffling, buffeting, and braiding of magnetic structures that presumably continues on up into the upper solar layers and the propagation of the higher-frequency-component[†] photospheric oscillations through the chromosphere and into the corona – all may contribute to heating and thus the production of a 1 MK or hotter corona. But in what proportion? And by how much? And what are the observational signatures of the possible heating sources?

The structure and topics of this chapter are summarized by Fig. 8.2. We are considering a photosphere in continual motion, threaded by magnetic fields and subject to 5 minute oscillations. The Swedish 1 meter Solar Telescope on La Palma and the Hinode satellite constituted observationally the state of the art in 2007, giving access to visible-wavelength observations of the photosphere and its attendant fields with a spatial resolution of roughly 0.1–0.2 arcsec, equivalent to 75–150 km on the solar surface. These dimensions approach the vertical scale height and the photon mean free path in the photosphere. Careful analysis of this wealth of observational data on the "lower boundary" of the solar atmosphere should eventually yield insight into the mechanism(s) heating the upper layers of the Sun. These upper layers comprise the chromosphere, a zone of gas at temperatures from 6000 to roughly 20 000 K, which extends some 2–5 Mm above the photosphere, and the corona, the Sun's tenuous "crown" of hot 1 MK plasma, which extends several hundred Mm above the photosphere.

[†] Waves with frequencies lower than the acoustic cutoff frequency, roughly 5 mHz, are evanescent above the photosphere and do not propagate energy into the higher solar atmosphere (unless the magnetic field geometry modifies the frequency at which the waves become evanescent).

Table 8.1. *Basic parameters for domains in the solar atmosphere, and their definitions. Note that all regions of the solar atmosphere are very inhomogeneous and that these values are only meant to give a rough idea of their magnitudes*

Region	n (m^{-3})	n_e/n_H	T (K)	B (gauss)	β
Photosphere[a]	10^{23}	10^{-4}	6×10^3	1–1500	> 10
Chromosphere[b]	10^{19}	10^{-3}	$2 \times 10^4 - 10^4$	10–100	10–0.1
Transition region[c]	10^{15}	1	$10^4 - 10^6$	1–10	10^{-2}
Corona[d]	10^{14}	1	10^6	1–10	$10^{-2} - 1$

[a] The *photosphere* is the layer from which the bulk of the electromagnetic radiation leaves the Sun. This layer has an optical thickness $\tau_\nu \lesssim 1$ in the near-UV, visible, and near-IR spectral continua, but it is optically thick in all but the weakest spectral lines;

[b] The *chromosphere* is optically thin in the near-UV, visible, and near-IR continua, but optically thick in strong spectral lines – it is often associated with temperatures around 10 000–20 000 K.

[c] The *transition region* is a thermal domain between the chromosphere and corona in which thermal conduction leads to a steep temperature gradient.

[d] The *corona* is optically very thin over the entire EM spectrum except for the radio waves and a few spectral lines – it is often used to describe the solar outer atmosphere out to a few solar radii with temperatures exceeding ~ 1 MK.

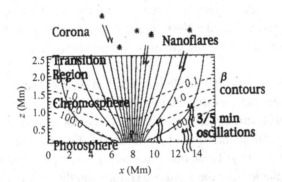

Fig. 8.2. Schematic view of the domains in the solar atmosphere (outside sunspots) of the magnetic field in it, and of the associated plasma β. In the photosphere the gas pressure is larger than the magnetic pressure, and photospheric motions are driven by convection – a mainly hydrodynamic phenomenon that brings up new magnetic flux and buffets, shreds, and reorganizes the existing field. Oscillations of somewhat higher frequency than the photospherically dominant 5 minute oscillations propagate into the chromosphere. Upon reaching the level where the exponentially decreasing gas pressure equals the magnetic pressure ($\beta = 1$), the surviving sound waves may be converted into other wave modes. It is likely that the processes heating the corona and magnetic chromosphere are episodic, inducing large temperature differences, flows, and other non-steady phenomena that in turn produce other wave modes.

On average the photospheric gas pressure $p_g = 10^4$ Pa is much greater than the pressure p_B represented by the observed average unsigned magnetic field strength, 1–10 gauss ($p_B = B^2/(2\mu_0) < 1$ Pa). However, in the largely isothermal chromosphere the gas pressure falls exponentially, with a scale height of some 200 km, while the magnetic field strength falls off much less rapidly even though the field is expanding to fill all space. Thus, depending on the actual magnetic field geometry, the magnetic pressure and energy density will overtake the gas pressure some 1500 km or so above the photosphere in the mid chromosphere. Another 1000 km, or 5 scale heights, above the level where $\beta = 1$ (see Eq. 3.11), the plasma's ability to radiate becomes progressively lesser while the dominance of the magnetic field becomes steadily greater. As we explain below, any given heat input in this region cannot be radiated away and will invariably raise the gas temperature to 1 MK or greater; thus a corona is formed. It is bound to follow the evolution of magnetic field as the field in turn is bound to be driven photospherically.

8.2 The photosphere

Solar convective motions continuously churn the solar atmosphere. Hot, high-entropy, gas is brought up to the photosphere where the excess energy is radiated away. Cool, low-entropy, material descends into the depths in steadily narrower lanes and plumes, as explained in the numerical models of Stein & Nordlund (1998). The result of these motions is the granular pattern shown in the left-hand panel of Fig. 8.3.

Recounting the wealth of knowledge gathered on the photospheric dynamics falls far outside the scope of this chapter. We will rather describe a limited set of high-resolution (100 km) observations made at the Swedish 1 meter Solar Telescope, paying special attention to observations of the structure and dynamics of photospheric bright points. These observations are also representative of the quality of data that is available from the Japanese Hinode satellite, from which images and spectra, including vector magnetograms and dopplergrams from the photosphere, are obtained with slightly lower spatial resolution (150 km) but much greater stability and duration in time than can be achieved on the ground.

In Fig. 8.3 we show typical images of the quiet to semi-quiet photosphere as well as a plage region. These images were observed in the so-called G-band, centered around 430 nm, which is formed some 100 km above the nominal photosphere. The observations were obtained with the Solar Telescope (Scharmer *et al.*, 2003) on La Palma. Examples can be seen of simple bright points in a fairly quiet region of the photosphere, i.e. a region in which the magnetic field is too weak to modify granular motions significantly. Isolated bright points are seen to be constrained to

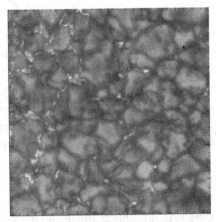 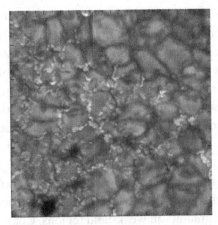

Fig. 8.3. The image on the left shows a typical quiet photospheric region (a network region) observed in the G-band with the Swedish 1 meter Solar Telescope. The image on the right includes a plage region, where the total amount of magnetic flux penetrating the photosphere is larger. The axes of both panels are numbered in arcseconds measured on the Sun; 1 arcsec is approximately 725 km. The G-band near 430 nm contains several spectral lines, notably the lines of the CH molecule, and is formed near the solar surface (the height where $\tau_{500nm} = 1$); the granulation and intergranular lanes are some 100 km above this height, and the bright points are some 200 km below, as explained in the text. The bright points are regions of enhanced magnetic field concentrated and contained by the granular motions. Notice also that the bright points are often pulled into ribbons and may fill the entire intergranular lane. In the plage region in the image on the right, notice the large number of phenomena showing complex structure, ribbons, "flowers", micropores, and isolated and seemingly simple bright points. The magnetic field in this right-hand image is in places strong enough to perturb the granulation dynamics and the granules appear "abnormal" while displaying a slower evolution than in the quieter regions of the photosphere.

intergranular lanes and do not seem to have any internal structure at this resolution. These isolated bright points are passively advected towards the peripheries of supergranular cells (Muller, 1983), where they gather and form the collection of bright points that define the supergranular network mentioned in Section 8.1.

Several studies of the statistical properties of the dynamical evolution of bright points have appeared in the literature. Berger and Title (1996) described a number of effects during the lifetime of bright points. These include shape modifications such as elongation, rotation, folding, splitting, and merging. They found that significant morphological changes can occur on time scales as short as 100 s and are strongly dependent on the local granular convective flow field. An average lifetime in excess of 9 minutes, with some bright points persisting much longer, has been found by several authors (Berger et al., 1998; Nisenson et al., 2003). The reported values for the mean lifetime are very similar to the evolution time scale of granulation.

Towards the left-hand edge of the left-hand panel in Fig. 8.3 we see examples of bright points forming ribbon-like shapes in the intergranular lanes. In places, these ribbons surround entire granules. Many examples are found also in the right-hand panel of Fig. 8.3, which shows a region of stronger average field strength, a *plage* region, than that in the left-hand panel. Flux concentrations with larger spatial extent are embedded in (micro-)pores with distinctly dark centers. Such small dark micropores may be the smallest manifestation of the phenomenon that produces pores and sunspots. The bright points in the plage region are not simple points but are seen to have structure and appear to modify the granular flow itself: the image shows that granules near bright points in the network or in plage regions are smaller and have lower contrast, and in addition display slower temporal behavior than granules in weak-field regions. Coalescing bright points in plage and network regions can form dark centers and thus become micropores if their density is large enough or, indeed, can brighten again if the granular flow breaks them apart into smaller flux elements.

In order for magnetic elements to become visible at all they must attain field strengths of sufficient amplitude to perturb the plasma in which they are embedded (even while being passively advected laterally). Thus, we expect that there also exists a whole hierarchy of magnetic structures which have not yet attained such a critical field strength, a conclusion that is supported by the first results from Hinode. What is the source of this field? One possible scenario is that it is brought up in granules at field strengths too low to leave an observational signature in the photospheric intensity. Granular flow then carries the field to the intergranular lanes and ultimately to the intergranular intersections, where it may become compressed and strong enough to leave an observational signature; such field strengths are of order 1500 gauss for a photospheric pressure of 10^4 Pa. This could be the reason why bright points always seem to appear in intergranular intersections.

8.2.1 *Why bright points?*

As explained above, photospheric bright points are correlated with regions of strong magnetic field and are subject to photospheric motions. Plage regions share similar characteristics but seem capable of modifying the background flow. Enhanced emission implies high temperatures but micropores, pores, and sunspots are dark. What is the relation between magnetic fields and the photospheric and lower chromospheric temperatures? Why are bright points bright?

One way to answer this question is through atmospheric modeling of the relevant phenomena using the MHD equations (3.1)–(3.4), where we include the divergence of the radiative flux $\mathbf{F}_r$ on the right-hand side of the energy equation (3.3).

Photospheric convection is driven by radiative losses from the solar surface. In order to model granulation correctly it is vital to construct a proper model of radiative transport. Nordlund (1982) devised a sophisticated treatment of this difficult problem in which spectral lines are grouped according to their opacity; in effect, one constructs wavelength bins that represent characteristic stronger or weaker lines and also the continuum, so that the radiation in all atmospheric regions is treated according to a certain approximation. If one further assumes that opacities are in local thermodynamic equilibrium (LTE) then the radiation from the photosphere can be modeled. After binning the opacities at all wavelengths into groups, the group mean opacities are used to calculate a group mean source function from which the emergent intensity, and thus the radiative losses, can be approximated.

This method of approximating solar radiative transport coupled with an MHD numerical code allows one to model solar convection with a high degree of realism (Nordlund, 1982). Carlsson *et al.* (2004) constructed convection simulations such as these in which an initial vertical magnetic field of magnitude 250 gauss was included. The equations were discretized on a numerical grid of $253 \times 253 \times 163$ points covering $6 \times 6 \times 2$ Mm^3. With the passage of time the magnetic field was carried along with the convective flow and formed quite complex topologies very similar to those observed in Fig. 8.3.

Once the model has evolved to a statistically steady state, high-fidelity synthetic spectra of the G-band (or any other wavelength band formed in the modeled region) can be computed a posteriori using a radiative transfer code containing more than the essentially four frequency points used in the MHD simulation. This sort of model contains essentially no free parameters (other than the initial magnetic field strength). The excellent agreement found between the computed and observed spectra is fairly strong evidence that most of the relevant physics is included in the model.

The advantage of having a model of the phenomena is clear: one can look at any given variable as a function of position and time. Armed with these results one is in a position to describe the emission from various photospheric regions, ranging over granules, intergranular lanes, bright points (magnetic flux tubes), and the region of strongest magnetic field concentration in the model.

The increased brightness in the magnetic elements is due to their lower density compared with the surrounding intergranular medium. One thus sees deeper layers where the temperature is higher. At a given geometric height, the magnetic elements are cooler than the surrounding medium because the magnetic fields prevent convective energy transport from deeper layers. At the edges of the flux concentrations the plasma is radiatively heated by the surrounding hotter, non-magnetic, plasma.

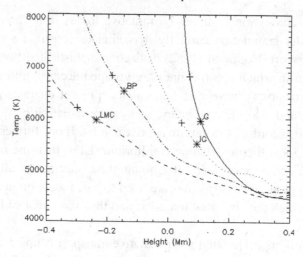

Fig. 8.4. Gas temperature as a function of height for four positions in the model of
Carlsson *et al.* (2004). The curves marked G and IG represent positions in a typical
granule center and in an intergranular lane. The asterisks on each curve show the
formation height of the G-band; the emergent radiation from the granule center,
curve G, is formed at a higher temperature and thus appears brighter than that
from the intergranular lane, curve IG. The curves marked BP and LMC represent
respectively a bright point and the region of greatest magnetic flux concentration;
the bright point appears bright because the opacity is low enough for gas at great
depths to be sampled, even though at any geometric point in the simulation the
temperature is higher in both the granules and the intergranular lanes. The same is
also true for the region of greatest magnetic flux, which is strong enough to hinder
convection and at the same time large enough in horizontal size to hinder radiative
heating from the "walls" of the flux tube. Heights are measured relative to the
solar surface ($\tau_{500nm} = 1$). The effective temperature of the Sun, T_{eff}, is 5780 K.

See Fig. 8.4 for examples of the temperature structure and heights of emergent
intensity formation for four different positions in the model.

8.3 The high-β chromosphere

With the observations and modeling of photospheric granulation and bright points
discussed in Section 8.2 in mind, let us turn to the chromosphere. What happens
as the perturbations implied in photospheric granulation propagate upward? How
does the magnetic field expand into the overlying regions? How does this change
the dynamics of the high atmosphere?

Originally the chromosphere – i.e. colored sphere – had its name from eclipse
observations. One can occasionally observe the strong intensely red light stemming
from the H-α line circumscribing the solar limb. The chromosphere is much thicker
than the photosphere; it contains roughly 10–12 scale heights, of some 200 km each,

and hence stretches for roughly 2500 km between the photosphere and the transition regions, where the temperature rises rapidly to coronal values.

8.3.1 Chromospheric oscillations

Much work has gone into elucidation of the thermal structure of the chromosphere and the heating needed to maintain this structure against radiative losses. Seminal in this regard was the work of Gene Avrett and co-workers in the 1980s (Vernazza et al., 1981). However, we would like to approach chromospheric structure from another perspective.

Assume that the chromosphere can be considered as an initially isothermal slab stratified by a constant gravitational acceleration g. Then the linearized equations of mass, momentum, and energy conservation can be written as (Sterling and Hollweg, 1988)

$$\frac{\partial}{\partial t}\delta\rho + \rho_0\frac{\partial}{\partial z}\delta v + \frac{\partial \rho_0}{\partial z}\delta v = 0, \tag{8.1}$$

$$\rho_0\frac{\partial}{\partial t}\delta v = -\frac{\partial}{\partial z}\delta p - g\delta\rho, \tag{8.2}$$

$$\frac{\partial}{\partial z}(\delta p - c_s^2\delta\rho) + \rho_0^\gamma\,\delta v\frac{\partial}{\partial z}\left(\frac{p_0}{\rho_0^\gamma}\right) = 0. \tag{8.3}$$

where δ signifies a perturbed quantity and the subscript 0 denotes an unperturbed quantity, the vertical velocity is denoted v, and $c_s = \sqrt{\gamma p_0/\rho_0}$, p, and ρ are the speed of sound, the gas pressure, and the density, respectively.

These equations may be combined into

$$\frac{\partial^2 Q}{\partial t^2} - c_s^2\frac{\partial^2 Q}{\partial z^2} + \omega_a Q = 0, \tag{8.4}$$

where $Q \equiv \rho_0(z)^{1/2}u$ and the acoustic cutoff frequency $\omega_a \equiv c_s/(2H_p) = \gamma g/(2c_s)$. This is a Klein–Gordon equation, the solution of which, after eventual initial transients have died down, is an oscillatory wake with a period close to the acoustic cutoff frequency. If we imagine photospheric dynamics as a driver – with typical driving frequencies in the 5 minute/3 mHz band, along with other excitations due to individual granule dynamics – then on the basis of Eq. (8.4) we expect the chromospheric response to be an oscillation with a frequency near 5 mHz, corresponding to a period of roughly 3 minutes. Note that this conclusion is reached without any consideration of effects such as the damping of waves through radiation and without close consideration of the driver spectrum.

In many cases a chromosphere dominated by 3-minute-period power is indeed what is found. In Fig. 8.5 we see continuum emission in the 104.3 nm band observed

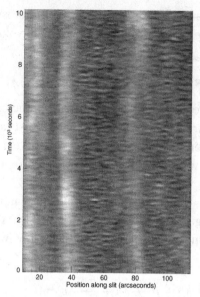

Fig. 8.5. Chromospheric oscillations in the "brightness temperature" of the C continuum formed near 104 nm at a height roughly 1100 km above the photosphere (the brightness temperature is the temperature that the radiating material would have if it were in local thermodynamic equilibrium). This composite image was obtained by pointing the SUMER slit at a given quiet-sun location on the Sun, making an exposure every 20 s or so, and then displaying consecutive exposures of the slit vertically. The persistent bright vertical bands represent areas of enhanced magnetic field. The horizontal structures clearly visible between the bright vertical bands can be explained by the presence of upwardly propagating 5 mHz oscillations (Wikstøl *et al.*, 2000).

with the SUMER instrument aboard the SOHO satellite. The image was made by stacking consecutive exposures of the slit vertically as a function of time. Note the horizontal bands of enhanced emission with horizontal extent of order tens of arcseconds (tens of thousands of kilometers), which recur with a period of roughly 3 minutes. These horizontal bands are omnipresent in the image presented here and can be explained as being due to upwardly propagating wave trains.

In fact, Carlsson and Stein (1995) showed that many aspects of the non-magnetic, so-called internetwork, chromosphere can be explained as a result of the presence of such photospherically excited, upward propagating, acoustic waves. In their simulations a photospheric driver, a piston taken from the doppler velocities measured in a Fe I line formed in the photosphere, was used to excite waves. These waves propagate upward, and as they do so their amplitude grows as the density decreases; consequently the wave steepens and forms shocks. A self-consistent radiative-transfer calculation produces energy transfer in the model as well as observational diagnostics, especially for the Ca II H line, a resonance line of Ca^+

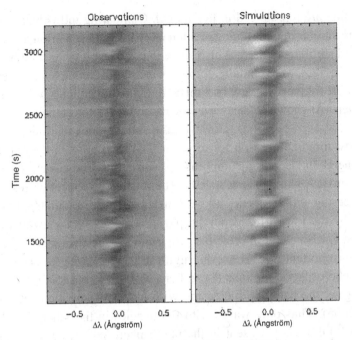

Fig. 8.6. The Ca II H-line profile as a function of time. Multiple repeated exposures are made with a slit pointed at a given position on the Sun. The observed line profile in a quiet, internetwork, position is shown in the left-hand panel. A one-dimensional radiation-hydrodynamic model due to Carlsson and Stein (1995), containing upwardly propagating acoustic waves driven by a piston computed from the doppler shift measured in an Fe I line in the blue wing, results in the Ca II H-line profiles shown in the right-hand panel. Waves appear first in the line wings and propagate towards the line center as the acoustic wave moves upward in the atmosphere. A peak brightening on the violet side of the line core, formed some 1000 km above the photosphere, indicates that the wave amplitude has grown and that the wave is non-linear, i.e. it has turned into a shock wave.

ions. The results of the model are shown in Fig. 8.6, which displays observed and computed Ca II H-line spectra as a function of wavelength and time. The figure shows the general behavior of the line emission in both observation and simulation: enhanced emission arises first in the line wings and thereafter moves in towards the line core. This emission is consistent with an upwardly propagating wave; the emission far out in the line wings is formed at lower heights than that near the line core, owing to the lower opacity at the off-center wavelengths. Upon reaching the line core, the perturbation causes the blue, or violet, peak to become very enhanced while a corresponding red peak is mostly absent. This anisotropy is a signature of a large velocity gradient in the region of emission formation, i.e. the wave has formed a shock at the height where the violet (K2V) peak is formed, roughly 1000 km above the photosphere. It is also worth noting that even the details of

Ca II emission are reproduced in this model. Comparing individual peaks shows that they are similar in both timing and intensity in the observation and the model.

These and similar simulations also show that the results indicated by the linear analysis in Eq. (8.4) is essentially correct: waves with frequencies lower than the acoustic cutoff frequency, roughly 5 mHz, decay exponentially with height in the chromosphere. As shown above, waves are naturally excited at the cutoff frequency of 5 mHz as wake oscillations. However, the simulations mentioned above show that the most important reason for a power peak at 5 mHz in the chromosphere is the exponentially decreasing amplitude of non-propagating evanescent waves of lower frequency. A photospheric spectrum dominated by 3 mHz waves, with wave power decreasing with frequency above 3 mHz, will change with height to become dominated by the lowest-frequency propagating waves. Waves with frequencies much above 5 mHz could in principle propagate up into the upper chromosphere and play an important role. However, high-frequency waves (a) are not strongly excited by photospheric motions (b) and are very strongly radiatively damped as they propagate, the damping increasing with increasing frequency (Fossum and Carlsson, 2005). Thus only waves with frequencies in the range 5 mHz to, say, 10 mHz that are already present in the photosphere propagate up and dominate internetwork chromospheric dynamics as they steepen and form shocks in the mid chromosphere.

There is therefore very strong evidence that the internetwork chromosphere is very dynamic and that the variations in physical quantities such as the temperature may be as large as the quantities themselves. In addition, these models show the grave danger posed by forming time-averaged models based on diagnostics that have a non-linear response to variations in the atmosphere. There is, for example, no need to assume a chromospheric temperature rise in the internetwork in order to explain the behavior of the Ca II line emission.

8.3.2 *The low-β chromospheric network*

There are several mysteries that remain regarding chromospheric emission. Let us return to the upper chromosphere, as imaged in Fig. 8.5. We now discuss the bright vertical bands with widths of 10 000 km or so. These regions coincide with regions of enhanced photospheric magnetic field. In addition, even the background emission in the dark internetwork bands is greater than that which can be explained solely by acoustic waves. Clearly, additional heating is required, a heating which presumably is connected in some way to the magnetic field.

As magnetic bright points are advected by the granular flow they tend, as discussed previously, to concentrate in a granular pattern, with typical horizontal dimensions of 20 000 km, that defines the supergranular flow field. The

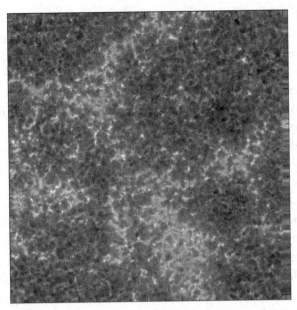

Fig. 8.7. The chromospheric network as observed with the Swedish 1 meter Solar Telescope in the Ca II H-line band in a region covering roughly 80×80 arcsec^2 (60×60 Mm^2). Most emission in this image is formed some hundreds of kilometers above the photosphere. Inverse granulation, an image of the "overshoot" of vertical granular flow with horizontal scales identical to that of regular granulation, can be seen. The magnetic elements (showing up here as bright points) are advected with the granular flow and are transported to the chromospheric network, which forms a honeycomb-like pattern with horizontal size scales of 20 000 km or so.

chromospheric network is the consequence of this flow field as seen in lines and continua formed in the chromosphere. In Fig. 8.7 we see the chromospheric network clearly outlined at greater scales than that of the the granular network seen in the form of inverse granulation in this image. This inverse granulation is a hydrodynamic phenomenon caused by the previously hotter granules rising, expanding, and cooling, whereupon the gas becomes compressed; it begins to fall back towards the lower photosphere and converges with the flow from other expanded granules. The network emission, however, seems to be caused by an enhanced heating due to the magnetic field, the nature of which is still unknown.

The magnitude of the magnetic field in the chromosphere (and corona) can be estimated by potential field extrapolations from the measured longitudinal field in the photosphere. This is done by solving the equation

$$\nabla \times \mathbf{B} = 0, \tag{8.5}$$

with boundary condition $B_z(x, y, 0) = B_{z0}(x, y)$, using an appropriate method (Seehafer, 1978). Alternatively, one can carry out force-free calculations with a constant field-line twist parameter α, or non-linear force-free calculations, by various methods as surveyed by Schrijver *et al.* (2006). Extrapolations show that the magnetic field will decay with a scale height that is roughly equal to the separation between the field sources on the solar surface. Thus, the granular-scale field will reach heights of order 1000 km, while fields on network scales will reach high into the corona, to about 20–30 Mm. Typical photospheric field strengths, which could vary from a few Gauss to 1000 Gauss or more, will dominate the gas pressure at heights varying from a few scale heights above the photosphere to a height of 1500 km or more. In other words, we can expect the surface at which plasma β equals unity to be very corrugated.

We have seen that we can expect a multitude of waves and wave modes to be excited in the solar convection zone. In most of the convection zone, these waves will be predominantly acoustic in nature. When acoustic waves reach the height where the Alfvén speed is comparable with the sound speed, i.e. roughly where $\beta = 1$, they undergo mode conversion, refraction, and reflection. In an inhomogeneous, dynamic, chromosphere this region of mode conversion will be very irregular and will change in time. We thus expect complex patterns of wave interactions that are highly variable in time and space.

What happens to the acoustic waves as they propagate from the photosphere into the chromosphere? McIntosh and co-workers found that there is a clear correlation between observations of wave power in SOHO SUMER observations and the magnetic field topology as found from potential field extrapolations based on from SOHO MDI observations of the longitudinal magnetic field (McIntosh *et al.*, 2001; McIntosh and Judge, 2001). An understanding of the basic phenomena can be built up by studying some simplified cases. Rosenthal and co-workers (Rosenthal *et al.*, 2002) made 2D simulations of the propagation of waves through a number of simple field geometries in order to obtain a better insight into the effect of differing field structures on the wave speeds, amplitudes, polarization, direction of propagation, etc. In particular, they studied oscillations in the chromospheric network and internetwork. They found that acoustic fast-mode waves in the photosphere become mostly transverse magnetic fast-mode waves when crossing a magnetic canopy where the field is significantly inclined to the vertical, as shown in Fig. 8.8. Refraction due to the rapidly increasing phase speed of the fast-mode waves results in total internal reflection of the waves.

This work was extended to other field geometries resembling a sunspot. Bogdan *et al.* (2003) studied four cases: excitation by either a radial or a transverse sinusoidal perturbation for two magnetic field strengths, either an "umbra" at the bottom boundary or a weak field. In the strong-field case, the plasma β is below unity at the location of the piston and the upward propagating waves do not cross a magnetic canopy. As the field is not exactly vertical at the location of the

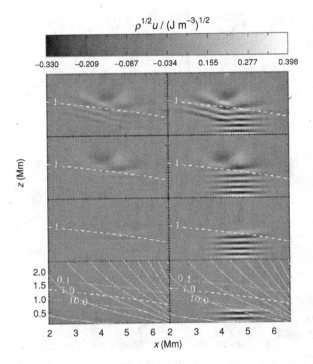

Fig. 8.8. Model results for horizontal (left-hand panels) and vertical (right-hand panels) wave velocities, scaled by the square root of the density. In this model the atmosphere is a schematic isothermal chromosphere with a magnetic field structure shown by the solid white lines; plasma β contours are shown by the white broken lines and dotted lines in the lower panels (cf. Fig. 8.2). A fast-mode (acoustic) wave is excited at the lower boundary by a vertically moving piston; its propagation is shown at four different times, from bottom to top. The fast-mode wave interacts with the magnetic field near $\beta = 1$ in such a way that the wave is refracted and essentially reflected back into the photosphere and towards regions of high plasma β. See Rosenthal *et al.* (2002) and Bogdan *et al.* (2003).

piston, both longitudinal and transverse waves are excited. The longitudinal waves propagate as slow-mode, predominantly acoustic, waves along the magnetic field. The transverse waves propagate as fast-mode, predominantly magnetic, waves. These waves are not confined by the magnetic field and they are refracted towards regions of lower Alfvén speed. They are therefore turned around and impinge on the magnetic canopy in the "penumbral" region. In places where the wave vector forms a small angle to the field lines, the waves are converted to slow waves in the lower region; in places where the attack angle is large there is no mode conversion and the waves continue across the canopy as fast waves. The simulations show that wave mixing and interference are important aspects of oscillatory phenomena in the solar chromosphere.

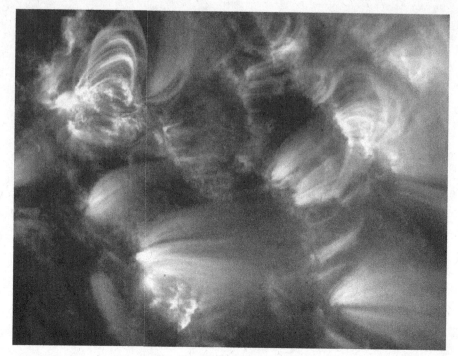

Fig. 8.9. The EUV corona as seen by the TRACE telescope in the 171 Å bandpass (TRACE is a satellite of the Stanford Lockheed Institute for Space Research and is part of the NASA Small Explorer program). The image shows plasma at around 1 MK.

8.4 Coronal heating

The problem of coronal heating has plagued solar physics since the discovery in the 1930s by Edlén and Grotrian that the corona has a temperature of order 1 MK. To this we can also add the problems of how the network chromosphere is heated and of the origin of the background emission in the internetwork chromosphere, neither of which can be explained by the action of 3 mHz or higher-frequency acoustic waves.

Clearly some "mechanical" form of heat input is necessary to raise the coronal plasma to a temperature much higher than the photospheric radiation temperature. The debate has raged in the decades that have followed: the convection zone produces more than enough mechanical energy flux, but how is this energy flux transported to the corona and how is it ultimately dissipated? Does the buffeting of magnetic flux tubes on relatively short time scales cause wave-heating or AC-heating? Or is the slower shuffling of flux tubes, causing stresses in the coronal magnetic field to build up, later to be episodically relieved in nano-flares as in the

DC-heating scenario, more important? (See e.g. Mariska (1993) and references cited therein for a summary of this debate.)

Apart from the question of energy transformation, other questions stand between us and an understanding of coronal heating. How is the energy flux ultimately thermalized? Do we need to continually inject new magnetic flux into the photosphere in order to sustain the corona? And if so, how much? Are there robust diagnostics that can separate the various heating mechanisms?

For reasons of personal preference we will unashamedly pursue the so-called nano-flare heating mechanism for the duration of this chapter, but please keep in mind that this set of problems is not by any means solved!

Why a million degrees? Before turning to a discussion of coronal modeling and coronal heating it is worth spending a few paragraphs on coronal temperatures. The densities inferred in the corona are generally high enough to ensure that velocity distribution functions are nearly Maxwellian and that electron and ion temperatures are nearly equal. But why is the coronal temperature of order 1 MK? Is achieving such a temperature a robust measure of the heating mechanism's success? To answer that question it is important to realize that the temperature of a plasma is set not only by the heat dissipated but also by the plasma's ability to lose energy.

The coronal plasma has essentially three possible ways to shed energy:

(i) through optically thin radiation, mostly from carbon, oxygen, nitrogen, iron, and neon (and at lower temperatures from effectively thin hydrogen Lyman-α transitions), described by

$$\Lambda(T_e) = n_e n_H f(T_e), \tag{8.6}$$

where n_e and n_H are the electron and total hydrogen densities and $\Lambda(T_e)$ is a function of temperature dependent mainly on line emission and, at higher temperatures, on thermal bremsstrahlung;

(ii) through thermal conduction along the magnetic field, with a conduction coefficient

$$\kappa(T_e) = -\kappa_0 T_e^{5/2} \nabla_\| T_e; \tag{8.7}$$

(iii) through the acceleration of a solar wind, in the case of the magnetically open corona. This is a very efficient energy-loss mechanism and it sets firm limits on coronal electron and ion temperatures, as described by Hansteen, Leer, and co-workers (Hansteen and Leer, 1995; Hansteen *et al.*, 1997a). It will be discussed in Chapter 9.

In short, if the plasma is dense then on the one hand $n_e n_H$ is large and variations in the heat input can be accommodated by small changes in the plasma temperature, which will remain at about 10^4 K or less (similar to the photospheric radiation temperature). Conduction, on the other hand, is very inefficient at these

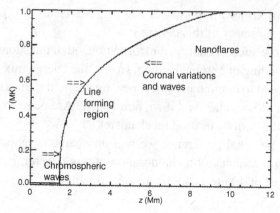

Fig. 8.10. Schematic stratification of the transition region between the chromosphere and corona. A persisting conductive flux from the corona ensures that the
temperature gradient must steadily increase with decreasing temperatures until
other terms (e.g. radiative losses) in the energy balance become significant. Optically thin spectral lines are formed over a relatively limited range in temperature
and can therefore give good diagnostics of how the transition region responds to
waves and other dynamic phenomena in the chromosphere below or the corona
above.

temperatures. However, the density drops exponentially with height, with a scale
height of only some hundreds of kilometers for a 10^4 K plasma. The efficiency of
radiative losses therefore drops very rapidly with height and *any* mechanical heat
input will raise the temperature of the plasma. The temperature will continue to rise
until thermal conduction can balance the energy input. Since thermal conduction
varies with a high power of the temperature this does not happen until the plasma
has reached 1 MK or so. Thus, we expect every heating mechanism to give coronal
temperatures of this order and we must conclude that the magnitude of the observed
coronal temperature is not a good guide to the mechanisms heating the corona.

8.4.1 The transition region

The argument used above necessarily implies that thermal conduction is the most
important energy-loss mechanism from the closed corona. This in turn means that
the energy flux carried away from the site of coronal heating will mainly be carried
by conduction and therefore will be roughly constant. As the temperature falls,
away from the heating site, the plasma's ability to carry a heat flux decreases
rapidly (as $T_e^{5/2}$, see Eq. 8.7) and the temperature gradient must become large to
compensate. This process sets the structure of the transition region; the interface
between the hot corona and the much cooler chromosphere is invariably sharp, as
shown schematically in Fig. 8.10, at least along individual flux tubes.

Owing to the very small spatial extent of the transition-region, line formation becomes particularly simple; the emission is optically thin and confined in space. Observations of transition-region lines could therefore potentially be both sensitive and understandable in terms of the processes heating both the chromosphere and corona.

Ions in the transition region will be in the ground state in general, excited occasionally by electron collisions that are followed immediately by a spontaneous de-excitation. Thus the intensity may be written

$$I = \frac{h\nu}{4\pi} \int_0^s n_u A_{ul} ds = \frac{h\nu}{4\pi} \int_0^s n_l C_{lu} ds, \qquad (8.8)$$

where the integration is carried out along the line of sight, n_u and n_l are the upper- and lower-level populations of the emitting ion, A_{ul} is the Einstein coefficient, and C_{ul} is the collisional excitation rate. The other symbols retain their usual meanings. The lower-level population may be rewritten as

$$n_l = \frac{n_l}{n_i} \frac{n_i}{n_H} n_H = \frac{n_l}{n_i} A_i n_H, \qquad (8.9)$$

where n_i is the total density of atoms of type i, n_l/n_i is the degree of ionization, n_H is the hydrogen number density, and A_i is the element abundance of atom type i relative to hydrogen. We may write the collisional excitation rate as

$$C_{lu} = n_e C_0 T_e^{1/2} \exp\left(-\frac{h\nu}{kT_e}\right) \Gamma_{lu}(T_e) \qquad (8.10)$$

where C_0 is a constant and $\Gamma(T_e)$ is a slowly varying function of the electron temperature.

Combining the above equations, and gathering the temperature-dependent parts of the intensity – including the ionization balance – into a rapidly varying function $g(T_e)$ that is sharply peaked around the region of maximum ion concentration n_l/n_i, allows us to rewrite the line intensity as

$$I = \frac{h\nu}{4\pi} A_i C_0 \int_0^s n_e n_H g(T_e) ds \qquad (8.11)$$

$$I \propto E(T_e) \equiv \int_{\Delta T_e} n_e n_H \frac{ds}{dT_e} dT_e. \qquad (8.12)$$

The intensity is, in other words, proportional to a quantity called the "emission measure" $E(T_e)$.

The emission measure – as well as the differential emission measure, which is essentially the integrand of the emission measure, Eq. (8.12) – may be observationally determined from measured line intensities. Likewise, given a coronal heating model it is relatively straightforward to construct the expected emission measure.

Comparisons of observed and predicted emission measures have met with little success; $E(T)$ has proven to be a difficult diagnostic for models to satisfy. In short, most if not all models predict much smaller line intensities for lines formed below 200 kK or so than those observed. To quote Athay (1982): *... On the other hand, the total failure of all models for $T < 2.5 \times 10^5$ K is a clear indication that the models have either a grossly incorrect geometry or they are omitting or misrepresenting a fundamental energy transport process.*

Another puzzling observation concerns the line shifts of lines formed in the transition region. Lines formed from the upper chromosphere and lower transition region (e.g. from singly ionized carbon, C II, the line at 133.4 nm) up to lines formed at temperatures of 500 kK are invariably red-shifted *on average*, the maximum red shift, of 10 km/s or greater, being found for lines formed at roughly 100 kK (e.g. the triply ionized carbon C IV 154.8 nm line). There is no reason to suppose that there is a net flow of material from the corona towards the chromosphere, so some preferential weighting mechanism is implied. This effect is stronger in regions where the magnetic field is assumed strong, such as in the network and in active regions. It is weaker or may be absent in the internetwork. Upward propagating sound waves, where plasma compression and the fluid velocity perturbation are in phase, would result in a net blue-shift. Does this imply that the red-shift is due to downward propagating waves formed, for example, as a result of nano-flare dissipation as suggested by Hansteen (1993)? Or is there some other active mechanism ensuring the preferential emission of downward moving plasma, as claimed by e.g. Peter *et al.* (2004)?

8.5 Forward modeling of the outer solar atmosphere

Analytical models, semi-empirical modeling, and close analysis of observations coupled with physical intuition have given researchers important insights into various aspects of the coronal heating question. However, it seems that this is not enough: the system including the convection zone and corona is of sufficient complexity to confound these methods of enquiry about the nature of coronal heating. Perhaps comprehensive interior-to-coronal ab initio numerical experiments can give insight into the problem?

It is only since the early 2000s that computer power and algorithmic develop-ments have allowed one even to consider taking on this daunting task. And still grave doubts remain on the validity of treating microscopic processes in the corona by the averaging methods inherent in the MHD approximation. The nano-flare scenario is based on photospheric shuffling and braiding resulting in the cre-ation of discontinuities in the coronal magnetic field. This implies that relatively large-scale photospheric dynamics drives the coronal field to steadily smaller scales, so that eventually the dissipation scale is reached and energy can be dissipated. Can we trust the results of calculations where this cascade is stopped by dissipation at scales many orders of magnitude larger than those presumably encountered in nature? And, if it happens that in fact we can create a coronal heating mechanism through our numerical modeling, how will we know that it is the right one? We will come back to this and other connected issues in the last section of this chapter.

Gudiksen and Nordlund (2002) showed that it is possible to overcome great numerical challenges, outlined below, to make initial attempts at modeling the photosphere-to-corona system. In their models, a scaled-down longitudinal magnetic field taken from a SOHO/MDI (Scherrer *et al.*, 1995) magnetogram of an active region was used to produce an initially potential magnetic field in the computational domain that covers $50 \times 50 \times 30$ Mm^3. This magnetic field is subjected to a parameterization of the horizontal photospheric flow on the basis of observations and, at smaller scales, on numerical convection simulations as the driver at the lower boundary. After a period of some 10–15 minutes solar time (and several months of cpu time!) stresses in the simulated corona accumulate and become sufficient to maintain coronal temperatures. Synthetic TRACE (Handy *et al.*, 1999) images constructed from these models show a spectacular similarity to images such as Fig 8.9. In addition, other synthetic diagnostics, e.g. of transition-region lines, show promising characteristics (Peter *et al.*, 2004).

There are several reasons why the attempt to construct forward models of the convection zone or photosphere-to-corona system has been so long in coming. We mention only a few.

The size of the simulation. We mentioned in Section 8.3.2 that the magnetic field tends to reach heights comparable with the distance between the sources of the field. Thus on the one hand if one wishes to model the corona to a height of, say, 10 Mm this requires a horizontal size close to twice that, i.e. 20 Mm, in order to form closed-field regions up to the upper boundary. On the other hand, resolving photospheric scale heights of 100 km or smaller and transition-region scales of some few tens of kilometers requires minimum grid sizes less than 50 km and preferably much smaller. (Numerical "tricks" can perhaps ease some of this difficulty, but will not help by much more than a factor 2). Putting these requirements together means that it is difficult to get away with computational domains of much less than 150^3 – a non-trivial exercise even on today's systems.

Thermal conduction. The "Courant condition", for a diffusive operator such as that describing thermal conduction, scales as $(\Delta z)^2$ instead of linearly with the grid size Δz as for the magnetohydrodynamic operator. This severely limits the time step Δt at which the code can be stably run. One solution is to vary the magnitude of the coefficient of thermal conduction when needed. Another is to proceed by operator splitting, such that the operator advancing the variables in time is $L = L_{\text{hydro}} + L_{\text{conduction}}$. Then the energy equation is solved by discretizing by the Crank–Nicholson method

$$\frac{\partial e}{\partial t} = \nabla \mathbf{F}_c = -\nabla \kappa_\parallel (\nabla_\parallel T),$$

where $\mathbf{F}_c$ is the energy flux transported by conduction, and then solving the system by for example the multi-grid method.

Radiative transport. The radiative losses from the photosphere and chromosphere are optically thick and will in principle require the solution of the transport equation. This can be done by the methods outlined in Section 8.2.1 for the case of the photosphere, which is close to local thermodynamic equilibrium. Modeling the much more transparent chromosphere may require that the scattering of photons is treated with greater care (Skartlien, 2000) or, alternatively that one uses methods assuming that chromospheric radiation can be tabulated a priori as a function of the local thermodynamic variables.

8.5.1 Convection zone to corona

With the proper tools in hand it is very tempting to attempt to model the entire solar atmosphere, from convection zone to corona. In Fig. 8.11 we show the result of such an experiment.

Into a box of dimension $16 \times 8 \times 16 \, \text{Mm}^3$ with well-established convection we insert a potential magnetic field generated by setting up six sources given by positive and a negative poles with magnitude 1000 Gauss and random locations at the lower boundary. The model is convectively unstable, driven by the radiative losses in the photosphere. The average temperature at the bottom boundary is maintained by fixing the entropy of the fluid entering through the bottom boundary. The bottom boundary, the properties of which are based on the extrapolation of characteristics, is otherwise open, allowing fluid to enter and leave the computational domain as required. The magnetic field at the lower boundary is advected with the fluid. As the simulation progresses the field is advected with the fluid flow in the convection zone and photosphere and individual field lines quickly attain quite complex paths through the model, as shown in Fig. 8.11.

To prevent immediate coronal cooling, the upper temperature boundary was initially set to 800 kK and the models were allowed to evolve from their potential state

Fig. 8.11. The thermal structure and magnetic field of a 3D model as it has developed after an hour's solar time evolution in a box of dimensions $16 \times 8 \times 16$ Mm. A horizontal slice of the temperature is shown near the photosphere; this surface shows regions where the temperature is in the range 15–400 kK. Selected magnetic field lines are plotted. Convection-zone and photospheric motions have deformed the originally quite simple potential field during the simulated hour. In the photosphere, the field is quickly concentrated in intergranular lanes. In the middle to upper chromosphere, β decreases below unity and the field expands to fill all space, constrained by its photospheric footpoints and forming loop-like structures. Note also that the field lines below the photosphere rarely breach the surface; rather, they become quite tangled as a result of convective buffeting (see also the top panels in Fig. 1.2).

for roughly 20 minutes. After this time the upper boundary was set to a zero temperature gradient, so that no conductive heat flux enters or leaves the computational domain. Aside from the temperature, the other hydrodynamic variables and the magnetic field were set using extrapolated characteristics.

The temperature structure in the model is also indicated in Fig. 8.11. It is found that the photosphere is some 1.5 Mm above the bottom boundary. The convection zone lies below, and has a temperature of some 16 kK at the lower boundary. Above the photosphere the chromosphere stretches upwards to the transition region over a span varying between 1.5 Mm and 6 Mm above the photosphere, depending on the local magnetic field configuration. The corona fills the remaining volume of the computational domain, with temperatures up to 1.5 MK in some locations but

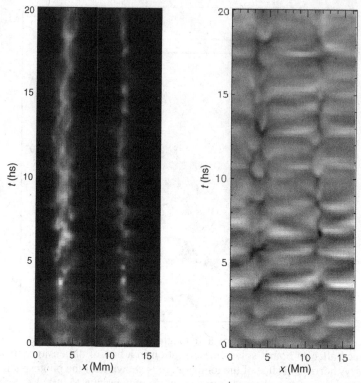

Fig. 8.12. Simulated observations of the O VI 103.2 nm line, which is formed in regions where the temperature is roughly 300 kK. The left-hand panel shows the total intensity in the line and the right-hand panel shows the average Doppler shift. These simulated observations are based on a 2D MHD model spanning a region 16 Mm × 10 Mm covering the convection to the lower corona. The magnetic topology in this model is similar to that shown in Fig. 8.11: a "loop" with footpoints close to $x = 4$ Mm and $x = 12$ Mm. The periodic oscillations visible are mainly due to upward propagating waves generated in or below the photosphere.

overall a fairly inhomogeneous temperature structure with temperatures as low as some few hundred thousand degrees even at large heights.

Transition-region diagnostics. There are several useful applications for a model such as the one described here. Of these, perhaps the most interesting lies in studying the generation and dissipation of magnetic field stresses in the corona, as described in Section 8.5.2 below. But there is also potential insight to be gained from studying the chromosphere and transition region in such models; 3D models of the magnetized chromosphere and transition region have been noticeably lacking. As an example let us consider the emission from the O VI 103.2 nm line, formed in regions where the temperature is roughly 300 kK. In Fig. 8.12 we show the average line intensity and average line doppler shift as a function of time obtained from a 2D model otherwise equivalent to the 3D model described above. The average magnetic field in this model has a fairly simple structure with a "loop" passing through the transition region having footpoints near $x = 4$ Mm and $x = 12$ Mm. The amplitude of the line emission is strong in regions where the magnetic field is nearly vertical, i.e. in the footpoints, and quite weak where the field is horizontal. The line shifts show that some signal of the chromospheric 3 minute oscillations reaches the model transition region, but at different times in regions of nearly vertical and nearly horizontal field. It is also interesting to note that the net line shift does not vanish in the footpoints but rather displays an average red-shift of 10 km/s. These images are remarkably similar to images constructed from SOHO SUMER observations of the same line and engender some confidence that this is a fruitful method for interpreting such observations.

8.5.2 Modeling nano-flares

As the stresses in the coronal field grow, so does the energy density of the field. This energy must eventually be dissipated, at a rate commensurate with the rate at which energy is pumped in. On the Sun, the magnetic diffusivity η is very small and gradients must become very large before dissipation occurs. In the models presented here, we operate with an η-value many orders of magnitude larger than on the Sun, and so dissipation starts at much smaller magnetic field gradients. The dissipated energy is

$$Q_{\text{Joule}} = \mathbf{E} \cdot \mathbf{j} \tag{8.13}$$

where $\mathbf{J} = \nabla \times \mathbf{B}$ is the current density and the resistive part of the electric field $\mathbf{E}$ is given by

$$E_x^{\eta} = \left[\frac{1}{2}(\eta_y^{(1)} + \eta_z^{(1)}) + \frac{1}{2}(\eta_y^{(2)} + \eta_z^{(2)}) \right] J_x \tag{8.14}$$

and similar for E_y^{η} and E_z^{η}. The diffusivities are given by

$$\eta_j^{(1)} = \frac{\Delta x_j}{P_{\text{m}}} (w_1 c_f + w_2 |v_j|), \tag{8.15}$$

$$\eta_j^{(2)} = \frac{\Delta x_j^2}{P_{\text{m}}} w_3 |\nabla_{\perp} \cdot \mathbf{v}|_{-} \tag{8.16}$$

where P_{m} is the magnetic Prandtl number, c_f is the fast-mode speed, w_1, w_2 and w_3 are dimensionless numbers of order unity, and the other symbols retain their usual meanings.

The working assumption in these models is that, although the artificial magnetic diffusivity used here and the diffusivity found on the Sun differ by many orders of magnitude, the total amount of energy actually dissipated in the chromosphere and corona should be similar (Galsgaard and Nordlund, 1996; Hendrix *et al.*, 1996).

The following conclusions are among the results of this modeling effort.

- A non-heated non-magnetic corona will cool significantly within 1200 to 1500 s. Even fairly strong hydrodynamic waves cannot maintain coronal temperatures.
- However, it seems that a corona threaded by even fairly weak fields can be maintained at temperatures of greater than 500 000 K by the stresses developed through convective and photospheric motions.
- The average coronal temperature (and heating) rises with increasing magnetic field strength. The structure of the field may also have some effect but perhaps mainly on the location (height) of the heating.
- Electromagnetic heating rates in the chromosphere are high, and we expect to see signatures of magnetic chromospheric heating in simulated emission.

8.6 The way forward

It seems that the modeling effort so far has had very promising results, as a number of observational characteristics are reproduced in these models. Starting from an observed magnetic field and a parameterization of the solar velocity field in the photosphere as boundaries and drivers, one can produce images that look remarkably like those observed in the coronal TRACE bands. In addition, synthetic spectral lines calculated on the basis of these models show characteristics that are very similar to those seen in SUMER and CDS spectra, as shown in Fig. 8.12 and Peter *et al.* (2004). And this congruence between observations and model is achieved with very few free parameters. Can we then conclude that the coronal heating problem is solved? And further, are these processes also relevant for the unanswered questions concerning the heating of the magnetic chromosphere? Can the methods and conclusions described here be used to understand the range of behavior we see in other stars?

Stars come in a huge variety. They have a large spread in mass, size, and surface temperature. Relatively small and cool stars of types F, G, K, and M have one characteristic that sets them apart from other stars. These stars are observable in X-ray emission in spite of their low surface temperatures. This X-ray emission is believed to originate in a stellar corona. Stars of types F, G, K, and M are stars similar to our Sun, with a convection zone and a hot corona like those of our Sun. The process that creates these stellar coronae is presumably identical to the one creating the corona of our Sun. It is thus puzzling that the X-ray luminosity within each of these star classes spans three orders of magnitude in a complete, volume-limited, sample (Schmitt and Liefke, 2004). Why is it that one star, almost identical to another on the basis of photosphere observations, produces a thousand times more energy in X-rays than the other? It is believed that the stellar rotation rate and resulting field strength are important parameters in this question.

Perhaps a word or two of caution is in order before we celebrate our successes and plan our next battles. We do have a promising hypothesis, but the following question remains. Are the tests to which we are subjecting it – e.g. the comparison of synthetic observations with actual observations – actually capable of separating a correct description of the Sun, or a stellar atmosphere, from an incorrect one? We have already demonstrated that we expect the solar corona to be heated to roughly 1 MK almost independently of the mechanism for such heating. Conduction along field lines will naturally make loop-like structures. This implies that reproducing TRACE-like "images" is perhaps not so difficult after all, and possible for a wide spectrum of coronal heating models. Transition-region and chromospheric diagnostics form a potentially much more discriminating test of the present heating model, but clearly it is still too early to say that the only possible model has been identified. It will be very interesting to see how these forward heating models stand up in the face of questions such as these: how does the corona react to variations in the total field strength or the total field geometry, and what observable diagnostic signatures do these variations cause?

Another issue is the fact that the treatment of the microphysics of dissipation is demonstrably wrong in these models. As stated in the previous section, the argument can be made that this does not matter: the only important factor is the amount of Poynting flux entering the corona. The way in which this is dissipated, and over which spatial and temporal scales, depends in part on the details of the microphysics, but the total amount of energy flux going into heating the corona remains the same and is independent of the exact physical process that thermalizes the Poynting flux. However, persuasive as this argument may seem, it would be a large step forward to find diagnostics that could confirm that the model described here is essentially correct.

One also wonders about the role of the emerging flux in chromospheric and coronal heating: how much new magnetic flux must be brought up from below in order to replenish the field dissipated in heating the outer atmosphere? And what, if any, is the role of an surface dynamo? The thermalization process is itself of great interest. Is it highly episodic, such as was found in the work of Einaudi and Velli (1999)? Do the electric fields built up by the churning of the magnetic field cause particle acceleration to large energies, as claimed by Turkmani *et al.*

(2005)? Is there a difference between the coronal heating mechanism and the process heating the network chromosphere? What happens in open magnetic field regions where the stresses built up by photospheric motions are free to propagate out into interplanetary space? There are certainly many open questions to be dealt with in the field of coronal heating, even if it should turn out that the basic scenario is correctly described by the current crop of forward models.

9

Stellar winds and magnetic fields

VIGGO H. HANSTEEN

The solar wind is responsible for maintaining the heliosphere and is the driving agent in the magnetospheres of the planets; furthermore, it provides the mechanism by which the Sun has shed angular momentum during the aeons since its formation. We can assume that other cool stars with active coronae have similar winds, which play the same roles for those stars (Wood *et al.*, 2005). The decade since the launch of SOHO has seen considerable changes in our understanding of thermally driven winds such as the solar wind, owing to theoretical, computational, and observational advances. Recent solar wind models are characterized by low coronal electron temperatures while proton, α-particle, and minor-ion temperatures are expected to be quite high and perhaps anisotropic. This entails an assumption that the electric field is relatively unimportant and that one is able to obtain in a quite natural way a solar wind outflow that has a high asymptotic flow speed while maintaining a low mass flux. In this chapter we will explain why these changes have come about and outline the questions now facing thermal wind astrophysicists.

The progress we have seen in the last decade is largely due to observations made with instruments on board Ulysses (McComas *et al.*, 1995) and SOHO (Fleck *et al.*, 1995). These observations have spawned a new understanding of solar wind energetics and the consideration of the chromosphere, corona, and solar wind as a unified system.

We will begin by giving our own, highly individual, "pocket history" of solar wind theory, highlighting the problems that had to be resolved in order to make the original Parker formulation of thermally driven winds conform with the obser-vational results. Central to this discussion are questions of how the wind's asymp-totic flow speed and mass flux are set, but we will also touch on higher-order moments such as the ion and electron temperatures and heat fluxes as well as the possible role of Alfvén waves and particle effects in driving the solar wind outflow.

9.1 A pocket history

There are many ways to begin a history of the solar wind, but, the author being Norwegian, it seems natural to him to begin with Kristian Birkeland. On the basis of laboratory studies and observations from stations in the Arctic, Birkeland concluded that the Sun continuously emits electric corpuscular radiation (Birkeland, 1908, 1913). The laboratory studies were performed using his "terella", designed to study the interaction of electrons from the Sun with the Earth's magnetic field: in this device cathode rays impinged on a magnetized sphere. The emission of particles from the Sun was studied also, by using the magnetized sphere as a cathode. Birkeland became so convinced of the correctness of his view that he assumed that *all* stars were emitting electrons and ions and that interstellar space was filled with "electrons and flying ions of all kinds". Unfortunately the work carried out by Birkeland went largely forgotten and it was first some 50 years later that the concept of a solar wind became universally known and accepted.

One could also begin such a history with the studies of Chapman and Ferraro in the 1930s (Chapman and Ferraro, 1929), in which they pointed out that magnetic storms could be explained by clouds of ions ejected from the Sun, or with Forbush's discovery (Forbush *et al.*, 1949) that the intensity of cosmic rays was lower when solar activity was high. The concept of an ionized gas outflow from the Sun was, however, first firmly established through the work of Biermann (1951, 1957) where it was pointed out that the light pressure from the Sun was not sufficient to account for the speed at which gases are blown away from the heads of comets.

By the 1950s, the other vital ingredient in the thermal solar wind – a hot, extended solar corona – had been established by the observational interpretative work of Grotrian (1933, 1939) and Edlén (1942) on emission lines such as Fe X and Fe XIV and the theoretical work of Alfvén (1941, 1947), Biermann (1946, 1948), and Chapman (1954, 1957). Much could and should be said about the fact that the Sun is surrounded by a corona of gas with temperature roughly 1 MK, as discovered by these authors. Here we will in the first instance simply note that a temperature of some millions of degrees implies that the sound speed c_s – essentially the mean ion speed – is much smaller that the Sun's escape speed v_g:

$$c_s \approx \sqrt{kT/m} \approx 100 \text{ km/s} \ll v_g = \sqrt{2GM_S/R_S} = 618 \text{ km/s}, \qquad (9.1)$$

where k is Boltzmann's constant, T the coronal temperature, m the mean particle mass, and G the universal gravitational constant; M_S and R_S are the solar mass and radius, respectively.

Mass and momentum balance radially away from the Sun along a flow tube with cross section A at heliocentric distance r can be written

$$\frac{d}{dr}(\rho v A) = 0,\tag{9.2}$$

$$\rho v \frac{dv}{dr} = -\frac{dp}{dr} - \rho \frac{GM_S}{r^2},\tag{9.3}$$

with ρ the mass density and v the flow speed. Then

$$p = 2nkT\tag{9.4}$$

is the gas pressure in an electron–proton plasma, n representing the electron or proton number density, so that $\rho = mn$ where $m \approx m_p/2$ is the mean particle mass for an electron–proton plasma.

Alfvén (1941) and Chapman (1957) showed that the consequence of thermal conduction in a million-degree corona is to extend the corona, i.e. the temperature falls off slowly with distance from the Sun. Thus, in a hypothetical *static* atmosphere, we find a pressure at infinity given by

$$\frac{d}{dr}(p) = -nm\frac{GM_S}{r^2},\tag{9.5}$$

$$p(r) = p_0 \exp\left[-\frac{GM_S m}{2k}\int_{R_S}^{r}\frac{dr}{r^2 T(r)}\right].\tag{9.6}$$

Thus, if the temperature falls off less rapidly than $1/r$, so that $\lim_{r\to\infty} p(r) > 0$, we expect a non-vanishing pressure at infinity when the corona is extended. In particular, we find that for reasonable temperatures and densities n_0, T_0 at the "coronal base" this pressure is much larger than any conceivable interstellar pressure.

Using this argument Parker (1958) deduced that the solar wind must expand supersonically into interstellar space, and he solved the equations of the solar wind under the assumption of a spherically symmetric, single-fluid, isothermal outflow. In this case the equations of mass and momentum conservation (Eqs. 9.2, 9.3) can be rewritten to give

$$\frac{1}{v}\frac{dv}{dr}\left(v^2 - \frac{2kT}{m_p}\right) = \frac{4kT}{m_p r} - \frac{GM_S}{r^2}.\tag{9.7}$$

Parker showed that the transsonic wind passes through a critical point

$$r_c = \frac{m_p GM_S}{4kT}\qquad\text{where}\qquad v_c = \sqrt{\frac{2kT}{m_p}}\tag{9.8}$$

and that such a flow could be made to match *any* pressure as $r \to \infty$.

Let us examine this trans-sonic wind solution in somewhat greater detail. If we integrate the force balance, Eq. (9.3), from the coronal base to the critical point r_c then we find a density ρ_c at the critical point given by

$$\rho_c = \rho(r_c) = \rho_0 \exp\left(-\frac{m_p G M_S}{2kT R_S} + \frac{3}{2}\right). \tag{9.9}$$

Note that this density is almost exactly the same as if there had been *no* solar wind flow: *the subsonic corona in the solar wind is essentially stratified as a* static atmosphere.

We can also find the resultant mass flux for the wind by examining the density and the velocity at the critical point:

$$(nv)_r = n_c v_c \frac{r_c^2}{r^2}$$

$$\propto \rho_0 T^{-3/2} \exp\left(-\frac{C}{T}\right) \tag{9.10}$$

where ρ_0 is the density at the coronal base. *The mass flux is proportional to the density at the coronal base and depends exponentially on the coronal temperature.*

For various reasons, Parker's solution did not at first find general acceptance, but debate quickly stilled after his prediction of a supersonic outflow were confirmed by the Mariner II spacecraft observations analyzed by Neugebauer & Snyder (1962).

9.2 The Parker spiral

The solar wind also carries the solar magnetic field with it. Let us briefly consider the outflow of ionized, magnetized, gas from a *rotating* star with a monopole magnetic field. The salient aspects of such a flow are found even when one considers only the equatorial plane and restricts attention to solutions where all variables are functions of r only. We make use of polar coordinates.

We find that then the equation of mass conservation can be written as

$$\frac{1}{r^2}\frac{\partial}{\partial r}\left(\rho v_r r^2\right) = 0, \tag{9.11}$$

while the ϕ component of the momentum equation is given by

$$\rho\left(v_r \frac{\partial v_\phi}{\partial r} + v_\phi \frac{v_r}{r}\right) = \frac{1}{\mu_0}\left(B_r \frac{\partial B_\phi}{\partial r} + B_\phi \frac{B_r}{r}\right) \tag{9.12}$$

or

$$\rho v_r \frac{1}{r}\frac{d}{dr}(r v_\phi) = \frac{1}{\mu_0} B_r \frac{1}{r}\frac{d}{dr}(r B_\phi). \tag{9.13}$$

Mass conservation implies that $\rho v_r r^2$ is constant, while a divergence-free field requires that $B_r r^2$ is constant. Multiplying Eq. (9.13) by r^3 we see that

$$r v_\phi - \frac{B_r r^2}{\rho v_r r^2} \frac{1}{\mu_0} r B_\phi = \text{constant} = L. \tag{9.14}$$

Under the assumptions above, the induction equation becomes

$$\frac{1}{r} \frac{d}{dr} \left(r(v_r B_\phi - v_\phi q B_r) \right) = 0. \tag{9.15}$$

For a star rotating with angular velocity Ω and radius R_s and having a monopole field, so that $B_{\phi s} \approx 0$, we find that the induction equation implies that

$$r(v_r B_\phi - v_\phi B_r) = \text{constant} \approx -R_s(R_s\Omega)B_{rs} = -\Omega r^2 B_r. \tag{9.16}$$

We can now solve Eqs. (9.14) and (9.16) for v_ϕ and B_ϕ and, using $M_A^2 = v_r^2/v_A^2$ with $v_A^2 = B_r^2/(\mu_0\rho)$, we find

$$v_\phi = \Omega r \frac{M_A^2(L/r^2\Omega) - 1}{M_A^2 - 1} \tag{9.17}$$

and

$$B_\phi = -\frac{B_r \Omega r}{v_r} \left[\frac{1 - L/(r^2\Omega)}{M_A^2 - 1} \right] M_A^2. \tag{9.18}$$

Both expressions show that we must have $1 - L/(r^2\Omega) = 0$ when $M_A^2 - 1 = 0$. We define $r \equiv r_a$ when $M_A^2 = 1$. Thus, we must have $L = r_a^2\Omega$. Notice that for large r we have that v_r tends to a constant and thus $M_A^2 \propto \Omega r^2$; as a result,

$$v_\phi \approx \frac{\Omega r_a^2}{r} \to 0 \tag{9.19}$$

and

$$B_\phi \approx -\frac{B_r \Omega r}{v_r}, \tag{9.20}$$

while for small r, close to the star, we have $v_\phi \approx r\Omega$ and $B_\phi \approx -B_r\Omega r/v_{Aa}$. In other words, the magnetic field and stellar wind rotate like a solid body out to the critical point r_a where the radial flow speed is equal to the "radial" Alfvén speed. Beyond this point the field is pulled along the wind into a spiral, the *Parker spiral*, as the flow becomes nearly radial far from the star (see Fig. 4.7 for the innermost part of such a spiral).

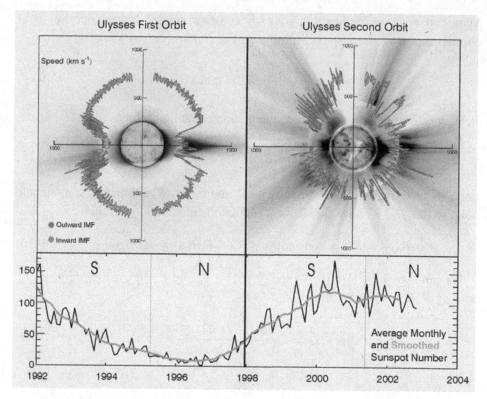

Fig. 9.1. Polar plots of solar wind speed as a function of latitude for Ulysses' first two orbits. The solar wind speed data was obtained by the SWOOPS (Solar Wind Observations Over the Poles of the Sun) instrument. The bottom pair of panels shows the sunspot number over the period 1992–2003. The first orbit occurred through the solar cycle's declining phase and minimum while the second orbit spanned the solar maximum (from McComas *et al.*, 2003).

9.3 Some solar wind properties

At this point it is useful to give a give a quick (and incomplete) overview of measured solar wind properties. A number of spacecraft have observed the solar wind in situ and from these measurements it is clear that at the very least one must divide the solar wind into two distinct states, the *fast* wind, with its origin in coronal holes near the Sun's poles and, irregularly, on the disk, and the *slow* wind, with its origin in or near the partly closed streamer belt associated with the solar equator. The two states and their spatial distribution are illustrated in Fig. 9.1, which also shows an overview of the wind speed and density measured in situ by the Ulysses spacecraft during two polar orbits around the Sun. Ulysses' orbit is highly inclined to the ecliptic and ranges from the Earth's distance, 1 AU, to Jupiter's orbit 5 AU from the Sun. The solar wind plots are plotted over solar images characteristic of a solar minimum (17 August 1996) and a solar maximum (7 December 2000). The solar images are, from the center out, composites from the Solar and Heliospheric Observatory (SOHO) extreme ultraviolet imaging telescope (Fe XII at 19.5 nm), the Mauna Loa K-coronameter (700–950 nm), and the SOHO C2 large angle spectrometric coronagraph (white light).

The Sun's coronal magnetic field is forced open into the heliosphere by the outward pressure of plasma as the latter is accelerated to form the solar wind. This opening of the field is represented well by models that impose a boundary condition of strictly radial field at a height some $1.5R_\odot$ above the solar surface. At that height, the higher-order multipoles that describe the coronal field on scales associated with individual active regions have decreased so much that the field entering the heliosphere is dominated by the low-order dipole and quadrupole components of the global photospheric field. As a result, the pattern of the radial field at a height of $1.5R_\odot$ mostly consists of two large patches of opposite polarity, separated by an evolving, undulating, neutral line where the radial field itself vanishes (see Schrijver, 2005). Mapping back from this surface using a potential field source surface (PFSS) (Neugebauer *et al.*, 1998; Liewer *et al.*, 2004) along potential fields to any height between the solar surface and $1.5R_\odot$ allows one to infer the source of various types of solar wind. Specifically it has been found that fast wind comes from polar coronal holes, while slow wind comes from near the boundaries between these holes and the low-latitude coronal streamer belt (the detailed origin of the slow wind remains a topic of active research).

The extension of this field pattern into the solar wind above the rotating Sun forms the spiraling heliospheric field, as described in Section 9.2, in which a warped current sheet separates fields of opposite radial polarities. This pattern sweeps by the Earth with each solar rotation, and as it does so, the magnetic polarities are observed to commonly form a two-sector pattern but with higher-order undulations occasionally leading to more than two current-sheet crossings per solar synodic rotation.

In general one finds the slow solar wind to be highly variable and filamentary while the fast wind is more uniform, displaying slower changes with time (see Table 9.1). We will mainly concentrate on the fast solar wind in the remainder of this chapter.

9.4 A pocket history, continued

As should be clear from Fig. 9.1, the fast solar wind regularly displays asymptotic velocities well in excess of 700 km/s. As first discussed by Parker (1965), such high speeds are very difficult to achieve by simply varying the parameters ρ_0 and T_0 at

Table 9.1. *Basic parameters of the fast and slow solar wind (based on Holzer (2005) and Feldman et al. (1977), and references cited therein). See also Section 11.3 and Table 11.1*

Property (1 AU)	Slow wind	Fast wind
Speed	430 ± 100 km/s	700–900 km/s
Density	$\simeq 10$ cm^{-3}	$\simeq 3$ cm^{-3}
Flux	$(3.5 \pm 2.5) \times 10^8$ cm^{-2}s^{-1}	$(2 \pm 0.5) \times 10^8$ cm^{-2}s^{-1}
Magnetic field	6 ± 3 nT	6 ± 3 nT
Temperatures	$T_p = (4 \pm 2) \times 10^4$ K	$T_p = (2.4 \pm 0.6) \times 10^5$ K
	$T_e = (1.3 \pm 0.5) \times 10^5$ K $> T_p$	$T_e = (1 \pm 0.2) \times 10^5$ K $< T_p$
Anisotropies	T_p isotropic	$T_{p\perp} > T_{p\parallel}$
Structure	filamentary, highly variable	uniform, slow changes
Composition	He/H$\simeq 1 - 30\%$	He/H$\simeq 5\%$
	low-FIP enhanced	near-photospheric
Minor species	n_i/n_p variable	n_i/n_p constant
	$T_i \simeq T_p$	$T_i \simeq (m_i/m_p)T_p$
	$v_i \simeq v_p$	$v_i \simeq v_p + v_A$
Associated with	streamers and transiently open field	coronal holes

the coronal base; thermal conduction is not capable of carrying sufficient energy into the corona to accelerate the wind to speeds of order 700 km/s or greater. Thus some additional acceleration, beyond the coronal base, must be assumed to operate. Leer and Holzer (1980) showed that this additional energy or push must be added beyond the critical point in order to be effective in accelerating the wind while maintaining mass fluxes within observational constraints.

We can understand the reasons for this result on the basis of our previous discussion. Equation (9.9) shows that the density at the critical point in a wind solution is essentially the same as that found in a static atmosphere. This means that the density of particles at the critical point will increase exponentially with increasing temperature or with some given "push" to the subsonic gas that effectively reduces g. In other words, if we try to accelerate the wind to high speed by increasing the coronal base temperature T_0 then the number of particles at the critical point will increase exponentially and the energy per particle that can be used to accelerate the flow to high speed may actually *decrease*, in the worst case leading to a *slower* wind. A guaranteed faster wind is only achieved if we add energy, or push the wind, beyond the critical point in which case the density is not perturbed and (most of) our effort goes into accelerating the wind.

The near-static nature of the subsonic corona has another consequence, which is closely related to the arguments given above. Equation (9.10) shows that, owing to the exponential dependence of the critical point density on the coronal temperature,

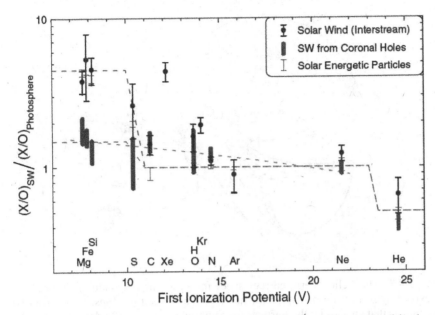

Fig. 9.2. Abundances of various elements (relative to oxygen) measured in the solar wind, relative to their photospheric abundances (relative to oxygen), ploted against an element's first ionization potential (FIP). The solid circles represent values typical of the slow solar wind (the "interstream" wind); the thick vertical lines represent values from the fast wind originating in coronal holes. The thin vertical lines represent data from solar energetic particles arising in flares and coronal mass ejections (from von Steiger and Murdin, 2000).

the solar wind mass flux at 1 AU will also depend exponentially on the coronal temperature. This argument leads to the following dilemma. On the one hand the observed mass flux at 1 AU does not vary much and can in fact be considered almost constant. On the other hand, in situ observations of coronal emission and freeze-in temperatures of ionization give one no reason to believe that the coronal temperature is essentially constant, as it should be if the theory above were correct. We must ask, what process ensures that the solar wind proton flux is invariant?

Below we will discuss ways to resolve this problem, and we will find that the answer probably lies in a consideration of solar wind energetics, but let us first consider a likely candidate for the "distant acceleration" of the solar wind.

9.5 An interlude with Alfvén waves

The realization that additional energy was required to ensure a fast solar wind flow was followed in the late 1960s and early 1970s by the detection of Alfvén waves (e.g. Belcher and Davis, 1971). Alfvén waves have the very interesting property

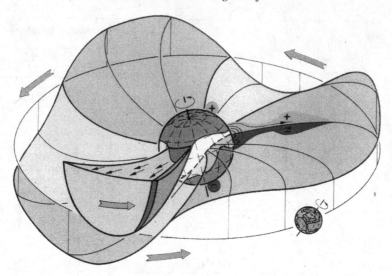

Fig. 9.3. The heliospheric current sheet forms a warped, undulating structure extending from the top ridge of the helmet streamer belt (cf. Figs. 1.3 (top panel) and 8.1) that sweeps by the Earth as the Sun rotates once per 27 days (the synodic period). The magnetic field changes direction across the current sheet. (From Alfvén, 1947.)

that, once they have been generated and have managed to propagate through the chromosphere and transition region, they are very difficult to dissipate and should propagate far into the wind without losing too much energy or momentum at a radial distance less than that of the critical point.

To see the effect of Alfvén waves on the flow we follow the excellent exposition of Hollweg (1973) by considering the steady-state conservation equations (see Eqs. 3.1–3.9)

$$\nabla \cdot (\rho \mathbf{v}) = 0, \tag{9.21}$$

$$\nabla \cdot (\rho \mathbf{v}\mathbf{v}) = -\nabla \left(p + \frac{B^2}{2\mu_0} \right) + \left(\frac{\mathbf{B}}{\mu_0} \cdot \nabla \right) \mathbf{B} + \rho \mathbf{g}. \tag{9.22}$$

We write each quantity as the sum of a fluctuating part, the wave, and an average part, i.e. as $\mathbf{B} = \mathbf{B}_0 + \delta\mathbf{B}$, and restrict our attention to incompressible Alfvénic fluctuations, for which $\delta\rho = \delta p = \delta\mathbf{B} \cdot \mathbf{B}_0 = \nabla \cdot \delta\mathbf{B} = 0$. We arrive at the desired equations by taking the time average of Eqs. (9.21) and (9.22) while neglecting any current $\mathbf{j}_0$. A particularly simple equation arises by noting that in the Wentzel–Kramers–Brillouin limit (WKB) limit, i.e. in the limit where the wavelength is smaller than the other characteristic lengths in the system, we have

$$\delta\mathbf{v} = \pm\delta\mathbf{B}(\mu_0\rho_0)^{-1/2}. \tag{9.23}$$

Dropping the subscript zero henceforth, and remembering that we are assuming no current, we find a momentum equation

$$\rho \mathbf{v} \cdot \nabla \mathbf{v} = -\nabla \left(p + \frac{\langle \delta B^2 \rangle}{2\mu_0} \right) + \rho \mathbf{g}_r. \tag{9.24}$$

In other words, the addition of Alfvén waves gives an additional component to the pressure that is proportional to the fluctuation in the squared magnetic field.

In order to estimate how the fluctuation in the magnetic field varies with distance from the Sun one could solve the azimuthal component of the gas momentum equation along with the time-averaged mass, momentum, and energy equations. Hollweg (1973) showed that in the WKB approximation one can find the time-averaged magnetic fluctuation, $\langle B^2 \rangle$, by instead considering the time-averaged energy equation

$$\rho \mathbf{v} \cdot \nabla \frac{3p}{2\rho} + p \nabla \cdot \mathbf{v} = -\nabla \cdot \left(\mathbf{q} + \mathbf{S} + \frac{1}{2}\rho \langle \delta v^2 \rangle \mathbf{v} \right) + \mathbf{v} \cdot \nabla \left(\frac{\langle \delta B^2 \rangle}{2\mu_0} \right). \tag{9.25}$$

Here $\mathbf{S} = (\langle \delta B^2 \rangle / \mu_0)(\mathbf{v} \pm \mathbf{v}_A)$ and $\mathbf{v}_A = \mathbf{B}(\mu_0 \rho)^{-1/2}$ is the Alfvén velocity. The heating due to Alfvén waves is

$$Q \equiv -\nabla \cdot \left(\mathbf{S} + \frac{1}{2}\rho \langle \delta v^2 \rangle \mathbf{v} \right) + \mathbf{v} \cdot \nabla \left(\frac{\langle \delta B^2 \rangle}{2\mu_0} \right). \tag{9.26}$$

Consider now Alfvén waves in a spiral magnetic field with intensity given by

$$B^2 = B_r^2 \left(1 + \frac{r^2 \Omega^2}{v^2} \right), \tag{9.27}$$

where $B_r \propto r^{-2}$ is the radial component of the field. In a spherically symmetric outflow the wave heating can be expressed as

$$Q = -\frac{1}{r^2} \frac{d}{dr} \left[r^2 \left(S_r + \frac{1}{2}\rho v \langle \delta v^2 \rangle \right) \right] + \frac{v}{2\mu_0} \frac{d}{dr} \langle \delta B^2 \rangle. \tag{9.28}$$

Alfvén waves are incompressible and hence difficult to damp, so there is no dissipation and the wave heating vanishes ($Q = 0$) for regions close to the Sun where the wave amplitudes are small. Using $S_r = (\langle \delta B^2 \rangle / \mu_0)(v + v_{Ar})$ and $\rho \langle \delta v^2 \rangle / 2 = \langle \delta B^2 \rangle / (2\mu_0)$ it is possible to rewrite Eq. (9.28) as

$$\langle \delta B^2 \rangle^{-1} \frac{d \langle \delta B^2 \rangle}{dr} - \frac{3}{2\rho} \frac{d\rho}{dr} + 2 \left(1 + \frac{v_{Ar}}{v} \right)^{-1} \frac{d}{dr} \left(1 + \frac{v_{Ar}}{v} \right) = 0. \tag{9.29}$$

This equation is readily integrated to obtain an expression for $\langle \delta B^2 \rangle$ as a function of heliocentric distance r:

$$\frac{\langle \delta B^2 \rangle}{\langle \delta B_0^2 \rangle} = \left(\frac{\rho}{\rho_0} \right)^{3/2} \left(\frac{1 + v_{Ar0}/v_0}{1 + v_{Ar}/v} \right)^2. \tag{9.30}$$

Thus, for distances near the Sun out to some tens of solar radii, Alfvén waves propagate outward without damping while $\delta B/B$ becomes steadily larger. When this ratio is sufficiently large, of order unity or so, the dissipation of Alfvén waves must take place. This limits $\delta B/B$ to some finite value, of order unity, while depositing wave energy in the outflowing plasma.

9.6 The coronal helium abundance and the proton flux

As noted by Leer and Holzer (1980), high-speed solar wind flow can be achieved by adding energy beyond the critical point, and in the previous section we showed that Alfvén waves are a possible candidate to fill this role. Let us now return to the question how the solar wind mass flux is determined, or, rather more specifically, to the question of whether it is possible to change the force balance in the sub-sonic corona in a manner which reduces the proton flux sensitivity to the coronal temperature.

Leer and Holzer (1992) found that the sensitivity of the proton flux to temperature could be reduced for coronal helium abundances as low as 10% (10% is the nominal photospheric abundance of helium; a 5% helium abundance is observed in the fast solar wind, see Table 9.1). To see why this is so, consider the momentum balance for particle species s (where s can represent protons, α-particles, or electrons). We write the species mass density $\rho_s = m_s n_s$ and denote the electric field by E, the charge of species s by Z_s, and the species pressure by $p_s = n_s k T_s$. The momentum equation for each particle species (see e.g. St-Maurice and Schunk, 1977) then takes the form

$$\rho_s \left[\frac{\partial v_s}{\partial t} + \left(v_s \frac{\partial}{\partial r} \right) v_s \right] + \frac{\partial p_s}{\partial r} - eZ_s n_s E + \rho_s \frac{GM_S}{r^2} = \frac{\delta M_s}{\delta t} + \rho_s D_s. \tag{9.31}$$

The term $\delta M_s/\delta t$ represents source and sink terms due to elastic and inelastic collisions with other particle species, and $\rho_s D_s$ is the force density on species s arising from an outward propagating mechanical energy flux.

We represent elastic collisions between particle species s and t by the collision frequency ν_{st}. In particular, Coulomb collisions between charged particles of type s and t are described by

$$\nu_{st} = 1.7 \left(\frac{\ln \Lambda}{20} \right) \left(\frac{m_p}{m_s} \right) \left(\frac{\mu_{st}}{m_p} \right)^{1/2} n_t T_{st}^{-3/2} Z_s^2 Z_t^2, \tag{9.32}$$

where ν_{st} is measured in s^{-1}, n_t in cm^{-3}, and T_{st} in K. The Coulomb logarithm is given by

$$\ln \Lambda = 23 + 1.5 \ln(T_e/10^6 \text{ K}) - 0.5 \ln(n_e/10^6 \text{ cm}^{-3}). \tag{9.33}$$

The reduced mass is $\mu_{st} = m_s m_t/(m_s + m_t)$, and the reduced temperature is $T_{st} = (m_t T_s + m_s T_t)/(m_s + m_t)$.

Inelastic reactions that destroy particles of type t while creating particles of type s are represented by the rate coefficient γ_{ts}.

With the definitions above we can write the momentum collision terms as

$$\frac{\delta M_s}{\delta t} = -\sum_t (n_s \nu_{st} + n_t \gamma_{ts}) m_s (u_s - u_t) + \sum_t \nu_{st} z_{st} \left(\frac{\mu_{st}}{k T_{st}}\right)\left(q_s - \frac{\rho_s}{\rho_t} q_t\right)$$

(9.34)

where the constant z_{st} equals $3/5$ for Coulomb collisions. The first sum represents the frictional force and the second sum is the so-called thermal force (see Section 9.10.2), which is seen to be dependent on the heat fluxes q_s and q_t. The expressions for the heat fluxes are lengthy and not necessary for the following discussion – full expressions can be found in Hansteen and Leer (1995).

A number of simplifications can be made for the electron equations. In the limit of small electron mass, only the electric, pressure gradient, and thermal forces need to be considered in the electron momentum equation (9.31); neglecting terms of the order of m_e/m_p and m_e/m_α in comparison with terms of order unity we obtain a polarization electric field

$$eE = -\frac{1}{n_e}\frac{dp_e}{dr} - \frac{15}{8}\frac{\nu_{ep} + \nu_{e\alpha}}{\nu_e'} k \frac{dT_e}{dr}$$

(9.35)

where the collision frequency ν_e', roughly equal to ν_{ee}, is defined in Hansteen and Leer (1995). The net charge needed to maintain the electric polarization field is very small, and the electron density is therefore given by

$$n_e = \sum_s Z_s n_s.$$

(9.36)

This relation allows us to write the electric field term in Eq. (9.31) in terms of the proton and ionized helium densities. We have also assumed that there is no current along the magnetic field, so that

$$j = \sum_s n_s Z_s u_s = 0.$$

(9.37)

Electrons are much lighter than ions and are therefore not gravitationally bound; rather, it is the electric field that hinders them from escaping the Sun. Conversely, the electric field represents an outward force on the much heavier protons and α-particles.

It turns out that when the coronal helium abundance A_{He} exceeds some 20%, the coronal scale heights in the low solar wind for helium and hydrogen are nearly the same. The balance of forces on the protons is then determined by gravity, the inward frictional force, the outward pressure gradient force, and the outward electric force. This leads to the following expression for the solar wind proton flux:

$$n_p u_p \approx \frac{GM_S}{r^2} \left(\frac{n_p}{\nu_{\alpha p}}\right) \frac{5}{4(3A_{He} + 2)}. \tag{9.38}$$

Note that, since $\nu_{\alpha p} \propto n_p T^{-3/2}$, the temperature sensitivity of the proton flux to the temperature is much reduced from the exponential dependence we found for a Parker-type proton–electron wind described by Eq. (9.7).

How has this change come about? It lies in the effect of helium: when the temperature increases, the α-particle scale height becomes larger and the number density of α-particles at a given height increases exponentially. This increases the frictional coupling between protons and helium, which reduces the increase in the proton flux. Conversely, when the temperature decreases, the α-particle scale height becomes smaller. This increases the electron pressure gradient (each α-particle contributes two electrons), which in turn increases the outward force from the electric field.

The mechanism for solar wind flux generation proposed here does actually work; it illustrates the role played by the polarization electric field and frictional interactions quite nicely. In addition a similar mechanism is responsible for the Earth's polar wind outflow of ionized hydrogen and helium (Axford, 1968; Demars and Schunk, 1994; Lie-Svendsen and Rees, 1996). However, we shall see in the next section that the explanation for the constancy of the solar wind mass flux probably lies in other processes.

9.7 The energy budget of the solar wind

Though we saw in the last section that a corona with a significant helium abundance could act as a regulator for the solar wind mass flux, that explanation of the observed proton flux invariance is rather convoluted and depends on several named and unnamed assumptions. Further insight into the formation of the solar wind mass flux can be gained by considering the energy budget of the solar wind.

We have so far constructed models defined by parameters set at the *coronal base*, or more specifically by the density ρ_0 and temperature T_0. The Parker-type thermally driven solar wind depends exponentially on temperature, and this means that given ρ_0 we can construct almost any model we desire by slightly varying T_0. But can these quantities be chosen independently of each other? To answer this, consider the steady-state integrated energy equation along the radial heliocentric

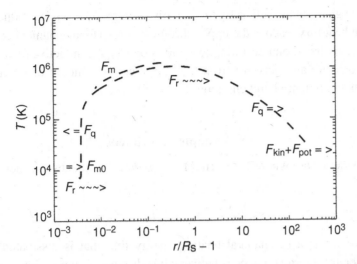

Fig. 9.4. Illustration of the solar wind energy budget from the chromosphere to 1 AU. An input mechanical energy flux F_{m0} is introduced at the bottom boundary and is dissipated in the lower corona and out to about $r = 1.3R_S$. Some of this energy flux is conducted (F_q) back to the chromosphere, where it is radiated away (F_r), and some is conducted outward and goes into the wind's kinetic and potential energy ($F_{kin} + F_{pot}$).

distance r,

$$F_{m0} + F_{q0} + M\left(-\frac{1}{2}v_g^2 + \frac{1}{2}v_0^2 + \frac{\gamma}{\gamma-1}\frac{p_0}{\rho_0}\right)$$
$$= F_m + F_q + M\left(-\frac{GM_S}{r} + \frac{1}{2}v^2 + \frac{\gamma}{\gamma-1}\frac{p}{\rho}\right) + F_r, \qquad (9.39)$$

where F_m is the input "mechanical" energy flux, F_q is the conductive heat flux, $M = \rho u A$ is the solar wind mass flux, and F_r is the radiative flux,

$$F_r = \int_{r_0}^{r} A n_e n_H f(T) \, dr, \qquad (9.40)$$

where $f(T)$ is a function only of T for optically or effectively thin radiation. The subscript 0 indicates some reference layer.

An interpretation of this equation is given by considering Fig. 9.4, which shows the temperature structure in a solar wind model and the positions of various energy fluxes. Choosing a reference level in the chromosphere we find that the energy budget is dominated by the input mechanical energy flux, in the form of MHD waves of high or low frequencies, presumably generated as a result of convective motions below the photosphere. Some of this energy flux may be converted to radiation as we move up through the chromosphere. In the transition region the energy flux is supplemented by a conductive heat flux down the transition-region temperature gradient as well as by the radiative energy flux that is produced as the conductive heat flux reaches the upper chromosphere. The mechanical energy flux is dissipated in the corona, resulting in a hot corona, heat conductive fluxes towards the chromosphere and outward from the Sun, a certain amount of radiative flux, and the kinetic, potential and enthalpy fluxes of the solar wind.

9.8 A simple experiment

Let us consider a case where we "turn off" the wind. This makes the energy budget very simple:

$$F_{m0} = F_m + F_q + F_r = F_{r\infty}. \tag{9.41}$$

Thus, if we insert a mechanical heating energy flux that is dissipated at some position then the energy balance is between this dissipated energy flux, the thermal heat conduction, and the radiative flux. The energy balance in this model is very similar to that described by Rosner *et al.* (1978).

In contrast, let us now compute a solution that has identical heating but where the plasma is allowed to flow into interplanetary space. In this case we can write the energy equation for species s as

$$\frac{\partial \frac{3}{2} p_s}{\partial t} + \frac{1}{A}\frac{\partial}{\partial r}(A\tfrac{3}{2}p_s v_s) + \tfrac{1}{A}p_s\frac{\partial A v_s}{\partial r} + \frac{1}{A}\frac{\partial A q_s}{\partial r} = \frac{\delta E_s}{\delta t} + Q_s - L_{r,s}, \tag{9.42}$$

where the term $\delta E_s/\delta t$ represents source and sink terms due to elastic and inelastic collisions with other particle species, Q_s is the heating rate from this energy flux for species s, and $L_{r,s} = \nabla \cdot F_{r,s}$ is the radiative loss rate.

The closed flux tube extends from the chromosphere out to some $2R_S$ above the solar surface, where a boundary enforcing $v = q = 0$ for all ion species and electrons is applied. This loop is heated by a mechanical energy flux F_m, where

$$F_m = f_m A = f_{m0} A_0 \qquad\qquad\qquad \text{for } r < r_m,$$
$$F_m = f_m A = f_{m0} A_0 \exp[-(r - r_m)/H_m] \quad \text{for } r \geq r_m; \tag{9.43}$$

here f_{m0} is the energy flux density at the base of the model. The energy flux remains constant until the height r_m, where it decreases exponentially with damping length H_m. This energy flux is consistent with a heat input for $r > r_m$ of $Q_m = F_m/(A H_m) = f_m/H_m$. All the energy input is in the form of heat, so that the force on the plasma associated with the coronal heating, D_s, equals zero for these models.

The open-flux-tube setup is essentially identical to the closed-flux-tube setup except for the fact that there is no upper boundary hindering the plasma from escaping the Sun.

In both models we describe a simple neutral H, H^+, e^- plasma but, in describing the radiative losses, which in large part are an energy sink in the electron fluid, we use loss rates consistent with a plasma of solar abundance.

The parameters describing the heating are identical in the open and closed models. We have chosen to deposit a mechanical energy flux $f_{m0} = 100$ W/m^2, which is dissipated in the region $r_m = 2R_S$ with dissipation scale $H_m = 0.1R_S$. The dissipated energy must be apportioned between electrons and protons in such a way that $Q_e = (1 - p)Q_m$ and $Q_p = pQ_m$, where p is a number in the range zero to one. In the models described here we have chosen equipartition, that is, we have set p equal to 0.5.

The results are shown in Fig. 9.5. In the upper two panels we show the closed model. On the left the electron and proton temperatures T_e and T_p and the total hydrogen number density n_H are shown for $1.0R_S \leq r \leq 2.5R_S$. The right-hand panel shows the electron and proton temperatures as well as the total gas pressure near the bottom 30 Mm of the model (equivalent to $h \approx 0.03\ R_S$). The bottom 2 Mm of the model contains the chromosphere, where $T_e = T_p \approx 7000$ K and the pressure scale height is some 200 km. The pressure at the top of the chromosphere is slightly less than 1 Pa. At 2 Mm the transition-region temperature rise occurs and the temperature increases quickly to 1 MK at $z = 13$ Mm, some 10 Mm above the top of the chromosphere.

The mechanical energy flux is deposited fairly far out into the corona where the particle density and the heating per particle are high, resulting in proton temperatures of order 4 MK. Some of this energy is carried away in the proton heat flux, but the main energy loss for the protons lies in collisional coupling to electrons, which attain a maximum temperature of some 3 MK. Particle densities are small so radiative losses in the corona are insignificant. However, electron heat conduction is quite efficient and as no energy is allowed to leave the system at the upper closed boundary this conductive flux is forced downwards into the transition region. The radiative losses scale roughly as the density squared, n_e^2, and become important near the top of the chromosphere where, in a steady state, they must balance the downflowing conductive fluxes exactly. It is this balance that determines the coronal density: increasing the heating rate F_{m0} will increase the amount of conductive heat flux reaching the upper chromosphere from above. If radiative losses there are too small then the plasma will heat up and evaporate into the corona, moving the position of the upper chromosphere to a location of higher density. Likewise, should the conductive heat flux be reduced owing to lower coronal heating rate, when the radiative losses become too large the material cools and the position of the upper chromosphere will move to a lower density commensurate with smaller radiative losses. The result is that coronal densities vary by orders of magnitude with heating rate, while the temperatures range over at most a factor 2–4.

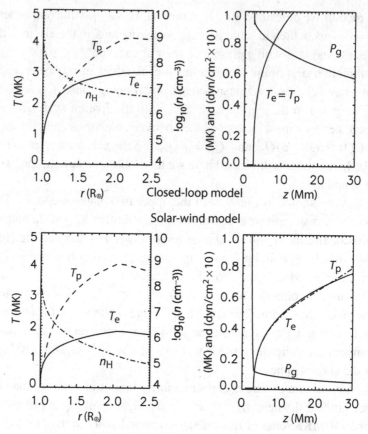

Fig. 9.5. Models for open and closed coronal fields illustrating heating at $2R_\odot$ with $H_m = 0.1R_\odot$. In the upper panels, the results of the closed model are shown: the left-hand panel shows the proton temperature (broken line), the electron temperature (solid line), and the total hydrogen density, i.e. neutral plus ionized (broken-and-dotted line); the right-hand panel shows transition-region details of the electron (solid) and proton (broken line) temperatures as well as the gas pressure p_g. The lower panels show the same variables for the open model.

This mechanism is also active in the open coronal model shown in the lower panels of Fig. 9.5. In the left-hand panel we see that the maximum proton temperature is roughly the same as that found in the closed model, some 4 MK. The electron temperature is much lower, however, showing a maximum of less than 2 MK. It is also clear that the proton and electron temperatures decouple at a lower height, $1.1R_S$ rather than at $1.2R_S$ as in the closed model. The total hydrogen density in the open model is much smaller than in the closed case: at $r = 2.5R_S$ the difference is greater than a factor 30 (or about a factor 1000 for radiative losses). This difference is also evident in the gas pressure (right-hand panels). It is an order of magnitude larger, at the top of the chromosphere, in the closed model than it is in the open one.

We are in a position to understand the differences between these models by considering the energy fluxes in the system. In the closed case essentially *all* the mechanical energy flux ends up as conductive heat flux down through the transition region. Hence the density and pressure at the top of the chromosphere are large. It follows that the coronal density and thus the collisional coupling between protons and electrons is strong. In the open case the mechanical energy flux can also go into electron conductive heat flux, into lifting the plasma out of the Sun's gravitational potential, and into accelerating the plasma to supersonic speed. The acceleration of the wind takes a substantial amount of energy, leaving less to flow back towards the chromosphere in the form of an electron (or proton) heat conductive flux. It follows that, in this experimental model, the gas pressure at the top of the chromosphere for the open field case is low, as is the coronal density and the collisional coupling between protons and electrons.

9.9 Solar wind models that include the chromosphere

We are now in a position to address the following question. Given an open coronal flux tube heated by some mechanical energy flux, what portion of this flux goes into accelerating the wind and what portion returns, via the electron conductive heat flux, to the top of the chromosphere to be radiated away? Or, returning to a form of Eq. (9.39), we can solve for the mass flux:

$$mnuA = \frac{F_{m0} + F_{q0} - F_{q\infty} - F_{r\infty}}{\frac{1}{2}(v_g^2 + u_\infty^2)}. \tag{9.44}$$

Now setting the bottom reference point in the chromosphere where $F_{q0} = F_{r0} = 0$ and the top reference point far enough out in the wind to have expended all the mechanical energy flux and all the conductive heat flux so that $F_{m\infty} = F_{q\infty} = 0$, we can ask the following question: what are the relative magnitudes of $F_{r\infty}$ and the kinetic and potential energy fluxes in the wind? (We have discarded the enthalpy flux at infinity $\propto p/\rho \simeq c_s^2$ since the asymptotic solar wind speed is highly supersonic, i.e. $u_\infty \gg c_{s,\infty}$.)

In order to answer this question, Hansteen & Leer (1995) computed a whole series of neutral-hydrogen proton–electron models, with lower boundary in the middle chromosphere and upper boundary at several tens of solar radii, in which the parameters describing coronal heating were varied. In particular the effect of varying the position of heat deposition r_m, the proportion of energy p put into protons versus that put into electrons, the proportional amount of direct push ρD_s versus heating, and, later (Hansteen *et al.*, 1997b) the effect of adding helium to the models.

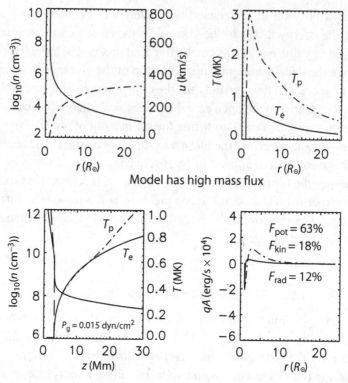

Fig. 9.6. Proton–electron solar wind model with heating close to the coronal base, at roughly $r_m = 1.1 R_S$ and $H_m = 0.5 R_\odot$. The upper left panel shows the particle density (solid line) and flow velocity (broken-and-dotted line), the upper right panel shows the proton (broken-and-dotted line) and electron (solid line) temperatures, the lower left panel shows details of the neutral-hydrogen (broken line) and proton (descending solid line) densities, as well as the electron (ascending solid line) and proton (broken-and-dotted line) temperatures in the upper chromosphere and transition region. The lower right panel shows the heat flux in the electron (solid line) and proton (broken and dotted line) fluids.

The most important findings of these studies are given in Figs. 9.6–9.8, where we show a set of hydrogen proton–electron models with heating deposited close to and far from the coronal base as well as a model that includes helium. Otherwise the models share the same input mechanical heat flux $f_{m0} = 100$ W/m² and the same geometry as the experiment described in Section 9.8.

Figure 9.6 shows the case where the mechanical heat flux is deposited, fairly close to the coronal base, at $r_m = 1.1 R_S$ with a scale height $H_m = 0.5 R_S$; the whole heat flux goes into the proton fluid ($p = 1$). The upper left panel shows the total hydrogen density and proton velocity. The particle density is some 10^{16} m⁻³ at the top of the chromosphere, a number that has decreased to 10^9 m⁻³ at $25 R_S$. The velocity increases rapidly and the acceleration is essentially complete by $5 R_S$,

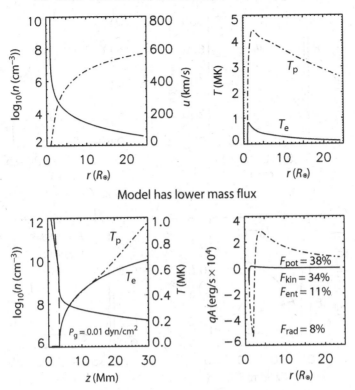

Fig. 9.7. Proton–electron solar wind model with heating far from the coronal base, at roughly $r_m = 2R_S$ and $H_m = 0.5R_\odot$. The plots show the same variables as in Fig. 9.6.

but the wind is fairly slow, 320 km/s, at $25R_S$. The maximum proton temperature is slightly below 3 MK at roughly $1.5R_S$, decreasing rapidly into the wind and down towards the transition region. The electron temperature is much lower, with a maximum equal to about 1 MK. The proton and electron temperatures are coupled up to $z = 15$ Mm where the temperature is roughly 700 kK. The electron heat flux dominates the downward conductive flux while it is the proton flux that is most important in the outward heat flux. In this model (which has a relatively high proton flux, of order 4.7×10^{12} m^{-2}s^{-1}) 63% of the deposited energy goes into potential energy, 18% into kinetic energy, and 12% goes into the radiative losses $F_{r\infty}$.

When we move the position of heat deposition out to $r_m = 2R_S$, leaving the other parameters unchanged, we arrive at the model illustrated in Fig. 9.7. In this model the coronal density is lower, while the wind velocity is some 550 km/s. The maximum proton temperature is higher than in the previous model, reaching 4.5 MK at $2.5R_S$. The electron temperature is much lower, its maximum value reaching only some 800 kK in the inner corona. Again the proton and electron

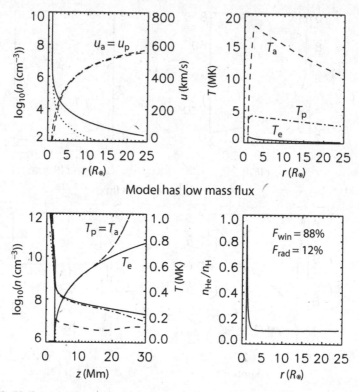

Fig. 9.8. Helium–hydrogen–electron solar wind model. The heating parameters are the same as those for the model shown in Fig. 9.7, but here 60% goes into heating the protons and 40% into heating the α-particles; the energy is deposited at roughly $r_m = 2R_S$. The upper left panel shows the total hydrogen (solid line) and helium (dotted line) densities as well as the proton (broken-and-dotted line) and α-particle (broken line) velocities and the upper right panel shows the electron (solid line), proton (broken-and-dotted line) and α-particle (broken line) temperatures. The lower left panel shows the neutral-hydrogen (short broken line), neutral-helium (dotted line), electron (nearly vertical solid line), proton (broken-and-dotted line), and α-particle (kinked descending solid line) densities as well as the electron, proton, and α-particle temperatures in the transition region and lower corona. The lower right panel shows the helium abundance n_{He}/n_H in the model.

temperatures are coupled up to $z = 15$ Mm where the temperature is roughly 550 kK. The proton heat flux dominates completely in the outer corona and beyond the critical point, but in the transition region it is the electron conductive flux that dominates and carries energy down towards the upper chromosphere. This model has a lower proton flux, 3.5×10^{12} m^{-2}s^{-1}, than the model with heating closer to the solar surface. The potential and kinetic energy fluxes in the wind are roughly equal, both claiming some 35% (there is also 11% going into the enthalpy flux at $25R_S$), while 8% goes into the radiative losses.

Finally, we examine the effect of adding helium to the model. Figure 9.8 shows a model where we deposit 60% of our mechanical energy flux into protons and 40% into α-particles at $r_m = 2R_S$. This model yields a velocity equal to 550 km/s for both protons and α-particles. In the upper right panel we see that the α-particle temperature is quite high, reaching almost 19 MK at maximum. The proton and electron temperatures are similar to those found in the previous proton–electron model with heating far from the solar surface (Fig. 9.7). The lower left panel shows that the ions and electrons decouple at $z = 15$ Mm, where all the temperatures are 650 kK; the proton and α-particle temperatures decouple at a greater height. The lower left panel also indicates that an overabundance of α-particles accumulates in the lower corona, starting roughly at $z = 25$–30 Mm. The lower right panel gives the helium abundance in the model; there is a large overabundance of α-particles in the lower corona. In this model 88% of the energy flux goes into the solar wind flow while 12% is conducted back to the upper chromosphere, where it is radiated away. Note that though this model is characterized by a large helium abundance in the lower corona it has essentially the same characteristics as the pure neutral-hydrogen proton–electron model with heating near $r_m = 2R_S$ shown in Fig 9.7.

It should be clear from the three examples given above, and it has been confirmed by the much more extensive set of models constructed by Hansteen and Leer (1995) and Hansteen *et al.* (1997b), a deposited energy flux in the open field corona goes mainly into the solar wind flow; only about 10% of the energy is conducted back towards the transition region and upper chromosphere. This entails that, to a very good approximation, the mass flux in the solar wind can be written as being proportional to the input mechanical heat flux F_{m0}:

$$mnA = \frac{F_{m0}}{\frac{1}{2}(v_g^2 + u_\infty^2)}. \tag{9.45}$$

Thus, we expect a more or less constant solar wind mass if the Sun produces roughly the same mechanical energy flux per unit surface and if magnetic flux tubes expand by roughly the same factor whatever the location of their origin on the surface. The same is true if, alternatively, the heating f_{m0} is proportional to the magnetic field strength on the surface B_0, in which case

$$(mnu)_E = \frac{A_0}{A_E} \frac{f_{m0}}{\frac{1}{2}(v_g^2 + u_\infty^2)} = \frac{B_E}{B_0} \frac{f_{m0}}{\frac{1}{2}(v_g^2 + u_\infty^2)}, \tag{9.46}$$

which is roughly constant since B_E is roughly constant at 1 AU.

In addition to explaining the lack of variation in the observed solar wind mass flux, the models described above show the following characteristics.

- While the mass flux in the wind is roughly set by the magnitude of the input mechanical heat flux, the ratio of kinetic and potential energy flux in the wind is set by the location of the heating; heating close to the sun implies a greater conductive flux down to the photosphere and thus a larger coronal particle density, smaller mechanical energy flux per particle, smaller kinetic energy flux, and greater mass flux. Heating far from the solar surface leads to a smaller coronal particle density, greater energy flux per particle, greater kinetic energy flux, and smaller mass flux. Note that heating must occur very close to the Sun in order to break the assumption on which Eq. (9.45) builds.

- The electron temperatures in the models described are fairly low; it seems difficult to achieve electron temperatures much larger than 1 MK. In fact, the models computed by Hansteen and Leer (1995) showed that roughly 80% of the input mechanical energy flux must be put into the electrons if the electron temperature is to equal the ion temperature. This is due to the highly effective nature of electron thermal conduction compared with the thermal conduction in the ion fluids. With low electron temperatures the electric field (Eq. 9.35)

$$eE \approx -\frac{1}{n_e}\frac{dp_e}{dr} \tag{9.47}$$

becomes much less important in driving the solar wind than in the traditional Parker-type thermally driven wind.

- Achieving high-speed flow with velocities greater than 700 km/s is difficult in the models presented here. The heat flux used in these models ensures that of order 10% of the input mechanical heat flux is lost as radiation from the upper chromosphere. This implies that it is difficult to construct models with arbitrarily low coronal densities and thus achieve high enough energy per particle to ensure high-speed flow. Alternative descriptions of the heat flux can ease this restriction.

- The α-particle temperatures found are very high and are roughly a factor m_α/m_p greater than the proton temperature. We will come back to this point in Section 9.10.3.

- The models show that the helium abundance in the lower corona can become very high. This is due in part to the thermal force, described below in Section 9.10.2, which hinders α-particles from settling down into the transition region, and in part to the difficulty in lifting the relatively heavy α-particles out of the Sun's gravitational well. Note, however, that the high α-particle density does not much impact the proton flux: the electric field set up by the α-particles is weak owing to the low electron temperatures, while the direct collisional coupling to protons is weak owing to the high proton and α-particle temperatures.

9.10 Discussion and conclusions

How do the models outlined above measure up against observations and what can they tell us about the processes heating the corona and accelerating the wind? Solar wind theory as described in the previous sections does a fairly credible job of describing and explaining why the Sun produces a transsonic outflow. In general one could state that the question of solar wind acceleration has become

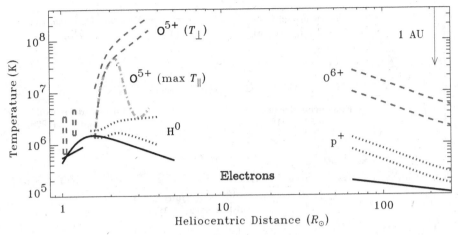

Fig. 9.9. Summary of the radial dependence of temperature in polar coronal holes and the fast wind at solar minimum: electron temperatures (solid line), neutral-hydrogen and proton temperatures (dotted line), and ionized oxygen temperatures (broken and broken-and-dotted lines). The paired sets of curves in the extended corona denote different empirical models derived from UVCS emission line properties. The narrow broken-line loops regions in the low corona ($r < 1.5R_S$) correspond to lower and upper limits on the O VI kinetic temperature deduced from SUMER line widths (from Kohl *et al.*, 2006).

subsumed by the question of how the (open) corona is heated. Yet there remain some interesting questions posed by the observations that are not obviously well explained by the current theory. In this final section we highlight some of these questions.

9.10.1 SOHO UVCS observations

Measurements made with the SOHO UVCS instrument, of the line profiles of the hydrogen Ly-α line at 121.6 nm and the oxygen O VI 103.2 nm and 103.7 nm lines (in addition to the Mg X 62.5 nm line), give a number of interesting clues; see Kohl *et al.* (2006) for a thorough review of methods and results.

The UVCS observations imply that proton temperatures are large, much greater than the electron temperatures, of order $T_p \simeq 3.5$ MK before possible non-thermal contributions have been removed. Accounting for the latter by assuming a model where Alfvén waves with a lower boundary amplitude, 30 km/s, are the cause of line broadening and assuming further that the Alfvén wave action is conserved still gives an estimated proton temperature $T_p > 2$ MK. A summary of these observations, including electron temperatures derived from SOHO SUMER observations as well as the measured freezing-in temperatures from Ulysses is shown in Fig. 9.9. The

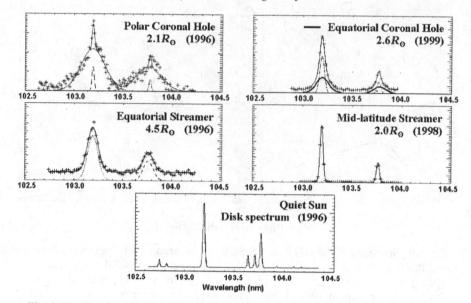

Fig. 9.10. (Top four panels) SOHO UVCS observations of the O VI 103.2, 103.7 nm doublet in four types of coronal structure. (Bottom panel) SOHO SUMER observations of the quiet solar disk in the same range of the spectrum. Figure from Kohl *et al.* (2006).

measured electron temperatures in coronal holes seem to be consistent with a model where $T_e \simeq 1$ MK and where $T_p > T_e$ is strongly implied.

Also shown in Fig. 9.9 are O VI temperatures as measured by line broadening (see Fig. 9.10 for O VI line widths measured for different types of coronal structures) and by the Doppler dimming technique (Kohl *et al.*, 2006). The former measurements should give an estimate of the temperature perpendicular to the magnetic field, $T_{O\perp}$, while the latter measurements give an estimate of the temperature $T_{O\parallel}$ (and flow velocity) along the magnetic field. The line-broadening results imply that in coronal holes the perpendicular temperature rises to greater than 10 MK at $r = 1.5R_S$ and exceeds 100 MK above $2R_S$–$3R_S$. These temperatures are high enough that they would not be substantially altered by subtracting any reasonable amount of non-thermal line broadening. The models of O VI Doppler dimming indicate that there is a large temperature anisotropy in the corona, such that $T_{O\perp}/T_{O\parallel} > 1$, which must be of order 3–10 in order to reproduce the range of observed line profiles above $2.5R_S$.

A word of warning is due before we continue our discussion of these findings. Raouafi & Solanki (2006) considered the effect of electron density stratification along the line of sight in the formation of Ly-α and the O VI and Mg X lines. They found that, while the width of the Ly-α line is relatively insensitive to the stratification, the widths of the oxygen and magnesium lines are strongly dependent.

The conclusion is drawn that the case for a large temperature anisotropy or, indeed, extremely high ion temperatures may not be as pressing as previously thought.

Nevertheless, the findings of what appear to be extremely high coronal ion temperatures and evidence for a large temperature anisotropy have led to a resurgence in interest in ion cyclotron resonance theories (again, see Kohl *et al.* (2006) for a thorough review) that can predict the observed ordering of temperatures ($T_{\text{ion}} \gg T_{\text{p}} > T_{\text{e}}$) as well as the temperature anisotropies ($T_\perp > T_\parallel$). In passing, we note the similarities between the UVCS near-Sun temperatures and those observed in situ in high-speed streams, as given in Table 9.1. In ion cyclotron heating there is a resonance between left-hand polarized Alfvén waves and the Larmor gyrations of positive ions. With a continuous distribution of wave energy emanating from the Sun, positive ions will increase their net energy in such a manner as to increase their perpendicular velocities and hence temperatures.

9.10.2 What is the thermal force?

The photospheric (and solar) helium abundance is measured to be of order 10% or less. Why then do we bother to construct models that contain coronal helium abundances of 30% or greater? The answer to this question lies in the large temperature gradient in the transition region and in the so-called thermal force that follows as a consequence in ionized systems with large heat fluxes. The thermal force is a result of Coulomb collisions between ions of differing masses in which ions heavier than protons experience a net force up the temperature gradient. This force is large and can, with numbers typical for the solar transition region, be several times larger than gravity, as first pointed out by Nakada (1969). The thermal force can therefore have a large effect on coronal abundances, including helium abundances as shown by Hansteen *et al.* (1997b) and discussed in Section 9.6.

The thermal force results from the net effect of Coulomb collisions (Eq. 9.32) when heavy and light ions, indicated by t and s, respectively, collide. The heavy ions will in general have a narrow velocity distribution function f_t; let us for the sake of simplicity assume that we can describe it as a delta function. The velocity distribution function for light ions is significantly wider, and will be slightly skewed when there is a temperature gradient. This skewed distribution is a result of the asymmetry of the surroundings, with, let us say, a warmer environment with a wider velocity distribution above and vice versa below. Then the number density of downward moving particles at a given absolute velocity coming from the warmer region above will be larger than that of the upward moving particles at the same absolute velocity coming from the cooler region below. If there is no net flux of particles, the core of the distribution must consequently be skewed in the opposite sense. Coulomb collision cross sections are very sensitive to the relative speeds of the colliding particles: they are proportional to $|u_t - u_s|^4$, where $u_t - u_s$ is the relative velocity of particles of types s and t. We have assumed that the heavier particles all share the same velocity; they will therefore experience more frequent collisions with light ions coming from below than with those from above, owing to the skewed distribution in the core of the lighter particles' distribution function f_s. The net result of all these collisions is a force up the temperature gradient.

There is a reason for concern here. The above argument implies that the description of the core of the distribution is important in order obtaining the correct description of the thermal force. The transport equations have their basis in the Boltzmann equation for the velocity distribution function f_s, i.e.

$$\frac{\partial f_s}{\partial t} + \mathbf{u} \cdot \nabla_r f_s + \frac{1}{m_s} \mathbf{F} \cdot \nabla_u f_s = \left(\frac{\delta f_s}{\delta t}\right)_{\text{coll}} \tag{9.48}$$

where r and v denote the radial position and velocity, $\mathbf{F}$ is the force acting on particle species s, ∇_r and ∇_u are the coordinate and velocity space gradient operators, and $(\delta f_s/\delta t)_{\text{coll}}$ represents the rate of change of f_s as a result of collisions. The transport equations are derived by multiplying Eq. (9.48) by a velocity moment u^n, where n ranges from zero up to some highest integer N, and then integrating the resulting equation over velocity space, as shown by e.g. Schunk (1977).

In order to close and complete the integrations, including the right-hand side collision terms, some approximation of the velocity distribution f_s must be made. The most common choice used in deriving the transport coefficients is a linear expansion about the Maxwellian such that $f_s = f_{s0}(1 - \phi)$ and

$$f_{s0} = n_s \left(\frac{m_s}{2\pi k T_s}\right)^{3/2} \exp\left(-\frac{m_s c_s^2}{2k T_s}\right) \tag{9.49}$$

where $c_s = u_s - v_s$ is the "random" velocity and v_s is the mean gas velocity for species s. The linear approximation is given by

$$\phi = \frac{m_s}{k T_s p_s}\left(1 - \frac{m_s c_s^2}{5k T_s}\right) \mathbf{q}_s \cdot \mathbf{c}_s \tag{9.50}$$

where $p_s = n_s k T_s$ is the species pressure. It turns out that this choice, which is a fairly good general approximation for many types of particle, is not optimal for particles that suffer Coulomb interactions. Killie *et al.* (2004) showed that fairly simple transport equations and collision terms are obtained if instead the following expansion is used:

$$\phi = \frac{m_s^2 c_s^2}{5k^2 T_s^2 p_s}\left(1 - \frac{m_s c_s^2}{7k T_s}\right) \mathbf{q}_s \cdot \mathbf{c}_s. \tag{9.51}$$

This expansion when used for a collision-dominated electron–proton plasma gives transport coefficients in good agreement with classical transport theory, in which

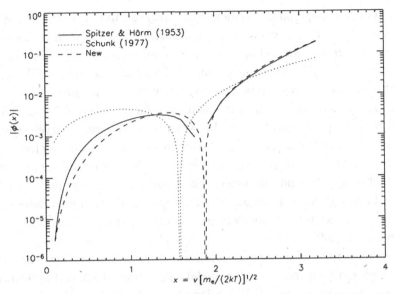

Fig. 9.11. Absolute value of the departure $\phi(x)$ (Eq. 9.51) from a Maxwellian, for electrons as a function of velocity measured in units of $m_e/(2kT)^{1/2}$; thus the distribution function is written as $f_s = f_{s0}[1 - \phi(u)]$ where f_{s0} is the Maxwellian distribution. The solid line shows the numerical solution produced by Spitzer and Härm (1953) for electrons and the dotted line shows the commonly used solution derived by Schunk (1977). The broken line shows the result derived by Killie *et al.* (2004), whence this figure is taken.

(Spitzer and Härm, 1953; Braginskii, 1965) the Boltzmann equation is solved numerically for electrons with the Fokker–Planck collision operator to obtain the correction term ϕ, as shown in Fig. 9.11.

An analysis of the new transport equations indicates that the thermal force on helium (and minor ions) may be reduced by as much as a factor 4 compared with the thermal force deduced from the transport equations based on Schunk's (1977) formulation. This a large enough difference to change substantially the rate at which helium is pushed up the transition-region temperature gradient into the lower corona, but this is still insufficient to preclude solar wind models with enhanced helium abundances in the lower corona (Janse *et al.*, 2007).

9.10.3 More questions

The large temperature anisotropies in the corona could be evidence that ion cyclotron waves are ubiquitous there. An open question is that of the origin of such waves. The Sun is expected to produce Alfvén waves with frequencies of order mHz as a result of convective motions and chromospheric flows buffeting the magnetic field in the photosphere and chromosphere; however, the cyclotron frequency is given by $\omega_c = Z_s e B / m_s$, which implies that Alfvén waves with frequencies of 10^4 Hz or greater are required in order to come into resonance for coronal conditions (say $B \simeq 0.001$ T). On the one hand, Axford & McKenzie (1992) suggested that high-frequency waves can be generated as a result of impulsive reconnection near the transition region. On the other hand Verdini & Velli (2007 and references cited therein) proposed mechanisms by which low-frequency Alfvén waves are converted to higher frequencies through various turbulent nonlinear processes.

The Sun is expected to produce Alfvén waves having periods commensurate with the photospheric and chromospheric dynamics. Evidence that such waves are indeed being generated and are able to propagate through the relatively short scale heights found in the chromosphere and transition region has been reported (Tomczyk *et al.*, 2007; De Pontieu *et al.*, 2007) but is still controversial.

Is markedly preferential heating, such as ion cyclotron heating, necessary to achieve high ion temperatures? The model calculations described in Section 9.9 imply that this may not be so. When ions are collisionally decoupled from the electron fluid their ability to lose energy is severely suppressed. When coupled to electrons, however, energy can be carried away very effectively by electron thermal conduction. Nevertheless the coefficient for *ion* thermal conduction is small and *any* energy source dissipated in the ion fluid will tend to raise the ion temperature until some loss mechanism can balance the input. A natural limit on this occurs when the ion achieves a kinetic temperature compatible with the escape speed at some height r (see Eq. 9.1), i.e.

$$T_s \simeq m_s \frac{2GM_S}{kr}. \tag{9.52}$$

In other words, in the absence of other loss mechanisms we would *expect* ion temperatures to scale with mass *whatever* the heating mechanism. And if temperatures greater than the ion-to-proton mass ratio m_s / m_p are observed, one could reasonably ask what is holding the ions back, preventing them from escaping the Sun's gravitational field.

A more compelling argument for preferential ion heating may come from the in situ measurement result that ion velocities in high-speed streams are larger than proton velocities by approximately the local Alfvén speed: $v_{\text{ion}} \approx v_p + v_A$. Does this prove that heavy ions are pushed or heated? Is it the two-stream instability that restricts their excess velocity to v_A?

Temperature anisotropies are also found in in situ measurements of high-speed solar wind streams near Earth for which $T_\perp > T_\parallel$. This is surprising, as in a collisional plasma the first adiabatic moment $u_\perp^2 / B$ is conserved; a high perpendicular temperature in the inner corona will very effectively be converted to a high *parallel* temperature with increasing distance from the Sun; this was also pointed out by Olsen & Leer (1999). They found that this problem could be alleviated by adding the wave pressure gradient from low-frequency Alfvén waves to the supersonic region of the flow, inducing the parallel temperature moment to cool adiabatically. The study of gyrotropic transport equations (Lie-Svendsen *et al.*, 2001; Janse *et al.*, 2007) may be able to shed light on this problem.

The slow wind, which has received scant attention in this chapter, seems in many respects to be fundamentally different from the more steady fast wind. What is the role of the opening of initially closed loops and their reconnection in driving the slow wind? Can the slow wind only be understood as a time-dependent phenomena? What is the slow wind's connection with the large-scale topology of the Sun's magnetic field, i.e. the streamer belt and coronal holes?

Finally, the first ionization potential (FIP) effect (see Fig. 9.2), clearly differentiates between fast and slow wind. An explanation of this difference and indeed the effect itself will require the understanding of processes in the chromosphere (where FIP elements are ionized) and in the wind. We leave the reader with the following questions. What is the role of gravitational settling in the upper chromosphere? Is the thermal force important? Is the magnetic topology important? What does all this tell us about the differences between the origin of the fast wind and the slow wind?

10

Fundamentals of planetary magnetospheres

VYTENIS M. VASYLIŪNAS

10.1 Introduction

The comparative study of the magnetospheres associated with various planets (and with some other objects within the solar system) aims to develop a unified general description of magnetospheric phenomenology and physics that is applicable to a variety of different systems. In this way concepts and theories, often developed in the first instance to fit specific phenomena in a particular magnetosphere, can be tested for correctness and applicability in a more general context.

The subject matter of magnetospheric physics, in general terms, deals with the configuration and dynamics of systems that result from the magnetic field interaction between isolated objects and their environments. Such systems exhibit a variety of fascinating physical phenomena, both visible (e.g. auroral light emissions) and invisible (e.g. charged-particle radiation belts), some of which have significant aspects of practical application (e.g. space weather effects at Earth, radiation dosage problems for spacecraft near Jupiter or Saturn). They also present a special challenge for physical understanding: as regions of transition between an object and its environment, magnetospheres are by their very nature spatially inhomogeneous systems, characterized by the inclusion of very different physical regimes, large ranges of parameter values, and the overwhelmingly important role of gradients.

The aim of this chapter is to present magnetospheric physics, at least in outline, as a self-contained discipline – a branch of physics described in logical sequence, as distinct from a study of individual objects (often presented in historical sequence). The detailed observational and theoretical study of the magnetospheres at individual planets (Chapter 13) remains, of course, the foundation stone of the general discipline, along with the detailed examination and modeling of the most accessible and best known of all magnetospheres, that of the Earth (see also Chapter 11).

10.2 Definitions and classifications

The term "magnetosphere" was introduced by Gold (1959) as the name for "the region above the ionosphere in which the magnetic field of the [Earth] has a dominant control over the motions of gas and fast charged particles". It quickly acquired a broader connotation as the region of space dominated, in a not always precisely defined sense, by the magnetic field of the Earth (replacing the term "geomagnetic cavity" introduced earlier by Chapman and Ferraro) and was soon being applied to analogous regions at other planets as well.

In the most general context, we consider a *central object*: a distinct well-defined body held together (in most cases) by its own gravity. It is immersed in a tenuous *external medium*, assumed to be sufficiently ionized that it behaves as a plasma. The *magnetosphere* is then the region of space around the central object within which the object's magnetic field has a dominant influence on the dynamics of the local medium. An alternative and in some ways more precise view is to regard the magnetosphere as the region enclosed by its bounding surface, the *magnetopause*, the latter being defined as the discontinuity of the magnetic field, where its direction changes; inside the magnetopause the controlling field is that of the central object, while outside it primarily the magnetic field of the distant external medium. This definition is particularly useful for the magnetospheres of planets in the solar wind: the continual variability of the interplanetary magnetic field direction in contrast with the relative constancy of the planetary magnetic dipole allows in most cases an easy observational identification of the magnetopause.

For a magnetosphere to exist, the central object must have a magnetic field of sufficient strength; if this magnetic field is dipolar, as in most cases of interest, the magnetic dipole moment must exceed a minimum value (discussed in Section 10.3.1 below). Even if the magnetic field of the central object is too weak to produce a true magnetosphere, the interaction of the central object with the magnetic field of the external medium may nevertheless create structures similar to those found in magnetospheres; in such cases the designation *magnetosphere-like system* is often used.

Within the solar system, magnetospheres have been observed at the planets Mercury, Earth, Jupiter, Saturn, Uranus, Neptune (external medium, the solar wind) and at Jupiter's moon Ganymede (external medium, the plasma of Jupiter's magnetosphere), magnetosphere-like systems at the planets Venus, Mars (external medium, the solar wind), at Jupiter's moon Io (external medium, the plasma of Jupiter's magnetosphere), and at Saturn's moon Titan (external medium, the plasma of Saturn's magnetosphere); all these systems are discussed in Chapter 13. The interaction of a comet with the solar wind may also be viewed as a magnetosphere-like system. On a much larger scale, the entire heliosphere may be viewed as being produced by the interaction of the Sun (the central object) with the local interstellar medium (the external medium) and described either as a magnetosphere or as a magnetosphere-like system, depending on what one chooses to emphasize; see Section 10.6 for further discussion. Finally, the concepts of magnetospheric physics have also been applied, albeit speculatively in most cases, to astrophysical objects such as pulsars (e.g. Michel, 1982), X-ray sources in binary systems (e.g. Vasyliūnas, 1979) and even to systems as large as radio galaxies (e.g. Miley, 1980).

In this chapter, as noted already, magnetospheric physics will be discussed as a general discipline, abstracted from the consideration of individual cases. Since the development of the discipline to date has been based to a large extent on studies of planetary magnetospheres, it is convenient to use the terminology "planet" and "solar wind" in place of the more general but somewhat colorless "central object" and "external medium". The presentation proceeds in three steps of increasing complexity. In Section 10.3 the interaction of the solar wind with just the dipolar magnetic field of the planet is discussed. In Section 10.4 the flow of plasma within the magnetosphere is added, which immediately brings in the coupling between the magnetosphere and the ionosphere. In Section 10.5 the plasma sources and the processes of plasma transport are considered. Finally, in Section 10.6 dimensional analysis is applied to determine the scaling relations of magnetospheric properties, by the use of which the various individual objects can be compared or contrasted.

10.3 Interaction of solar wind with a planetary magnetic field

10.3.1 A closed magnetosphere

The basic configuration of a prototypical planetary magnetosphere is shown in Fig. 10.1. Many of its characteristic structures can be understood on the basis of a simple model that takes into account only the two ingredients indispensable for the formation of a magnetosphere: the solar wind, with mass density ρ_{sw} and bulk velocity $\mathbf{v}_{sw}$, and the planetary magnetic field, with dipole moment $\mu = B_p R_p^3$ where B_p is the surface magnetic field strength at the equator and R_p is the radius of the planet. As a consequence of the constraints imposed by the magnetohydrodynamic (MHD) approximation (discussed under various aspects in Chapters 3–7 and briefly in Section 10.4 below), the boundary surface between the solar wind and the planetary magnetic field – the magnetopause – is nearly impermeable both to plasma and to magnetic field, resulting in a clear separation between two distinct regions of space: the magnetosphere itself, within which the magnetic field lines from the planet are confined and from which the solar wind plasma is excluded, and

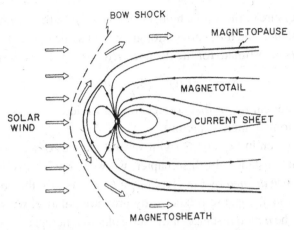

Fig. 10.1. Schematic view of a magnetically closed magnetosphere, cut in the noon–midnight meridian plane. Open arrows, solar wind bulk flow; solid lines within magnetosphere, magnetic field lines (directions appropriate for Earth).

the exterior region beyond the magnetopause, to which the plasma that comes from the solar wind is confined. This simple *closed magnetosphere* is only a first-order approximation (in reality the magnetopause is not completely impermeable but allows, under certain conditions, some penetration of plasma and of magnetic field to produce the *open magnetosphere* described in Section 10.3.3); it does, however, describe fairly accurately the size and shape of the main structures.

The solar wind flow, initially directed away from the Sun, is diverted around the magnetosphere, as indicated in Fig. 10.1. Since the initial flow speed is supersonic and super-Alfvénic (faster than both the speed of sound and the Alfvén speed v_A), the solar wind is first slowed down, deflected, and heated at a detached *bow shock* situated upstream of the magnetopause (the bow shock is analogous to the sonic boom in supersonic aerodynamic flow past an obstacle). The region between the bow shock and the magnetopause, within which the plasma from the solar wind is flowing around the magnetosphere, gradually speeding up and cooling, is called the *magnetosheath*, a term introduced by Dessler and Fejer (1963) replacing the term *transition region* used in the older literature.

The location of the magnetopause is determined primarily by the requirement of pressure balance: the total pressure (plasma plus magnetic) must have the same value on each side of the discontinuity. In the simple closed magnetosphere considered here, the plasma pressure inside the magnetopause and the magnetic pressure outside are both neglected. The exterior pressure then scales as the linear momentum flux density in the undisturbed solar wind, $\rho_{sw} v_{sw}^2$ (often called the *dynamic pressure* of the solar wind), and is a maximum in the sub-solar region, where the plasma near the magnetopause is almost stagnant. The interior pressure scales as the magnetic pressure of the dipole field, $(1/8\pi)(\mu/r^3)^2$ with μ the magnetic dipole moment of the planet, and thus varies strongly with distance from the planet. Equating the two gives an estimate for the distance R_{MP} of the sub-solar magnetopause:

$$R_{MP} = (\xi\mu)^{1/3}(8\pi\rho_{sw}v_{sw}^2)^{-1/6} \tag{10.1}$$

where ξ is a numerical factor that corrects for the added field from magnetopause currents ($\xi \simeq 2$ to first approximation).

The distance given in Eq. (10.1) (with various choices of ξ) is often called the Chapman–Ferraro distance. In this chapter, we will consistently use the symbol R_{CF} for the distance *defined* by Eq. (10.1) with $\xi = 2$, i.e. for the nominal distance of the sub-solar magnetopause predicted by pressure balance; we will reserve the symbol R_{MP} for the *actual* distance of the sub-solar magnetopause in any particular context. Thus, $R_{MP} \simeq R_{CF}$ in the present case of a simple closed magnetosphere but this is not necessarily so in the case of more general models.

The pressure balance condition, combined with assumptions about the sources of the magnetic field within the magnetosphere, may be used to calculate not only the distance to the sub-solar point but also the complete shape of the magnetopause surface; for a discussion of such models at Earth, see Chapter 11 and also reviews by e.g. Siscoe (1988). Typically the magnetopause is roughly spherical on the dayside of the planet, facing into the solar wind flow (the effective center of the sphere being located behind the planet, very roughly at a distance $0.5\,R_{MP}$), and is elongated in the anti-sunward direction.

The magnetopause distance R_{MP} may be regarded as the characteristic scale for the size of a magnetosphere. Being equal to R_{CF} in the case of negligible plasma pressure and no magnetic field sources other than the planetary dipole inside the magnetosphere, R_{MP} can be readily calculated from Eq. (10.1) given only a few basic parameters of the system. In the case where the plasma pressure or a non-dipolar field in the outer regions of the magnetosphere is not negligible, the qualitative effects on R_{MP} can still be estimated from the pressure balance, as illustrated in Fig. 10.2: (a) the actual distance R_{MP} is larger than the nominal distance R_{CF} (the value $\xi = 2$ instead of $\xi = 1$ is in fact a consequence of the non-dipolar field from the magnetopause currents); (b) a change in the solar wind dynamic pressure produces a larger change in the magnetopause distance – the magnetosphere is less "stiff" if the plasma pressure in the interior is significant.

10.3.2 Stress considerations

From a more general physical viewpoint, determining the magnetopause from the pressure balance is an instance of a recurrent theme in magnetospheric physics: the shaping of the magnetic field configuration by the mechanical stresses of the

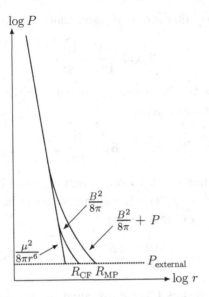

Fig. 10.2. Variation in the total pressure (magnetic plus plasma) with distance from the planet and its relation to the radial distance of the sub-solar magnetopause. The relationship for R_{MP} in Eq. (10.1) may be compared with the schematic representation of a more realistic plasma-filled, non-dipolar planetary magnetic field.

plasma. The dipole magnetic field as such exerts no stresses outside the planet, its magnetic pressure being exactly balanced by its magnetic tension; the pressure balance at the magnetopause arises solely because the planetary dipole magnetic field, which by itself would extend to infinity, has been pushed back by the solar wind and confined to a finite space. Since the magnetosphere presents an obstacle to the solar wind flow, it is subject to forces as the flow is deflected. Locally, at any point of the magnetopause, the pressure exerted by the exterior plasma is balanced by the pressure of the interior magnetic field. Globally, these questions arise: what is the total force on the magnetosphere from the exterior medium, and on what is it really exerted?

A powerful tool for investigating such questions is the well-known momentum equation of the plasma, expressed in the standard conservation form (the partial time derivative of the density of a conserved quantity plus the divergence of the flux density of that quantity):

$$\frac{\partial \rho \mathbf{v}}{\partial t} + \nabla \cdot (\rho \mathbf{v}\mathbf{v} + \mathbf{P} - \mathbf{T}) = 0 \tag{10.2}$$

where ρ, $\mathbf{v}$, and $\mathbf{P}$ are, respectively, the mass density, bulk flow velocity, and pressure tensor of the medium; $\mathbf{T}$ is the Maxwell stress tensor $\mathbf{T}_m$ plus the stress

Fundamentals of planetary magnetospheres

tensors representing any other forces, if significant:

$$\mathbf{T}_m = \frac{\mathbf{BB}}{4\pi} - \frac{\mathbf{I}B^2}{8\pi} \tag{10.3}$$

($\mathbf{I}$ is the unit dyad, $I_{ij} = \delta_{ij}$; the units are Gaussian). The divergence of $\mathbf{T}_m$ is equal to the Lorentz force per unit volume:

$$\nabla \cdot \mathbf{T}_m = (\nabla \times \mathbf{B}) \times \frac{\mathbf{B}}{4\pi} = \frac{1}{c}\mathbf{J} \times \mathbf{B}. \tag{10.4}$$

The total force $\mathbf{F}$ acting on any given volume is given by the volume integral of the divergence terms in Eq. (10.2) and therefore can be written as the surface integral over the boundary of the volume:

$$\mathbf{F} = \int \left(\frac{\mathbf{BB}}{4\pi} - \frac{\mathbf{I}B^2}{8\pi} - \rho \mathbf{vv} - \mathbf{P} \right) \cdot d\mathbf{S}. \tag{10.5}$$

By Newton's second law, this force must equal the net acceleration of the total mass enclosed within the volume, i.e. the integral of the time-derivative terms in Eq. (10.2); quantitatively, for a planetary magnetosphere (Vasyliūnas, 2007),

$$\mathbf{F} \simeq M \left(\frac{d^2\mathbf{R}}{dt^2} - \mathbf{g} \right) \tag{10.6}$$

where $\mathbf{g}$ is the gravitational acceleration due to the Sun, M is the total mass and $\mathbf{R}$ is the center of mass of the entire enclosed system, magnetosphere plus planet. Siscoe (1966) first recognized (in a terrestrial context) that, because the mass contained in the magnetosphere is smaller by many orders of magnitude than the mass of the planet, the entire force from the solar wind interaction must be exerted on the massive planet, with a negligible net force on the magnetosphere itself (for further discussion of the forces on the magnetosphere and the planet, see Siscoe (2006) and Vasyliūnas (2007) and references therein).

While the pressure from the external medium thus accounts for the formation and shape of the magnetosphere on the dayside of the planet, it cannot by itself explain the formation of the *magnetotail* on the nightside. This structure, shown also in Fig. 10.1, is a region of magnetic field lines pulled out into an elongated tail in the anti-sunward direction, with the magnetic field reversing direction between the two sides of a *current sheet* or *plasma sheet* in the equatorial region. To form this structure one needs an appropriate stress: a tension force pulling away from the planet. If we choose a closed volume bounded by a surface just outside the magnetopause plus a cross section of the magnetotail (a vertical cut at, say, the right-hand edge of Fig. 10.1) and evaluate the force by applying Eq. (10.5), the total tension force F_{MT} is given by the integral over the cross section and the total pressure force F_{MP} by the integral over the magnetopause:

$$F_{MT} \simeq \frac{B_T^2}{8\pi} A_T, \qquad F_{MP} \simeq \rho_{sw} v_{sw}^2 A_T \qquad (10.7)$$

where B_T is the mean magnetic field strength and A_T the cross-sectional area of the magnetotail (typically, A_T exceeds πR_{MP}^2 by a factor 3–4). Both F_{MP} and F_{MT} are directed away from the Sun and, as pointed out by Siscoe (1966), are exerted ultimately on the planet.

There are several possible physical mechanisms that could produce this tension force, and which of them is the most important may well differ between magnetospheres. The most straightforward mechanism (and, at least at Earth, the most widely assumed) is that for an open magnetosphere (discussed below in Section 10.3.3): magnetic field lines connect directly from the magnetotail into the solar wind flow and are dragged along with it. Alternatively, and less specifically, the solar wind flowing along the magnetopause is assumed to exert a *tangential drag* on the plasma adjacent to the interior of the magnetopause. Other mechanisms, as discussed in Section 10.5, rely on internal sources of magnetospheric plasma: a plasma pressure within the magnetosphere sufficiently strong to inflate the magnetic field lines, or a pronounced outflow from interior sources that pulls out the magnetic field, somewhat analogously to the action of the solar wind on the interplanetary magnetic field. In both cases the day–night asymmetry of the pressure outside the magnetosphere tends to confine the extended fields to the night-side of the planet.

10.3.3 The open magnetosphere

At the locations of the planets, the interplanetary magnetic field is weak in the sense that its energy density is very small in comparison to the kinetic energy density of the solar wind bulk flow: $v_A^2 \ll v_{sw}^2$. The flow of solar wind plasma past the magnetospheric obstacle deforms the magnetic field lines within the magnetosheath and drapes them around the magnetopause, a process well modeled at Earth (e.g. Spreiter *et al.*, 1968; see Section 11.4). The magnetic field is amplified and may become dynamically no longer negligible as the magnetopause is approached, but the total pressure is in general not greatly modified, an increase in magnetic pressure often being offset by a decrease in plasma pressure. One might therefore anticipate that the effect of the interplanetary magnetic field on planetary magnetospheres should be minimal.

What is overlooked in the above discussion is the possibility that, through the process of *magnetic reconnection* (the physics of which is discussed in detail in Chapter 5), the magnetic field lines from the planet may become connected with

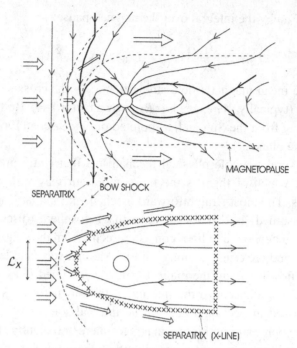

Fig. 10.3. Schematic representation of a magnetically open magnetosphere. The upper panel shows a cut in the noon–midnight meridian plane; the bold lines are magnetic field lines within the separatrix surfaces separating open lines from closed lines or open lines from interplanetary field lines; the other conventions are the same as in Fig. 10.1. The lower panel shows a cut in the equatorial plan; the perimeter line of × symbols represents the intersection with the two branches of the separatrix (compare with Fig. 4.3 for the hypothetical case with zero wind speed); the solid lines are streamlines of magnetospheric plasma flow, and $\mathcal{L}_X$ represents the projection of the dayside magnetic reconnection region along the streamlines into the solar wind (see Section 10.4.3).

those of the interplanetary magnetic field, to produce a magnetically *open magnetosphere*, shown Fig. 10.3 for the simplest case in which the interplanetary magnetic field is parallel to the planetary dipole moment. The magnetopause is now no longer impermeable to the magnetic field and, as a consequence, it need no longer be impermeable to plasma either.

The modifications of the magnetospheric system implied by an open character of the magnetosphere are in some ways minor and in other ways very far-reaching. The location and shape of the dayside magnetopause is for the most part not greatly modified (in agreement with our expectations above). The component B_n of the magnetic field normal to the magnetopause is in general small compared with the magnitude of the field, $|B_n| \ll B$ (so much so that it is often difficult to establish by direct observation that $B_n \neq 0$, and much of the evidence for an open magnetosphere has been indirect). However, the total amount of open magnetic flux Φ_M of one polarity can (at least at Earth) become comparable with the maximum amount that could reasonably be expected to be open, estimated as $\sim \mu / R_{\text{MP}}$, the dipole flux beyond the distance of the sub-solar magnetopause; despite the fact that $|B_n| \ll B$, this is possible if the effective length of the magnetotail $\mathcal{L}_T \gg R_{\text{MP}}$. As noted in Section 10.3.2, that the magnetosphere is open provides immediately an explanation of the tension force needed to maintain the magnetotail, as well as a mechanism for magnetospheric convection (discussed below in Section 10.4.3). Finally, the efficiency of the reconnection process depends greatly on the relative orientation of the magnetic fields on the two sides of the magnetopause, one result of which is that the open character of the magnetosphere is most pronounced when the interplanetary magnetic field is parallel to the planetary dipole moment (i.e. antiparallel to the dipole magnetic field in the equatorial plane), $\mathbf{B}_{\text{sw}} \cdot \hat{\mu} > 0$. Since the direction of the interplanetary magnetic field is highly variable on all time scales, this can lead to pronounced time-varying changes of magnetospheric configuration as well as energy input and dissipation; such processes have been extensively observed and modeled at Earth.

From the point of view of comparing magnetospheres (see Chapter 13) an open magnetosphere embodies primarily the effects of the solar wind acting on the planetary magnetic field, and one major question is the relative importance of these effects in comparison with the effects of internal processes arising from rotation and interior sources of plasma (Sections 10.4 and 10.5.1).

10.4 Plasma flow and magnetosphere–ionosphere interaction

10.4.1 Fundamental principles

A well-known consequence of the MHD approximation is a constraining relation between the plasma bulk flow and the magnetic field: plasma elements that are initially on a common field line remain on a common field line as they are carried by the bulk flow. Since magnetic field lines in the magnetosphere of a planet connect to the ionosphere of the planet, any discussion of plasma flow in the magnetosphere immediately involves questions of magnetosphere–ionosphere interaction. The field lines extend in fact into the interior of the planet, which in many cases is highly conducting electrically; hence it might seem that magneto-spheric flow is constrained by the planet itself. This does not happen, however, because, as pointed out by Gold (1959) in the same paper in which he introduced the term "magnetosphere", most planets possess an electrically neutral (and effectively non-conducting) atmosphere sandwiched between the ionosphere and the planetary interior. Although very thin in comparison with the radius of the planet, this layer suffices to break the MHD constraints and thus allow the plasma in the ionosphere and the magnetosphere to move without being necessarily attached to the planet; without such an insulating layer, much magnetospheric dynamics as we know it would not be possible.

While the plasma in the ionosphere can thus move relative to the planet, it remains constrained to move more or less together with the plasma in the magnetosphere. The conventional formulation, however, describes the plasma flow rather differently in the two regions. On the one hand the magnetosphere is treated, to a first approximation at least, as an MHD medium, with the electric field $\mathbf{E}$ related to the plasma bulk flow $\mathbf{v}$ by the MHD approximation and with the electric current $\mathbf{J}$ related to the plasma pressure by stress balance. On the other hand the ionosphere is treated as a moving conductor (the conductivity results primarily from collisions between the ions and the neutral particles, planetary ionospheres being for the most part weakly ionized), with $\mathbf{J}$ related by a conductivity tensor to $\mathbf{E} + \mathbf{v}_n \times \mathbf{B}/c$, where $\mathbf{v}_n$ is the bulk velocity of the neutral medium. The magnetosphere and the ionosphere are coupled by current continuity,

$$\nabla \cdot \mathbf{J} = 0, \tag{10.8}$$

which implies a connection between the currents in the two regions, and by Faraday's law

$$\frac{\partial \mathbf{B}}{\partial t} = -c\nabla \times \mathbf{E}, \tag{10.9}$$

which implies, in this context, the continuity of the tangential electric field.

The resulting scheme of calculation is illustrated in Fig. 10.4 in the form in which it was first proposed by Vasyliūnas (1970) and Wolf (1970), generalizing earlier work by Fejer (1964). Since then it has been elaborated considerably in detail, but neither the logic nor the equations have changed appreciably (for a review see e.g. Wolf (1983) and references therein). What has changed to some extent in recent years is the physical understanding of the equations: stress and flow are emphasized for the ionosphere as well as for the magnetosphere, and the limitation of the theory to quasi-steady equilibrium situations is better appreciated. Motivated in part by discussions on the relative role of $\mathbf{B}$ and $\mathbf{v}$ versus $\mathbf{J}$ and $\mathbf{E}$ in plasmas (Parker, 1996, 2000, 2007; Vasyliūnas, 2001, 2005a,b; and references therein), this new understanding is particularly relevant when discussing the magnetospheres of the giant planets (see Chapter 13): within these, rotational stresses are of obvious importance, and the long wave-travel times imply correspondingly long time scales for establishing equilibrium. Accordingly, before describing corotation and other plasma flows of the various magnetospheres, we will summarize the new aspects in the understanding of magnetosphere–ionosphere coupling; a more conventional treatment of the theory, as applied at Earth, is given in Chapter 11.

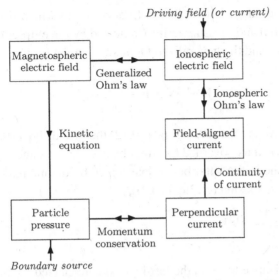

Fig. 10.4. Schematic diagram of magnetosphere–ionosphere coupling calculations (after Vasyliūnas, 1970).

The principal insights are the following.

(1) As long as $v_A^2/c^2 \ll 1$ (i.e. the inertia of the plasma is dominated by the rest mass of the plasma particles, not by the relativistic energy-equivalent mass of the magnetic field), $\mathbf{v}$ produces $\mathbf{E}$ but $\mathbf{E}$ does not produce $\mathbf{v}$ (Buneman, 1992; Vasyliūnas, 2001). The primary quantity physically is thus the plasma bulk flow, established by appropriate stresses. The electric field is the result of the flow, not the cause; its widespread use in calculations is primarily for mathematical convenience (one example is mentioned below in the discussion of Eq. (10.14)).

(2) The electric current in the ionosphere is not an Ohmic current in the physical sense, and its conventional expression by the "ionospheric Ohm's law"

$$\mathbf{J}_\perp = \sigma_P \mathbf{E}_\perp^* + \sigma_H \hat{\mathbf{B}} \times \mathbf{E}^*, \qquad \text{where} \qquad \mathbf{E}^* \equiv \mathbf{E} + \frac{1}{c}\mathbf{v}_n \times \mathbf{B} \qquad (10.10)$$

has a merely mathematical significance. Here σ_P and σ_H are the Pedersen and the Hall conductivities (e.g. Matsushita, 1967), $\mathbf{E}^*$ is the electric field in the frame of reference of the neutral atmosphere, and $\hat{\mathbf{B}}$ is a unit vector.

Physically, the current is determined by the requirement that the Lorentz force should balance the collisional drag between the plasma and the neutral atmosphere when their bulk flow velocities differ (Song *et al.*, 2001; Vasyliūnas and Song, 2005). The governing equations (giving horizontal components only) are the momentum equation of the ionospheric plasma

$$\frac{\partial \rho \mathbf{v}}{\partial t} + \cdots = \frac{1}{c}\mathbf{J} \times \mathbf{B} - \nu_{in}\rho(\mathbf{v} - \mathbf{v}_n) \qquad (10.11)$$

(ν_{in} is the ion–neutral-particle collision frequency), in which the time derivative term and the inertial and pressure terms (indicated by the ellipses) on the left-hand side are neglected, and the generalized Ohm's law in the form

$$0 = \mathbf{E} + \frac{1}{c}\mathbf{v} \times \mathbf{B} - \frac{\mathbf{J} \times \mathbf{B}}{n_e e c} + \cdots. \tag{10.12}$$

Eliminating $\mathbf{v}$ between Eqs. (10.11) and (10.12) then yields Eq. (10.10). The current in the ionosphere is thus governed by stress balance in the same way as the current in the magnetosphere, the latter being determined by the momentum equation for the magnetospheric plasma (cf. Eq. (10.2)),

$$\frac{\partial \rho \mathbf{v}}{\partial t} = -\nabla \cdot (\rho \mathbf{v}\mathbf{v} + \mathbf{P}) + \frac{1}{c}\mathbf{J} \times \mathbf{B}, \tag{10.13}$$

with the time derivative term on the left-hand side neglected.

(3) Underlying this neglect of the time derivative (i.e. acceleration) terms in the momentum equations is the implicit assumption that any imbalance between the mechanical and the magnetic stresses (which, fundamentally, is what determines the acceleration of the plasma) produces a bulk flow that acts to reduce the imbalance, which then becomes negligible over a characteristic time scale that is easily shown to be of the order of the Alfvén wave travel time across a typical spatial scale, e.g. along a field line. The theory is thus applicable only to systems that are stable and evolve slowly, on time scales $\tau \gg \mathcal{L}/v_A$.

(4) For phenomena that are large-scale in the sense that their time and length scales are long in comparison with those of plasma oscillations ($\tau \gg 1/\omega_p$ and $\mathcal{L} \gg \lambda_e \equiv c/\omega_p$ where ω_p is the electron plasma frequency and λ_e is the electron inertial length, also known as the collisionless skin depth), $\mathbf{J}$ adjusts itself to become equal to $(c/4\pi)\nabla \times \mathbf{B}$, not the other way around (Vasyliūnas, 2005a, b). Although $\mathbf{B}$ is in principle determined from a given $\mathbf{J}$ by Maxwell's equations (on a time scale of the light travel time $\sim \mathcal{L}/c$), in a large-scale plasma any $\mathbf{J} \neq (c/4\pi)\nabla \times \mathbf{B}$ is immediately (on a time scale $\sim 1/\omega_p$) changed by the action of the displacement-current electric field on the free electrons in the plasma. The current continuity condition $\nabla \cdot \mathbf{J} = 0$ is thus satisfied automatically; there is no separate requirement of current closure – what is often discussed under that rubric is in reality the coupling of the Maxwell stresses along different portions of a field line.

10.4.2 Corotation

Corotation with the planet is the simplest pattern of plasma flow in a planetary magnetosphere and one that plays a major role particularly in the magnetospheres of the giant planets (see Chapter 13). Understanding the conditions that are needed for corotation to exist as well as those that result in significant departures from corotation is an essential aspect of magnetospheric studies.

If the planet possesses an insulating atmosphere, the rotation of the planet itself has no *direct* effect on plasma flow in the magnetosphere, as discussed in Section 10.4.1. What does affect plasma flow is the motion of the neutral upper atmosphere (the thermosphere) at altitudes of the ionosphere where the neutral and the ionized components coexist and interact. "Corotation with the planet" is therefore not quite an accurate description. What really is meant is co-motion with the upper atmosphere, which in turn is then assumed to corotate with the planet, for reasons unrelated to the magnetic field, i.e. the vertical transport of horizontal linear momentum from the planet to the neutral atmosphere (e.g. by collisional or eddy viscosity or similar processes) together with an assumed small relative amplitude of the neutral winds.

Any difference between the bulk flow of the neutral medium and the ionized component of the plasma in the ionosphere results in a collisional drag that must be balanced by the Lorentz force; without it, the drag force would soon bring the plasma to flow with the much more massive neutral medium. The Lorentz force in the ionosphere is coupled to a corresponding Lorentz force in the magnetosphere, which in turn must be balanced locally by an appropriate mechanical stress. The net result is that departure from corotation requires a mechanical stress in the magnetosphere to balance the plasma–neutral-particle drag in the ionosphere; conversely, the plasma will corotate if the stress in question is negligibly small. (It is fairly obvious that the direction of the stress must be more or less azimuthal, opposed to the direction of rotation.) Quantitatively, the requirements for the corotation of magnetospheric plasma may be expressed by four conditions.

(1) **Planet–atmosphere coupling**   This is simply the assumption, discussed above, that the upper atmosphere effectively corotates with the planet.

(2) **Plasma–neutral-particle coupling in the ionosphere**   The collisional drag of the neutral medium on the plasma must be sufficiently strong to ensure that $\mathbf{v} \simeq \mathbf{v}_n$. The quantitative condition is derived in principle from Eq. (10.11), with the left-hand side set to zero, but with one complication: what is relevant for the interaction with the magnetosphere is not the local current density $\mathbf{J}$ but the current per unit length integrated over the extent of the ionosphere at altitude z, i.e. the height-integrated current $\mathbf{I} \equiv \int \mathbf{J} \, dz$. A direct integration of Eq. (10.11) over height, however, is not simple because $\mathbf{v}$ varies strongly with z even when $\mathbf{v}_n$ is independent of z, as usually assumed. The horizontal electric field, however, is essentially constant over the entire (relatively thin) height range of the ionosphere, from continuity of tangential components implied by Faraday's law. It is thus convenient first to express $\mathbf{J}$ by Eq. (10.10) and then integrate over height

to obtain

$$\mathbf{I}_\perp = \frac{B}{c} \left[\Sigma_P \hat{\mathbf{B}} \times (\mathbf{v}_0 - \mathbf{v}_n) - \Sigma_H (\mathbf{v}_0 - \mathbf{v}_n)_\perp \right] \qquad (10.14)$$

where $\mathbf{v}_0$ is the plasma flow at the topside of the ionosphere and is related to the electric field by

$$\mathbf{E}^* = -\frac{1}{c} (\mathbf{v}_0 - \mathbf{v}_n) \times \mathbf{B} \qquad \text{or equivalently} \qquad \mathbf{E} = -\frac{1}{c} \mathbf{v}_0 \times \mathbf{B}. \qquad (10.15)$$

In Eq. (10.14) $\Sigma_P \equiv \int \sigma_P dz$ and $\Sigma_H \equiv \int \sigma_H dz$ are the Pedersen and the Hall conductances (height-integrated conductivities), Σ_P being the more important for magnetosphere–ionosphere interactions (Hall currents close just within the ionosphere, to a first approximation).

To ensure that $\mathbf{v}_0 \simeq \mathbf{v}_n$, the ionospheric conductance Σ_P must be sufficiently large in relation to the height-integrated current $\mathbf{I}$, which scales as the current per unit length in the magnetosphere and hence ultimately as the mechanical stresses in the magnetosphere. For a more precise criterion, one must consider a specific process; some examples are discussed in Section 10.4.4.

(3) **Magnetohydrodynamic coupling from ionosphere to magnetosphere along magnetic field lines** Conditions (1) and (2) merely ensure that the plasma corotates at the topside of the ionosphere, at the foot of a magnetic flux tube within the magnetosphere. For corotation to extend into the magnetosphere itself, the MHD constraining relation between the flow and the magnetic field must hold. In the steady state, one may simply invoke Eq. (10.12) (the third term on the right-hand side may often be neglected) and calculate the electric field by treating the field lines as equipotentials. The description of the physical process, however, by which the initial corotation of the ionosphere is *communicated* to the magnetosphere is considerably more complex. The MHD approximation (10.12), the plasma momentum equation (10.13), and Faraday's law (10.9) combine to show that the initial transverse flow will launch an Alfvén wave propagating along the field line to set the entire flux tube into motion.

(4) **A stress balance maintaining centripetal acceleration in the magnetosphere** If conditions (1)–(3) are satisfied, the plasma will be corotating at least as far as the components of $\mathbf{v}$ perpendicular to $\mathbf{B}$ are concerned, but the flow parallel to $\mathbf{B}$ remains unconstrained. For the entire flow to be corotational, one further condition must be satisfied: there must exist a radial stress to balance the centripetal acceleration of the corotating plasma. In most cases, this stress is produced by the corotation itself, as the magnetic field lines are pulled out until their tension force becomes sufficiently strong to balance the centripetal acceleration.

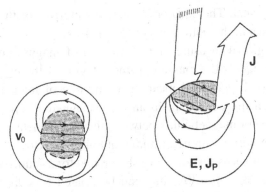

Fig. 10.5. Schematic diagram of magnetospheric convection over the Earth's north polar region (after Vasyliūnas, 1975a). (Left) Streamlines of the plasma bulk flow; the Sun is on the left. (Right) Electric field lines and associated Pedersen currents, and the magnetic-field-aligned current (also known as the Birkeland current; see the large arrows). Section 10.4.3 gives a detailed description and Section 11.6 and Fig. 11.9 give corresponding MHD model results for the electric potential. For a schematic representation of magnetospheric convection throughout the magnetosphere, see Fig. 13.4.

10.4.3 Magnetospheric convection

Magnetospheric convection may be considered as the other canonical pattern, besides corotation, of plasma flow in a planetary magnetosphere, one that plays an overwhelmingly important role in the magnetosphere of Earth (see Chapter 11 for a detailed discussion). The basic concept is that the flow of solar wind plasma past the magnetosphere imparts some of its motion to plasma in the outermost regions of the magnetosphere, either directly by MHD coupling along open field lines (Dungey, 1961) or through an unspecified tangential drag near the magnetopause (Axford and Hines, 1961).

By continuity of mass and magnetic flux transport, the flow then extends into the region of closed field lines or the interior of the magnetosphere, setting up a large-scale circulation pattern (which has some superficial resemblance to, but no real physical commonality with, what is called convection in ordinary fluid dynamics). Figure 10.5 illustrates the pattern, projected onto the topside ionosphere of the planet: shown on the left-hand side are the streamlines of the plasma bulk flow v_0, which are also the equipotentials of the electric field according to Eq. (10.15). The electric field lines and the associated Pedersen currents are shown on the right-hand side, along with a sketch of the implied Birkeland (i.e. magnetic-field-aligned) currents. The shaded region is the polar cap, identified with the region of open field lines in the open magnetosphere; it represents the mapping (along field lines) of the boundary region where the solar wind motion is being imparted to magnetospheric plasma. The equatorial-plane counterpart of the flow outside the polar cap was shown in the bottom panel of Fig. 10.3.

A quantitative global measure of the strength of magnetospheric convection is the emf (which equals the maximum line integral of the electric field) across the polar cap, Φ_{PC}. Its physical meaning is the rate of magnetic flux transport (advection) through the polar cap. In an open magnetosphere, $c\Phi_{PC}$ equals the rate of reconnection of magnetic flux between the interplanetary and the planetary magnetic fields. Numerous empirical studies at Earth (e.g. Boyle *et al.*, 1997; Burke *et al.*, 1999; and references therein) have shown that for a southward interplanetary magnetic field (i.e. $\mathbf{B}_{sw} \cdot \hat{\boldsymbol{\mu}} > 0$), Φ_{PC} can be related to solar wind parameters approximately as follows:

$$c\Phi_{PC} \simeq v_{sw}(\mathbf{B}_{sw} \cdot \hat{\boldsymbol{\mu}})\mathcal{L}_X \qquad (10.16)$$

where $\mathcal{L}_X$ is a length that typically is a fraction (~ 0.2 to ~ 0.5) of the magnetopause radius R_{MP}. When comparing different magnetospheres, one often supposes that the ratio $\mathcal{L}_X/R_{MP}$ is a more or less universal constant. Physically, $\mathcal{L}_X$ may be looked at as the length of the reconnection X-line on the magnetopause at which the magnetic field lines from the solar wind and from the planet first become interconnected, projected along the streamlines of the magnetosheath flow back into the undisturbed solar wind, as illustrated in Fig. 10.3 (see Section 13.1.4 for applications to other planets).

Although magnetospheric convection is often talked about as if it were simply a mapping-plus-continuation of the solar wind or boundary layer flow, it does involve mechanical stresses in an essential way. The Birkeland currents shown in Fig. 10.5 indicate that the magnetic field is deformed both at the near-planet end and also at the distant end, with Lorentz forces that must be balanced by mechanical stresses at both ends. The Lorentz force in the ionosphere is balanced by the collisional drag of the neutral atmosphere; the total force integrated over the entire ionosphere points approximately anti-sunward (to the right in Fig. 10.5), in agreement with the preponderance of anti-sunward plasma flow and hence sunward drag force. The Lorentz force in the distant region is sunward and acts to slow down the anti-sunward flow of the solar wind plasma. In the presence of the strongly converging field lines of the dipole, all these forces interact, also in a complicated way, with the interior of the planet, as recently elucidated by Siscoe (2006) and Vasyliūnas (2007).

An alternative and in some ways physically more accurate view is to consider magnetospheric convection as the result not directly of an imposed flow but of an imposed stress on the magnetic field, in this case an anti-sunward pull on the field lines by the solar wind flow. Once this point of view is adopted, mechanisms for producing flows of the magnetospheric convection pattern other than by the solar wind are immediately apparent. A mechanical stress in the magnetosphere will deform the magnetic field to reach local equilibrium; the deformation of the field extends necessarily into the ionosphere, where it must be balanced in turn by a mechanical stress; however, the only one available is collisional drag and hence plasma flow relative to the neutrals. Thus any quasi-localized unidirectional mechanical force in the magnetosphere can produce a flow pattern similar to that of Fig. 10.5, the anti-sunward direction being replaced by the direction of the force and the polar cap being replaced by the region mapped along field lines from the location of the force in the magnetosphere.

The most significant mechanisms for such internally driven convection flows in planetary magnetospheres are one-sided (azimuthally non-symmetric) enhancements of plasma pressure or, in the case of corotating plasmas subject to centripetal acceleration, of mass density; these play an important role in transport processes, as discussed below in Section 10.5. If these enhancements are fixed relative to the rotation, e.g. if they are associated with asymmetries or anomalies of the planetary magnetic field (Dessler and Hill, 1975), the convection pattern remains fixed in the corotating frame of reference and is known as *corotating convection*; see Hill *et al.* (1981) and references therein.

10.4.4 Limits to corotation

The conditions (1)–(4) for corotation discussed in Section 10.4.2 are all in essence local conditions at a given magnetic flux tube. Deviations from corotational flow when one or other of these conditions is no longer satisfied need not, therefore, be global but can be confined to limited regions. Typically, plasma flow in any particular magnetosphere may follow corotation in the inner regions, out to a critical radial distance in the equatorial plane, and then deviate significantly from corotation at larger distances. The critical distance depends on which of conditions (1)–(4) given in Section 10.4.2 is violated and by which process.

Violation of (1): Deviations of $\mathbf{v}_n$ from the corotation velocity $\mathbf{\Omega} \times \mathbf{r}$ may occur for two reasons: (a) the presence of a neutral wind velocity field, which in most cases is not directly associated with the magnetosphere and is not connected with any particular distance (but which can have effects in the magnetosphere since the plasma tends to follow the combined flow field of corotation plus winds); (b) when the plasma flow is already strongly non-corotational because of stresses applied from the magnetosphere, the collisional drag on the neutrals may become non-negligible in comparison with the stresses from the planet, modifying the flow of the neutrals to reduce the difference from the plasma.

Violation of (2): The mechanical stresses associated with magnetospheric convection (Section 10.4.3) may act to produce non-corotational flows. The importance

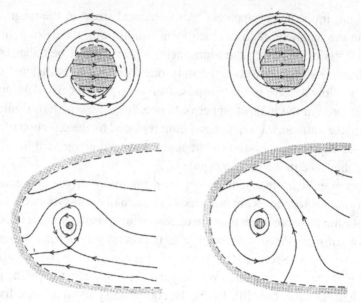

Fig. 10.6. Streamlines of plasma flow in a magnetosphere: (upper panels) looking down on the topside ionosphere, and (lower panels) projected along magnetic field lines to the equatorial plane of the magnetosphere. On the left magnetospheric convection is dominant and on the right corotation is dominant (after Vasyliūnas, 1975); compare with the observation-based model for Jupiter's magnetospheric configuration in Fig. 13.6.

of solar-wind-driven magnetospheric convection in comparison with corotation may be gauged by comparing the polar cap potential given by Eq. (10.16) with the corresponding corotational potential between the colatitude θ_{pc} of the polar cap boundary and the pole:

$$c\Phi_{\mathrm{CR}} \simeq \Omega R_{\mathrm{P}}\theta_{\mathrm{pc}}{}^2 B_{\mathrm{P}}. \qquad (10.17)$$

By conservation of magnetic flux,

$$\Phi_M = \pi R_{\mathrm{P}}\theta_{\mathrm{pc}}{}^2 (2B_{\mathrm{P}}) \approx \mathcal{L}_{\mathrm{X}}\mathcal{L}_{\mathrm{T}} (\mathbf{B}_{\mathrm{sw}} \cdot \hat{\boldsymbol{\mu}}) \qquad (10.18)$$

(although perhaps confusing, the use of Φ both for potential and for magnetic flux is traditional; for $\mathcal{L}_{\mathrm{X}}$ and $\mathcal{L}_{\mathrm{T}}$ see Section 10.3.3). From Eqs. (10.16)–(10.18) we obtain

$$\frac{\Phi_{\mathrm{CR}}}{\Phi_{\mathrm{PC}}} \approx \frac{\Omega \mathcal{L}_{\mathrm{T}}}{2\pi v_{\mathrm{sw}}}. \qquad (10.19)$$

The plasma flow patterns for the two extreme cases of small and large values for the ratios in Eq. (10.19) are shown in Fig. 10.6. When $\Phi_{\mathrm{CR}}/\Phi_{\mathrm{PC}} \gg 1$ (right-hand panels), the magnetosphere is said to be corotation-dominated; the corotation extends almost to the magnetopause. When $\Phi_{\mathrm{CR}}/\Phi_{\mathrm{PC}} \ll 1$ (left-hand panels), the magnetosphere is convection-dominated; corotation is confined to the inner region, which extends to a distance R_{cr} given approximately by

$$\left(\frac{R_{cr}}{R_{MP}}\right)^{\nu} \sim \frac{\Omega \mathcal{L}_T}{v_{sw}} \tag{10.20}$$

where the exponent ν varies between 1.5 and 2 depending on the details of the model for the convection flow. The two extreme cases, convection-dominated and corotation-dominated, were first distinguished and applied to the magnetospheres of Earth and Jupiter by Brice and Ioannidis (1970) (see also Vasyliūnas, 1975a).

Nevertheless, even when $\Phi_{CR}/\Phi_{PC} \gg 1$ the corotation of plasma does not necessarily extend to the magnetopause but may be self-limited by another mechanical stress that arises when the magnetosphere has sources of plasma deep within its interior (see Section 10.5.1 below). The plasma supplied from such a source must be transported outward and, as long as corotation is maintained, the angular momentum per unit mass of the plasma increases as Ωr^2. This requires an appropriate torque to be applied to the plasma by the azimuthal component of the Lorentz force, which is then balanced in the usual way by an azimuthal collisional drag in the ionosphere, constituting a departure from corotation. Physically, an outward-moving plasma element initially lags behind corotation (by the conservation of angular momentum), pulling the magnetic field line with it; the resulting deformation of the magnetic field in the ionosphere reduces the corotation of the plasma until the drag force needed to balance the torque in the magnetosphere is reached. This process of *partial corotation* was first described and quantitatively modeled by Hill (1979), who found that the departure from corotation is very slight at small equatorial radial distances and increases gradually with increasing required torque; the flow becomes strongly sub-corotational beyond a distance R_H (often called the Hill radius), given for a dipole field by the simple expression

$$R_H{}^4 = \left(\frac{\pi \Sigma_P}{c^2}\right)\frac{\mu^2}{S} \tag{10.21}$$

where S is the net outward mass flux from the internal source (recall that μ is the magnetic dipole moment of the planet).

Violation of (3): By the very nature of the MHD approximation, departures from it occur for the most part only for small-scale structures. Hence one does not expect a direct influence on the mapping of large-scale flows from non-MHD effects (which can, however, have a very significant influence on related phenomena such as the aurora or particle acceleration).

Violation of (4): The radial stress balance needed to maintain the centripetal acceleration of corotating flow can exist only as long as the Lorentz force, or equivalently the tension of the stretched-out field lines, is sufficiently strong to match or exceed $\rho\Omega^2 r$ in the equatorial region of the magnetosphere. This is no longer possible beyond a critical distance R_0, often estimated by the rough argument of setting the Alfvén speed equal to the corotation speed. A more precise estimate can easily be made if the plasma source is located at a radial distance r_s, beyond which the plasma is assumed to be confined to a planar sheet of thickness $h_s \ll r$, with magnetic flux tube content (the mass per unit magnetic flux)

$$\eta \equiv \int \frac{dl\rho}{B} \approx \frac{\rho h_s}{B_z} \tag{10.22}$$

being assumed approximately constant for $r > r_s$; then R_0 is given by (Vasyliūnas, 1983, and references therein)

$$R_0{}^4 \simeq \frac{\mu^2}{\pi \rho_s h_s r_s{}^3 \Omega^2} \tag{10.23}$$

where ρ_s is the plasma mass density at the source location. Under the same assumptions about the geometry of the plasma distribution, the total mass M from the internal source can be estimated as

$$M \approx 2\pi \rho_s h_s r_s{}^2; \tag{10.24}$$

hence R_0 can also be written as

$$R_0{}^4 \simeq \frac{2\mu^2}{M r_s \Omega^2}. \tag{10.25}$$

Obviously the above estimate of R_0 and indeed the entire concept of the breakdown of radial stress balance is applicable only if the plasma is still corotating at distances as far out as R_0, which requires that $R_0 \ll R_H$; when comparing the two distances, one may relate M and S by $M = S\tau_{tr}$, where τ_{tr} is the global transport time to be discussed below in Section 10.5.2. If this is the case, the plasma flow beyond R_0 is expected to change from corotation to a more nearly outward motion and ultimately to a general quasi-radial outflow, which then also stretches out the magnetic field lines until they break, via the reconnection process. Such a rotationally driven radial outflow of a corotating plasma from an internal source (possibly channeled into the magnetotail by the external pressure on the magnetopause) was proposed for the magnetosphere of Jupiter by Hill *et al.* (1974) and Michel and Sturrock (1974), who named it the *planetary wind*; the term *magnetospheric wind* is also used, sometimes in the more general sense of a radial outflow without reference to a specific physical origin. A widely cited but rather speculative diagram of the associated changes in the magnetic field topology (Vasyliūnas, 1983) is shown in Fig. 10.7.

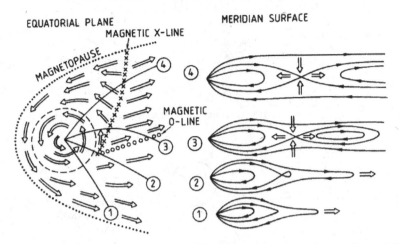

Fig. 10.7. Schematic of planetary wind flow and magnetic topology (after Vasyliūnas, 1983).

10.5 Plasma sources and transport processes

Planetary magnetospheres exhibit an enormous variety; almost every observed magnetosphere differs significantly and qualitatively from all the others (see Section 13.1.3). A primary reason for such diversity, in the face of the few structures of more or less universal character predicted by the interaction of the solar wind with the planetary magnetic field (Section 10.3), lies in the existence of several different types of source for the plasma within the magnetosphere, as well as several different types of transport process that determine the motion and distribution of the plasma.

10.5.1 Sources of magnetospheric plasma

Three types of plasma source can be distinguished on the basis of their location: (1) at the outer boundary of the magnetosphere, i.e. the interface with the solar wind; (2) at the inner boundary of the magnetosphere, i.e. the interface with the ionosphere and atmosphere of the planet; (3) within the interior volume of the magnetosphere, at interfaces with objects within the magnetosphere, e.g. the moons and the rings of the planet. Figure 10.8 shows schematically the various locations and the associated plasma source regions.

(1) **Outer-boundary source: solar wind.** In a magnetically open magnetosphere (Section 10.3.3), plasma from the solar wind can flow along open magnetic field lines directly into the magnetotail, forming a *plasma mantle*. Although much of this plasma remains on open field lines and continues to flow down the magnetotail, some plasma may, as a result of magnetic reconnection in the magnetotail,

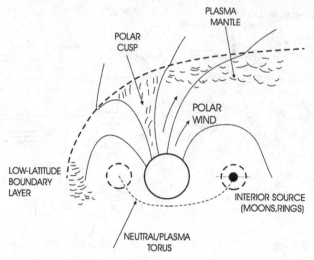

Fig. 10.8. Schematic of magnetospheric plasma source locations (not to scale, schematic, only northern hemisphere shown). See Fig. 13.3 for a diagram for Earth's magnetosphere in particular.

be transported into the closed-field-line region of the magnetosphere. In addition, even when the magnetopause ideally should be impenetrable to plasma, as in a closed magnetosphere (Section 10.3.1) or in those parts of the open magnetosphere boundary where the normal component of the magnetic field $B_n = 0$, some solar-wind plasma crosses as the result of non-ideal MHD processes; this is evidenced by the existence of the *low-latitude boundary layer*, where plasma with proper-ties typical of the magnetosheath is found on closed field lines just inward of the magnetopause. The *polar cusp* (see e.g. Fig. 1.3) is another prominent region of plasma entry, but whether it represents flow along open or transport across closed field lines (or a combination of both) is still disputed.

(2) **Inner boundary source: ionosphere.**   Plasma from the ionosphere can flow directly along the magnetic field lines upward into the magnetosphere, pro-vided only that there is sufficient energy to overcome the gravitational binding to the planet. The process may be simply a plasma analog of Jeans escape, involving those particles with thermal speeds in excess of the gravitational escape speed (but with the added complication that, since electrons are much lighter than ions, plasma escape also involves electrostatic effects). There may be a systematic upward bulk flow, the *polar wind*, formed by processes similar to those that lead to the outflow of the solar wind from the Sun. Finally, magnetospheric processes, espe-cially those connected with the aurora, may lead to acceleration of both elec-trons and ions in the ionosphere to speeds well above the gravitational escape speed.

(3) **Interior source: moons and planetary rings**. Solid bodies traveling on orbital paths that lie wholly or partly within the volume of the magnetosphere may act as sources of plasma, in most cases by an indirect process: the release of neutral gas, which is subsequently ionized. Such a source is significant particularly in the case of the giant planets. Jupiter and Saturn both possess a number of moons, some of size comparable with or larger than the Earth's moon, located deep within the magnetosphere; Saturn has in addition the well-known rings, the constituents of which may range from dust to boulder size. The neutral gas may be released from the surface of the body or from its atmosphere (in the case of moons large enough to keep one), by evaporation, or by the sputtering produced by impacting plasma particles from the magnetosphere; it may be ionized by photoionization or by electron-impact ionization. For a more detailed discussion of specific objects, see Chapter 13.

The strength of any particular plasma source is generally parameterized by the rate S, i.e. the mass added per unit time; the role of the mass added from an interior source in limiting corotation was discussed in Section 10.4.4. For the solar wind source, S is given roughly by the incident solar wind mass flux (which is reasonably well known) multiplied by the efficiency of entry into the magnetosphere (which is poorly known, but probably small). For an S from the ionospheric source the efficiency may be taken as effectively 1, but the upward mass flux is very uncertain in most cases. For an interior source, S depends on the rate of injection of neutral particles and on the rate of ionization, both of which (if sputtering and electron impact ionization are important) depend on the parameters of magnetospheric plasma, of which S itself is a source; thus a complex nonlinear interaction determines S. As a result of all these uncertainties, at present the values of S for the different sources in various magnetospheres can, for the most part, only be estimated empirically (see Section 10.6 for further discussion).

An important marker for identifying the plasma from the different sources is the composition of the plasma ions, which reflects the ionic composition of the source. The solar wind contributes primarily H^+ (protons), with a small admixture of He^{++} (alpha particles). The contribution of the ionosphere depends on the atmosphere of the planet. For the giant planets, the atmospheres consist primarily of hydrogen, hence the contribution of their ionospheres is not easily distinguishable from that of the solar wind, except for some subtle effects, e.g. the absence of He^{++} and the possible presence of molecular hydrogen ions; for Earth, the presence of O^+ is an indicator of the ionospheric contribution. The contribution of planetary moons and rings varies with the object but in almost all cases is dominated by heavy ions – e.g. sulfur and oxygen ions from the moon Io at Jupiter and water-group ions from most of the other (ice-covered) moons or rings at Jupiter and Saturn.

10.5.2 What moves the plasma?

Plasma in the magnetosphere from the various sources described in Section 10.5.1 is located initially on magnetic field lines that thread the respective source regions. For a solar wind source, they are field lines near the magnetopause: open (plasma mantle, possibly a part of polar cusp) or closed (low-latitude boundary layer, possibly the other part of polar cusp). For a interior source, they are the field lines crossing the orbits of the moons or rings. An ionospheric source does, in principle, populate all the field lines, but its contribution is, at least initially, heavily concentrated toward low altitudes. If the plasma is to be redistributed over wide regions of the magnetosphere, as observed in many magnetospheres for plasma from different sources, transport processes that move the plasma away from the source regions must be present.

Plasma from an ionospheric source may be redistributed by direct flow along the magnetic field lines (its density, however, becomes greatly reduced as the cross-sectional area of a flux tube increases with increasing altitude). This is the simplest transport process, and one not impeded by any MHD constraints. To redistribute the plasma from other types of source, transport processes are required that necessarily involve flow across magnetic field lines, including in particular flow in the radial direction.

Plasma bulk flows that are transverse to the magnetic field are subject to the generalized Ohm's law, e.g. Eq. (10.12), and hence are predominantly $\mathbf{E} \times \mathbf{B}$ drifts, with some contribution from gradient and curvature drifts in the azimuthal direction only; thus they transport magnetic flux along with the plasma. The plasma sources described in Section 10.5.1, however, are sources only of plasma and not of magnetic flux: they add a net amount of plasma to the magnetosphere, while the total amount of magnetic flux threading the magnetosphere remains fixed (it is set by the number of field lines of one polarity emerging from the planet). The added plasma must ultimately be removed from the magnetosphere, predominantly by transport out of it. Of the other loss processes, recombination is negligible in most cases; precipitation of plasma onto the planet, although potentially of appreciable local impact (on e.g. the formation of the aurora or the heating and ionization of the atmosphere), is of very little significance for the global mass budget simply because the surface area of the planet is usually very small in relation to the magnetosphere.

The quintessential problem of plasma transport in planetary magnetospheres, therefore, is how to achieve a flow that carries both plasma and magnetic flux yet removes plasma but not magnetic flux. The almost universally accepted solution is to consider circulating flow patterns that do not appreciably change the magnetic field (they are often referred to as *interchange motions* (Gold, 1959), at least in a qualitative sense – the precise denotation of that term is sometimes disputed) but in which the outflowing segments have a larger plasma content than the inflowing segments. The prototype of such circulating interchange motions is magnetospheric convection, in the generalized sense discussed at the end of Section 10.4.3; the various models of the transport process may be viewed as applications of magnetospheric convection, driven by different physical processes and with different assumed spatial and temporal scales. There are two basic dichotomies of the adopted model: (I) magnetospheric convection is either (a) driven by the solar wind or (b) driven by the internal dynamics of the magnetosphere; (II) magnetospheric convection is either (1) quasi-steady (systematic) or (2) fluctuating (random or chaotic). Each of the four resulting possibilities has found application in some magnetosphere or other.

(1a) **Quasi-steady magnetospheric convection driven by the solar wind**. This is expected to be the primary transport process in magnetospheres that have no significant interior sources and that are convection-dominated in the sense defined by Eq. (10.19), e.g. the magnetospheres of Mercury and Earth. In magnetospheres that are corotation-dominated, however, magnetospheric convection may still be present but produces no net radial transport; the latter is effectively suppressed since each plasma element is carried by the rotation between inbound and outbound segments of the convection pattern.

(1b) **Quasi-steady magnetospheric convection driven by internal dynamics**. As pointed out in Section 10.4.3, for the internal dynamics to give rise to magnetospheric convection, a necessary condition is an azimuthally asymmetric configuration of the stresses in the magnetosphere; for the convection to be quasi-steady, the asymmetry should be permanent for a given location. An example of transport by this process is corotating convection, proposed as a possibility for the magnetosphere of Jupiter (Hill *et al.*, 1981, and references therein), the permanent asymmetry being provided by an observed asymmetry (the "magnetic anomaly") of Jupiter's internal magnetic field. Very recently, corotating convection has also been proposed as a transport process for the magnetosphere of Saturn (Goldreich and Farmer, 2007; Gurnett *et al.*, 2007); since no azimuthal asymmetry of Saturn's internal magnetic field has been observed to date, however, it is not clear what could hold an asymmetric stress configuration in place.

(2a) **Fluctuating magnetospheric convection driven by the solar wind**. Since the strength of solar-wind-driven magnetospheric convection depends on the direction of the interplanetary magnetic field (Eq. 10.16), which is highly variable, a fluctuating state may be viewed as the normal condition. As far as the plasma transport process is concerned, the main difference from the quasi-steady state is that the suppression of radial transport by the effects of rotation, discussed under (1a) above, can be overcome by fluctuations that have time scales appropriately related to the rotation period. The principal application of this concept has been to the inner regions of the magnetosphere of Earth, roughly inside the distance given by Eq. (10.20), where corotation is the dominant flow and also the azimuthal flows associated with gradient and curvature drifts may be large. Fluctuating magneto-spheric convection has been invoked, sometimes under different designations, to explain how radial transport (needed to account for e.g. buildup of the ring current, erosion of the plasmasphere, and formation of the radiation belts; see Chapter 11) can be maintained despite the large azimuthal motions.

(2b) **Fluctuating magnetospheric convection driven by internal dynamics**. Almost universally, this is assumed to be the primary transport process in the corotation-dominated magnetospheres of the giant planets, Jupiter and Saturn. If the plasma comes predominantly from interior sources deep within a corotating magnetosphere, its distribution is unstable to the formation of localized mass concentrations that move outward, alternating with adjacent mass dispersions that move inward; this is the *rotational* (or centrifugal) *interchange instability*, which is somewhat analogous to the Rayleigh–Taylor instability of a heavy fluid on top of a light one. As the primary mechanism of radial transport, it was first proposed for Jupiter by Ioannidis and Brice (1971) and has been extensively developed, e.g. by Siscoe and Summers (1981), Southwood and Kivelson (1989), Pontius and Hill (1989), Vasyliūnas (1989b), Hill (1994), Ferrière *et al.* (2001), Vasyliūnas and Pontius (2007), and others.

To describe quantitatively the mean transport resulting from random fluctuations, a diffusion equation for the mass per unit magnetic flux (i.e. the flux tube content) η, defined in Eq. (10.22), is commonly employed. It is most simply derived by a method (adapted from work by G. I. Taylor in 1921 on fluid turbulence) first applied in the magnetosphere of Earth by Cole (1964) and independently in an astrophysical context by Jokipii and Parker (1968). If each quantity Q is represented as a mean $\langle Q \rangle$ plus a fluctuation δQ, the mean transport of flux tube content is then given by

$$\langle \eta \mathbf{v} \rangle = \langle \eta \rangle \langle \mathbf{v} \rangle + \langle \delta \eta \delta \mathbf{v} \rangle. \tag{10.26}$$

The first term on the right-hand side, giving the plasma transport by the mean flow, is *advection*; the second term, giving the transport by the correlated density and velocity fluctuations without any mean flow, is *diffusion*. In the absence of sources and sinks, the flux tube content is conserved by the flow:

$$\frac{\partial \eta}{\partial t} + \mathbf{v} \cdot \nabla \eta = \text{sources} - \text{sinks}. \tag{10.27}$$

Now, if $\delta \eta$ is calculated in the quasi-linear approximation by integrating Eq. (10.27) this allows the second term in Eq. (10.26) to be rewritten in standard diffusion form:

$$\langle \delta \eta \delta \mathbf{v} \rangle \simeq -\mathbf{D} \cdot \nabla \langle \eta \rangle \tag{10.28}$$

with diffusion coefficient

$$\mathbf{D} = \int_{-\infty}^{t} dt' \langle \delta\mathbf{V}(t')\delta\mathbf{V}(t) \rangle \tag{10.29}$$

(it has been assumed that the autocorrelation of the velocity field $\langle \delta\mathbf{V}\delta\mathbf{V} \rangle$ is non-negligible only for separations short in comparison with the evolution time scale of the mean configuration). Note that the concept of diffusive transport is not restricted to flux tube content but can also be derived for any quantity that is conserved by the flow, i.e. obeys an equation of the same form as Eq. (10.27); it has been extensively applied in studies of trapped particles and radiation belts (e.g. Schulz and Lanzerotti, 1974, and references therein).

The diffusion coefficient (when not simply taken as an empirical parameter to be determined by fitting to observations) can be expressed in various ways: as the mean square amplitude of the velocity fluctuations multiplied by a correlation time (Cole, 1964) or as the power spectrum at zero frequency (Jokipii and Parker, 1968). In the present context of fluctuating magnetospheric convection, it is most conveniently represented in terms of the characteristic linear dimension λ of the circulating flow pattern and of the circulation time τ_c as

$$D \simeq \lambda^2/\tau_c. \tag{10.30}$$

The net outward mass flux S is given by

$$S = r \oint d\phi \, B_e \langle \delta\eta\delta v \rangle \simeq -r \oint d\phi \, B_e D \frac{\partial \eta}{\partial r} \tag{10.31}$$

where B_e is the equatorial magnetic field at radial distance r and the integration is over all azimuthal angles around the planet. With the total mass M given by

$$M = \int r dr \oint d\phi \, B_e \eta, \tag{10.32}$$

the global transport time defined by $\tau_{tr} = M/S$ can be estimated to an order of magnitude as

$$\tau_{tr} \sim \tau_c \left(\frac{r_s}{\lambda}\right)^2 \tag{10.33}$$

where r_s is the radial distance of the source region.

The circulation time τ_c can be estimated relatively reliably from considerations of magnetosphere–ionosphere coupling (Siscoe and Summers, 1981), but the detailed spatial geometry of the transport process is still largely unknown. Not only the values of λ and their variation with distance but even the basic flow patterns – nearly circular, or radially elongated? – remain uncertain, with no firm, generally accepted, conclusions to date from the extensive observational, theoretical, and computational studies that have been made (e.g. Krupp *et al.*, 2004; Vasyliūnas and Pontius, 2007; Yang *et al.*, 1994; and references therein).

10.5.3 Plasma structures

The great variety of magnetospheric structures that can be formed by the interplay of plasma source and transport processes are surveyed in Chapter 13 and (for Earth) in Chapter 11. Here we simply make a few remarks on some generic relationships.

(1) *Neutral* and *plasma tori* are the direct result of an interior source from a moon of the planet. A neutral torus around the orbit of the moon forms if the neutral particles leave the moon with speed high enough to escape its gravitational field but not the gravitational field of the planet (as first suggested by McDonough and Brice, 1973); the corresponding plasma torus then forms by ionization of the neutral torus. If the ionization time is appreciably shorter than the time for the neutrals to spread over the orbit of the moon, the neutral torus will not be complete but will cover only a segment of the orbit near the moon and will move with its Keplerian speed; the plasma torus, however, will still be complete (except in the case of a moon at synchronous orbit) as long as it is carried around by corotation.

(2) *Plasma sheets* (or, equivalently, *current sheets*) can be formed by two different processes. With an interior source, plasma is transported outward, stretching out the magnetic field; the resulting plasma or current sheet is also called the *magnetodisk*, particularly if the stretching is by rotational stresses. With a solar wind source, plasma is transported inward, preventing the field lines stretched by the solar wind from collapsing back to a quasi-dipolar form.

(3) *Ring currents* in the traditional terrestrial sense of plasma populations with energy content high compared with that supplied by local sources (as distinct from merely a synonym for the magnetodisk) and *radiation belts* are structures that arise from strong adiabatic compression of the plasma and suprathermal particles, respectively. They therefore presuppose inward transport over appreciable distances and are most readily produced by an exterior boundary source. They can, however, be produced by an interior source if the plasma or particles are first transported outward to the boundary regions, heated or accelerated, and then transported (recirculated) back inward. A unique feature of radial diffusion is that it transports always in the direction of decreasing content; thus it is capable of such bidirectional transport if the radial gradients at low and high energies are opposite.

(4) The *plasmasphere* is usually understood as the region filled with the plasma coming primarily from the ionosphere. Because of the unfavorable area to volume ratio of the ionospheric source, high densities in the magnetosphere can be reached only if the source acts for a long time; hence the outward transport from the region

must be negligible (if this is the case, the reason is usually the local predominance of corotation).

10.6 Scaling relations for magnetospheres

The comparative study of magnetospheres is a very broad topic, both qualitatively (regarding the diversity of phenomena) and quantitatively (regarding the wide range of the numerical values of parameters). For a systematic comparison and classification, one would like to identify a few key parameters characterizing each magnetosphere and its environment; from these, one would like to be able to decide which differences among magnetospheres are essential, in the sense that they arise from different physical processes, and which differences result merely from the same process acting at different scales. Also, if such distinctions are to be made on the basis that some particular parameter is large or small, the parameter in question must be expressed in a dimensionless form, for the designation "large" or "small" is meaningful only when applied to a dimensionless quantity (whenever something is described as large or small, the question must always be asked: large or small compared to what?).

The parameters may be separated into two classes. The *input* or given parameters define the system; in principle they may be specified arbitrarily. The *derived* parameters describe the properties of the given system; they are determined by the relevant physical processes.

10.6.1 Input parameters

The physical parameters that may be assumed given for a particular magnetosphere fall naturally into three groups: those describing the solar wind or, more generally, the external medium, those describing the planet or central object, and those describing the given – as distinct from derived – properties of the magnetosphere itself.

(1) The **solar wind parameters** important for the interaction are the following (their dependence on radial distance r from the Sun is shown, with values at $r = 1$ AU designated by the subscript a):

mass density, $\rho_{sw} \simeq \rho_a (1 \text{ AU}/r)^2$,

bulk flow speed, $v_{sw} \simeq v_a$,

magnetic field transverse to the radius vector from the Sun, $B_{sw-t} \simeq B_a (1 \text{ AU}/r)$,

radial magnetic field, $B_{sw-r} \simeq B_a (1 \text{ AU}/r)^2$

(the radial magnetic field is generally of minor significance for studies of magnetospheres, because $B_{sw-r} \ll B_{sw-t}$ for magnetospheres located beyond $r = 1$ AU and, more importantly, because the radial field is aligned with the bulk flow and thus does not contribute to the $-\mathbf{v} \times \mathbf{B}$ electric field of the MHD approximation).

Not included in the above list is the temperature of the solar wind, which has been largely "forgotten" at the bow shock of the magnetosphere; the post-shock temperature, equal to some fraction of the solar wind bulk-flow kinetic energy per particle, is essentially independent of the pre-shock temperature value, and the latter therefore plays no significant role in the interaction (except possibly in the distant downstream regions). Note that an important dimensionless parameter, the Alfvén Mach number based on the transverse magnetic field,

$$M_A = \left(\frac{4\pi \rho_{sw} v_{sw}^2}{B_{sw-t}^2} \right)^{1/2} \approx \left(\frac{4\pi \rho_a v_a^2}{B_a^2} \right)^{1/2} \tag{10.34}$$

is independent of the distance from the Sun and hence has the same value, on average, at all the planets.

(2) The **planetary parameters** relevant to the magnetosphere and the interaction with the solar wind include

- radius of the planet, R_p,
- rotation frequency, Ω,
- surface magnetic field strength at the equator, B_p,
- magnetic dipole moment, $\mu = B_p R_p^3$,
- typical value of Pedersen conductance in the ionosphere, Σ_P,
- mass of the planet, M_p (important only for gravitational effects).

For describing most aspects of the magnetosphere, the parameters B_p and R_p appear only in the combination $B_p R_p^3 = \mu$; the values of B_p and R_p taken separately are of little or no significance (e.g. Vasyliūnas, 1989a, 2004). Also, M_p hardly ever appears explicitly in any magnetospheric calculation.

(3) **Magnetospheric parameters** are not always easily classified as given or derived. Most of the simple parameters one can construct to represent the characteristic or global properties of a magnetosphere are to a large extent determined by the interaction with the solar wind and can hardly be considered as specifiable a priori. If the magnetosphere has an interior source of plasma associated with a single moon, e.g. Io at Jupiter, the

- radial distance of the interior source, r_s,

is one obvious given parameter. Aside from that, almost the only other magnetospheric parameter that is generally treated as given is the

- strength of the interior source of plasma, S

(Section 10.5.1). The theoretical arguments for this are rather inconclusive: on the one hand, it may be argued that the interior source is governed primarily by the injection of neutral particles and is independent of the magnetosphere to first order, while on the other hand one may point to plasma effects on the ionization rate and even (via sputtering) on the injection rate. The real motivation for treating S as given appears to be primarily the availability of reasonably robust empirical estimates of S for the magnetospheres of Jupiter and Saturn.

In addition to the magnitudes of the various parameters listed above, one must also include the directions of the vector quantities: the solar wind velocity $\mathbf{v}_{sw}$, the interplanetary magnetic field $\mathbf{B}_{sw}$, the planetary rotation axis $\mathbf{\Omega}$, and the magnetic moment $\boldsymbol{\mu}$.

10.6.2 Derived parameters

Despite the almost boundless complexity in detail of magnetospheres, several simple parameters that prove useful in characterizing their essential gross properties can be defined; some have been mentioned in earlier sections. Specifically, we now consider the following *derived parameters*:

- distance to the sub-solar magnetopause, R_{MP} (Section 10.3.1),
- effective length of the magnetotail, $\mathcal{L}_T$ (Section 10.3.3),
- amount of open magnetic flux, Φ_M (Section 10.3.3),
- EMF across the polar cap, $c\Phi_{PC}$ (Section 10.4.3),
- rate of energy dissipation in the magnetosphere, $\mathcal{P}$,
- mass input from the solar wind, S_{sw} (Section 10.5.1),
- limiting distance of corotation, R_{cr} (Section 10.4.4).

The question now is whether scaling relations can be found that allow the derived parameters to be expressed in terms of the input parameters.

10.6.3 Constraints from dimensional analysis

Two guiding principles constrain the search for scaling relations. The first is the principle of dimensional similitude: if a derived (i.e. dependent, calculated) parameter is to be expressed as a combination of input (independent, given) parameters, that combination must have the same physical dimensions as the dependent parameter. The second is the principle of physical relevance: parameters associated with a particular physical process should appear only if that process plays a significant role.

The first principle imposes a specific functional form for each derived parameter: it must be proportional to a dimensionally correct combination of the given parameters, multiplied by a function of any or all dimensionless quantities formed from the given parameters, including the relative angles between the vectors (the form of the function is not constrained by dimensional analysis). The second principle, less specific than the first, is of help in deciding which of the many dimensionless quantities that can be formed are to be included.

The basic length scale defined by a combination of solar wind and planetary parameters is the Chapman–Ferraro distance R_{CF} (see Eq. (10.1) and the following discussion). The dependences of the derived parameters of Section 10.6.2 can then be expressed as follows:

$$R_{MP} \sim R_{CF}\Psi_{MP}, \tag{10.35}$$

$$\mathcal{L}_T \sim R_{CF}\Psi_T, \tag{10.36}$$

$$\Phi_M \sim B_{sw}R_{CF}{}^2\Psi_M, \tag{10.37}$$

$$S_{sw} \sim \rho_{sw}v_{sw}R_{CF}{}^2\Psi_S, \tag{10.38}$$

$$c\Phi_{PC} \sim v_{sw}B_{sw}R_{CF}\Psi_{PC}, \tag{10.39}$$

$$\mathcal{P} \sim \rho_{sw}v_{sw}{}^3R_{CF}{}^2\Psi_{\mathcal{P}}, \tag{10.40}$$

$$R_{cr} \sim R_{CF}\Psi_{cr}, \tag{10.41}$$

where the Ψs are (dimensionless) functions of the appropriate dimensionless quantities. Not all the Ψs are independent: Eqs. (10.16) and (10.18) imply that

$$c\Phi_{PC} \sim \frac{v_{sw}}{\mathcal{L}_T}\Phi_M, \quad \text{which is equivalent to} \quad \Psi_{PC}\Psi_T \sim \Psi_M. \tag{10.42}$$

Equations (10.35)–(10.41) represent the derived parameters as dimensioned scale factors multiplied by dimensionless proportionality constants; the form of the equations has been chosen so that the dimensioned scale factors depend only on the given solar wind parameters and on the Chapman–Ferraro distance R_{CF}, which has been taken as the scale factor for the parameters with the dimension of length. For all the other derived parameters, the scale factors are simply the corresponding solar wind values: for $c\Phi_{PC}$, the solar wind electric potential over a distance R_{CF}; for Φ_M, S_{sw}, and $\mathcal{P}$, the interplanetary magnetic flux, the solar wind mass flux, and the solar wind bulk-flow kinetic energy flux, respectively, through an area $R_{CF}{}^2$. Everything to do with the magnetosphere is thus contained in the dimensionless proportionality constants, the Ψs, which are functions of the relevant dimensionless quantities.

With the exception of the angles and the Alfvén Mach number M_A, most of the dimensionless quantities have no universally used symbols; we will designate them simply as Q_1, Q_2, ... and choose between a quantity and its inverse in such a way as to make the dimensionless quantity small when the associated physical process is not important.

Of the angles between the vectors, the most important is the angle θ between $\mathbf{B}_{sw}$ and $\boldsymbol{\mu}$; empirically, θ is found to be the dominant factor for determining the values of Ψ_M, Ψ_{PC}, and Ψ_P in the magnetosphere of Earth. Aside from the angles, the only dimensionless quantity that can be formed from the solar wind parameters alone is the Alfvén Mach number, Eq. (10.34); for magnetospheric purposes, the inverse quantity

$$M_A^{-1} = \left(\frac{B_{sw}^2}{4\pi \rho_{sw} v_{sw}^2}\right)^{1/2} \equiv Q_1 \qquad (10.43)$$

is more informative: M_A^{-1} represents the influence of the interplanetary magnetic field and is thus, despite its typically small numerical value, $\sim O(10^{-1})$ (see Table 9.1), always important if the magnetosphere is open.

Of the planetary parameters, the only one that appears explicitly in Eqs. (10.35)–(10.41) is the magnetic moment μ, through its role in determining the Chapman–Ferraro distance R_{CF}. Any other planetary parameters that have magnetospheric effects must therefore appear as factors in the dimensionless arguments of the Ψs. No dimensionless quantity that is relevant to the magnetosphere, however, can be formed from the planetary parameters alone: they must be taken together with solar wind parameters. The electrodynamic influence of the ionosphere can be represented by the dimensionless quantity

$$\frac{4\pi \Sigma_P v_{sw}}{c^2} \equiv Q_2 \qquad (10.44)$$

and that of the planetary rotation by

$$\frac{\Omega R_{CF}}{v_{sw}} \equiv Q_3 . \qquad (10.45)$$

Finally, a dimensionless quantity that involves, of the planetary parameters, only the dipole moment is

$$\left(\frac{e}{mc} \frac{B_{sw} R_{CF}}{v_{sw}}\right)^{-1} \equiv Q_4 \qquad (10.46)$$

where e and m are the charge and mass of the proton. In essence, this is the ratio of the kinetic energy per unit charge of the solar wind bulk flow and the solar wind electric potential across the size of the magnetosphere, but it can also be related (Vasyliūnas *et al.*, 1982) to the comparative importance of the gradient and curvature as against $\mathbf{E} \times \mathbf{B}$ drifts, as well as to other manifestations of finite gyroradius effects, in particular an effective viscosity. The essentially new aspect of the quantity Q_4 is the explicit appearance of the particle mass and charge, implying that Q_4 is important only if non-MHD processes play a significant role (the MHD equations do not contain any reference to individual particle properties). Its numerical value is generally quite small ($\sim O(10^{-2})$ for Earth and much smaller for Jupiter and Saturn); its possible importance for Mercury was suggested by Ogilvie *et al.* (1977).

From the magnetospheric given parameters, three dimensionless quantities can be formed. The first, somewhat trivial, is the ratio of r_s and R_{CF} (or some other length scale),

$$\frac{r_s}{R_{CF}} \equiv Q_5. \tag{10.47}$$

The second is the ratio

$$\frac{S}{\rho_{sw} v_{sw} R_{CF}^2} \equiv Q_6 \tag{10.48}$$

of the interior source strength and the solar wind mass flux through an area roughly comparable with the cross section of the magnetosphere; it has proved useful in describing the relative importance of interior sources (e.g. Vasyliūnas, 1989a). The third (little known, recently introduced in Vasyliūnas, 2008) is

$$\frac{S \Omega r_s^5}{\mu^2} \equiv Q_7 \tag{10.49}$$

(in this form it may look rather incomprehensible, but writing it as

$$S \left\{ \left[(\mu/r_s^3)^2 / (\Omega r_s)^2 \right] (\Omega r_s) (r_s^2) \right\}^{-1} \tag{10.50}$$

should make its derivation apparent); it contains magnetospheric and planetary but no solar wind parameters and, as discussed in Section 10.6.4 below, is closely related to the limits on corotation.

10.6.4 Interrelationships of parameters

We have now six independent dimensionless quantities, Q_1–Q_6, defined by Eqs. (10.43)–(10.48); Q_7, defined by Eq. (10.49), can be shown to satisfy $Q_7 \sim Q_6 Q_3 (Q_5)^5$. Obviously, if the proportionality constants were to depend on all six quantities at comparable levels of importance, dimensional analysis would not impose any significant constraints on the derived parameters. Some dimensionless quantities may be neglected, however, if the associated physical processes play only a minor role.

In the simplest case (at the level of approximation discussed in Section 10.3) of an MHD interaction between the solar wind and the planetary magnetic dipole, the only dimensionless quantity (aside from angles) that is relevant is $Q_1 = M_A^{-1}$. The set of all magnetospheres can be approximated in this limit as a two-parameter family: the proportionality constants in the scaling relations (10.35)–(10.41) depend only on M_A^{-1} and on $\hat{\mathbf{B}}_{sw} \cdot \hat{\boldsymbol{\mu}}$ (for simplicity, any dipole tilt away from perpendicular to the solar wind flow is ignored here). Plausible rough dependences are suggested by the empirical results for Earth's magnetosphere (see e.g. Chapter 11); for the derived parameters that can be discussed at this level of approximation, they are

$$\Psi_{MP} \sim 1 - O(10^{-1})F, \tag{10.51}$$

$$\Psi_{T} \sim O(M_A), \tag{10.52}$$

$$\Psi_{M} \sim O(M_A)F, \tag{10.53}$$

$$\Psi_{S} \sim O(M_A^{-3}) \tag{10.54}$$

where F (here and in the following equations) represents $F(\hat{\mathbf{B}}_{sw} \cdot \hat{\mu})$, the function for the angle dependence, which is frequently approximated as follows:

$$F(x) = \begin{cases} x, & x > 0, \\ 0, & x < 0. \end{cases} \tag{10.55}$$

The remaining derived parameters, $c\Phi_{PC}$, $\mathcal{P}$, and R_{cr}, are meaningful only if plasma flow and ionospheric effects are included (at the level discussed in Section 10.4), which brings in Q_2 and Q_3 as additional dimensionless arguments on which the proportionality constants depend; the set of magnetospheres is now a four-parameter family. The derived parameter primarily affected by this extension is the emf across the polar cap, $c\Phi_{PC}$, which is observed to approach a limiting value as $v_{sw}B_{sw}$ increases (Siscoe *et al.*, 2002c, and references therein), a saturation effect predicted as a result of ionospheric currents by Hill *et al.* (1976); the model developed by Siscoe *et al.* (2002c) is equivalent to setting

$$\Psi_{PC} \sim O(1)\frac{F}{1 + 1/(Q_1 Q_2 F)}; \tag{10.56}$$

note that

$$Q_1 Q_2 = \frac{4\pi \Sigma_P v_A}{c^2}. \tag{10.57}$$

For the energy dissipation rate $\mathcal{P}$, the two most widely used empirically based models are the Burton–McPherron–Russell equation (Burton *et al.*, 1975) and the so-called ϵ parameter (Perrault and Akasofu, 1978), both first put into a dimensionally correct form by Vasyliūnas *et al.* (1982). The equivalent values of $\Psi_\mathcal{P}$ are given by Eqs. (10.58) and (10.59), respectively:

$$\Psi_\mathcal{P} \sim O\left(M_A^{-1}\right) F \tag{10.58}$$

$$\sim O\left(M_A^{-2}\right) F', \tag{10.59}$$

F being taken from Eq. (10.55) for the first form and a somewhat modified, smoothed, function F' for the second. Both values of $\Psi_\mathcal{P}$ depend (in slightly different ways) on M_A^{-1} and on angles only, with no explicit reference to ionospheric parameters.

As the final step of increasing complexity, one may add the interior plasma sources (at the level of approximation discussed in Section 10.5), bringing in Q_5 and Q_6 as additional dimensionless quantities. With the large number of dimensionless quantities now included, dimensional analysis as a predictive tool may be of limited power; however, it is also useful as a classifying tool, for encompassing the great variety of observed magnetospheres and magnetosphere-like systems that result from the diversity of plasma sources and transport processes.

One significant result of a strong interior source is to increase the size of the magnetosphere, scaled by the R_{MP}, to well above the value R_{CF} set by the pressure balance between the solar wind and the magnetospheric magnetic field, as the pressure of plasma inside the magnetosphere becomes larger; this effect is discussed in Section 10.3.1 and illustrated there in Figure 10.2. The increase in the ratio R_{MP}/R_{CF} as the result of internal plasma pressure can be considerably larger ($R_{MP}/R_{CF} \sim 2$–3 is observed at Jupiter and Saturn) than the decrease associated with an open magnetosphere ($\sim O(10^{-1})$ is observed at Earth). To represent the effect quantitatively in dimensionless form, one may therefore neglect the influence of M_A^{-1} and replace Eq. (10.51) by

$$\frac{R_{MP}}{R_{CF}} \sim \Xi \left(\frac{S}{\rho_{sw} v_{sw} R_{CF}^2} \right) \tag{10.60}$$

where the function $\Xi(x)$ increases monotonically with increasing x from $\Xi(0) \sim 1$. (In this context, S need not be limited to the interior source from moons and planetary rings but may be taken as including any sources from the ionosphere.)

One may envisage a situation in which S becomes so large that the plasma pressure inside the magnetosphere is much larger than the magnetic pressure almost everywhere, to the point that the planetary magnetic field no longer plays any significant role in balancing the external pressure from the solar wind; according to the definitions of Section 10.2 we then have a magnetosphere-like system rather than a true magnetosphere. We may still designate the distance to the interface with the solar wind as R_{MP}, but its value as given by Eq. (10.60) obviously must be independent of R_{CF} (which in this limit plays no role). This requires that

$$\Xi(x) \to \sqrt{\frac{x}{\zeta}} \quad \text{as} \quad x \to \infty \tag{10.61}$$

where ζ is a constant, equal to the limiting value of $S/(\rho_{sw} v_{sw} R_{MP}^2)$. It is easily shown that, for a pressure balance between the solar wind and plasma flowing outward with speed v from a localized source, $\zeta \sim v_{sw}/v$.

It is convenient to rewrite Eq. (10.60) by normalizing S to the solar wind mass flux through the observed distance R_{MP}^2 instead of the theoretical distance R_{CF}^2

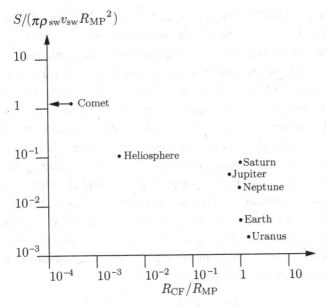

Fig. 10.9. Normalized interior mass source vs. R_{CF}/R_{MP}. The parameter values are mostly taken from Chapter 13. The values for a comet and for the heliosphere are only very rough estimates. The distance R_{CF} for the Sun (admittedly a somewhat artificial concept) was calculated by balancing a dipole field of 1 G at the solar surface against an interstellar pressure that stops the solar wind at ~ 100 AU.

and by inverting to give S as a function of $1/R_{MP}$:

$$\frac{S}{\rho_{sw} v_{sw} R_{MP}^2} \sim \Upsilon\left(\frac{R_{MP}}{R_{CF}}\right) \tag{10.62}$$

where the function $\Upsilon(x)$ has the following limiting values:

$$\Upsilon(x) \to \zeta \quad \text{as} \quad x \to 0, \qquad \Upsilon(x) \to 0 \quad \text{as} \quad x \to 1. \tag{10.63}$$

A plot of the two variables of Eq. (10.62) for various observed or inferred magnetospheres and magnetosphere-like systems is shown in Fig. 10.9. The true magnetospheres are clustered near $R_{CF} \simeq R_{MP}$ and spread upward and to the left ($R_{CF} < R_{MP}$ as the interior source strength increases). On the far left ($R_{CF} \ll R_{MP}$) are magnetosphere-like systems of the type exemplified by a cometary interaction with the solar wind. (Magnetosphere-like systems of the other type, exemplified by Venus and Mars, in which the solar wind interacts directly with the ionosphere or atmosphere, may be included in this diagram at the far right, where $R_{CF} \gg R_{MP}$.)

The heliosphere may be viewed as a magnetosphere-like system in which a central object (the Sun) interacts with an external medium (the interstellar gas and plasma); the solar wind now plays the role of the internal plasma of the system. The interaction is in some ways similar to that of a comet with the solar wind:

the pressure to hold off the external medium is provided primarily by the outflow of plasma from the central object. The Sun differs from a comet in possessing a magnetic field, but the stresses of the interplanetary magnetic field are negligible in comparison with the dynamic pressure of the solar wind. From Fig. 10.9, the heliosphere is in fact intermediate between Jupiter and a comet: the interaction with the interstellar medium may be viewed either as *Jupiter-like* but with the interior source greatly enhanced (to the point of dominating dynamically over the magnetic field) or as *comet-like* but with the addition of a magnetic field from the central object (a field not strong enough, however, to have a significant dynamical effect).

The final topic in which the interior plasma source plays a major role is that of providing a limit to corotation. This was extensively discussed in Section 10.4.4, and the result obtained was that the limiting distance of corotation R_{cr} is the smallest of the distances R_c (Eq. 10.20), R_H (Eq. 10.21), and R_0 (Eq. 10.24); here we simply recast these distances in the appropriate dimensionless form. In terms of the dimensionless quantities $Q_1, Q_2, \ldots$ we have,

$$\frac{R_c}{R_{CF}} \sim \left(\frac{Q_3}{Q_1}\right)^{1/\nu}, \qquad (10.64)$$

$$\frac{R_H}{R_{CF}} \sim \left(\frac{Q_2}{Q_6}\right)^{1/4}, \qquad (10.65)$$

$$\frac{R_0}{R_{CF}} \sim (Q_3 Q_5 Q_6 \Omega \tau_{tr})^{-1/4} \qquad (10.66)$$

where ν is the exponent introduced in Eq. (10.20) and τ_{tr} is the global transport time (Section 10.5.2). The distances R_H and R_0 can also be expressed in dimensionless form in another way, with the use of only the planetary and magnetospheric given parameters:

$$\frac{R_H}{r_s} \sim \left[\left(\frac{4\pi \Sigma_P \Omega r_s}{c^2}\right)\left(\frac{1}{Q_7}\right)\right]^{1/4}, \qquad (10.67)$$

$$\frac{R_0}{r_s} \sim (Q_7 \Omega \tau_{tr})^{-1/4}. \qquad (10.68)$$

This was to be expected, since the physical processes associated with these distances are purely internal to the magnetosphere and do not involve the solar wind.

11

Solar-wind–magnetosphere coupling:
an MHD perspective

FRANK R. TOFFOLETTO AND GEORGE L. SISCOE

11.1 Introduction

In Chapter 10 the basic principles that govern magnetospheric structure and dynamics in general were explained. There it was made clear that the magnetosphere is a time-dependent, three-dimensional, multi-component, interacting system. The point of the present chapter is that to obtain a quantitative description of so complex a system, one must resort to numerical models. Moreover, to check the self-consistency of the assumptions behind a conceptual model, one must rely on global numerical simulations. In this chapter we use numerical simulation models that are generally available to illustrate their use in giving global-scale depictions of the structure and dynamics of the terrestrial magnetosphere and we compare the results with our understanding of these subjects developed by other means.

We consider the flow of mass, momentum, energy, and magnetic flux through and around the magnetosphere. These topics involve the macro-scale coupling between the solar wind and the magnetosphere and so are well suited for treatment by global MHD simulations. We start by explaining where numerical models are positioned in the magnetospheric scientist's armory and by briefly describing the illustrative models that we will use.

It comes as no surprise that of the known magnetospheres Earth's is the best observed and understood. What we know about its structure, dynamics, and interaction with the solar wind has been gleaned from decades of ground- and space-based observations, each of which gives us a snapshot of one or more physical processes in the magnetosphere. Despite years of accumulating data, the huge volume of the magnetosphere means that data coverage has nonetheless been sparse, although recent efforts to use fleets of spacecraft and imagers have resulted in significant advances (Burch, 2005; Goldstein *et al.*, 2006). Data paucity can lead to differing physical interpretations, for example, those pertaining to magnetospheric substorms (e.g. Kennel, 1995). That said, it is still the case that the many observations that have been made and their interpretations have produced a fairly comprehensive picture of the Earth's magnetosphere, which, as mentioned, shows it to be both complex and dynamic. The lower panel of Fig. 1.3 is a conceptual synthesis of the magnetosphere made from an accumulation of data snapshots and by applying the general principles discussed in Chapter 10. It shows various plasma regions in an open-magnetosphere version of Fig. 10.1. The boundaries of this system – the solar wind and the ionosphere – are also complex and variable. It is well established that the solar wind plays a pivotal external role in driving and controlling the magnetosphere and that the ionosphere acts as an internal mass source and sink and as a regulator of energy and momentum coupling. So we are dealing with a complex object, the magnetosphere animated and constrained by complex boundary conditions and, in the face of this complexity, we are trying to construct a global picture of it out of a collection of observations that greatly undersamples the system.

The advent of powerful numerical simulations has helped to compensate for incomplete data coverage. They have become the tool of choice for sorting out cause and effect relationships in the vast magnetospheric system and for clarifying the magnetosphere's coupling to the solar wind driver and to ionospheric sources and sinks. Just as increased sophistication and abundance of data on the magnetosphere have enabled increased sophistication in magnetospheric theory, so advances in numerical methods and computer power have elevated numerical simulations to the point where now they play a role complementary to data acquisition in advancing the understanding of magnetospheric structure and processes.

We should acknowledge that analytical theory continues to be an important tool in the theorist's toolbox, both as a way of confirming numerical solutions and as a means of providing fundamental understanding of the basic physics. Chapter 10 demonstrated the power of the analytical approach to discussing the properties of magnetospheres in general. In addition, many excellent articles (e.g. Hill, 1983; Walker and Russell, 1995) have been written that describe in detail the basic physics of how the solar wind and Earth's magnetosphere and ionosphere interact. As mentioned earlier, this chapter focuses on the numerical modeling part of the magnetospheric scientist's armory.

11.2 Global MHD models

Our goal here is to demonstrate the value of global MHD simulations as a resource for acquiring an understanding of the structure of the whole magnetosphere and of the processes that take place in and around it. This demonstration should also make clear that MHD simulations are powerful tools for magnetospheric research. But one should also be made aware at the outset that numerically based magnetospheric models have reached the status of valuable resources and powerful research tools only after a long development and testing process. At first the space physics community received the advent of such models with some skepticism. Much of this was directed towards global MHD models in which the validity of the MHD approximation when applied to the whole magnetosphere was assumed. It is well known that there are regions of the magnetosphere where MHD is not applicable, such as the inner magnetosphere (Wolf *et al.*, 2007) and reconnection regions on the magnetopause and in the tail (Birn *et al.*, 2001). However, efforts have been made to augment MHD with other physics models and approximations, and these have met with some success (e.g. Raeder *et al.*, 2001b; Wiltberger *et al.*, 2004; De Zeeuw *et al.*, 2004; Taktakishvili *et al.*, 2007). A second issue related to numerical models is the validity of their solutions, since in all cases one must make some compromise between accuracy and robustness, which could in some cases introduce unexpected or non-physical results (Raeder, 2000; Gombosi *et al.*, 2000). Then there is the difficulty factor; constructing numerical codes to solve the MHD equations poses challenges for the modeler, especially codes for the magnetosphere with its vast variations in temporal and spatial scales. Nonetheless, the significant accomplishment of the modelers has been to develop these codes into valuable research tools, as we shall see.

In the following references numerical methods for solving the MHD equations and the validity of the results, which for some is still an open question, are discussed (Huba and Lyon, 1999; Tóth, 2000; Tóth *et al.*, 2006; Sokolov *et al.*, 2006; Lankalapalli *et al.*, 2007). But comparisons between model results and data, which started in the 1980s with the International Solar Terrestrial Program (Papadopoulos *et al.*, 1999), coupled with ongoing model improvements have led to the general acceptance and use of global MHD models within the community (e.g. Raeder *et al.*, 1998; Slinker *et al.*, 1999; Wiltberger *et al.*, 2003; Siscoe *et al.*, 2004; Papadopoulos *et al.*, 1999; Guild *et al.*, 2004; Ohtani and Raeder, 2004; Ridley and Kihn, 2004; Rastatter *et al.*, 2005).

We will use two community-accessible global MHD numerical simulation models to examine various large-scale physical properties of Earth's magnetosphere and to compare their results with traditional ideas about global magnetospheric behavior. One reason for using two models is to separate results that are common to both from results that depend on detailed computational properties of the individual codes (the grid structure, numerical techniques, etc.). In addition, we will use one model or the other according to which is better tailored for a particular application. For simplicity, only cases of steady solar wind input will be treated here. Moreover, space limitations allow us to consider only a few specific physical examples. In particular, we will consider only the normal range of solar wind conditions and omit any discussion of extreme conditions such those producing magnetic storms. The latter topic will be treated in Vol. II of this series. The output of the models is web-accessible through a graphics interface at the Community Coordinated Modeling Center (CCMC, http://ccmc.gsfc.nasa.gov), which allows the user to select parameters and views to suit his or her interest. Thus, the reader can reproduce and multiply any example given here.

The models are the University of New Hampshire OpenGGCM code (Raeder *et al.*, 2001b) and the University of Michigan BATS-R-US Global MHD code (Gombosi *et al.*, 2004). The OpenGGCM code was first developed in the early 1990s and the BATS-R-US in the late 1990s. As mentioned, the OpenGGCM and BATS-R-US model results were obtained from the Coordinated Community Modeling Center (CCMC) education runs, which are available on the world wide web. The codes use different grids and numerical schemes, which are described in detail in the references above. All model results presented here were interpolated onto a uniform Cartesian grid of spacing $0.5 R_E$; so some of the differences may be attributable to issues related to this interpolation rather than differences in the codes themselves. Some other differences between the model's results are likely to be due to the different assumptions and numerical methods used. However, as will be shown in this chapter, the models do in fact agree reasonably well.

People have wondered whether collisionless magnetic reconnection, which is a non-MHD process, might invalidate MHD simulations of magnetospheric dynamics and, if so, to what extent they would be invalidated. This is an important consideration since collisionless magnetic reconnection appears to be an essential process in the coupling between the solar wind and the magnetosphere and in the phenomenon of substorms. Collisionless magnetic reconnection involves physical processes that are not included in the MHD equations and that operate on spatial scales smaller those resolved by MHD codes. There are three possibilities, that the microphysics of collisionless reconnection determines the rate of reconnection, or that macro-scale dynamics determine the rate of reconnection, or that some combination of microphysics and macrophysics determines the rate of reconnection. If the first case applies, to obtain valid MHD simulations it would be essential to incorporate collisionless reconnection microphysics correctly in the MHD codes or to parameterize it correctly in terms of MHD variables as a sub-gridscale process.

How to parameterize correctly the microphysics of collisionless reconnection in an MHD code was the subject of a campaign in the late 1990s involving theorists and modelers from many institutions. The results were reported in the March 2001 issue of the *Journal of Geophysical Research*. The campaign established that the Hall effect, which is not included in the usual MHD equations, is a critical

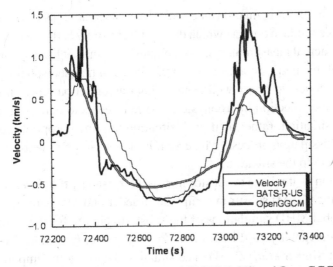

Fig. 11.1. Measured (DMSP) and computed (BATS-R-US and OpenGGCM) iono-spheric plasma velocities over the polar cap, showing a particularly good match. The figure was prepared as part of a space-weather metrics campaign.

ingredient in determining collisionless reconnection rates (Birn *et al.*, 2001). The question, therefore, is this. In the absence of the correct collisionless reconnection parameterization, how well do global MHD codes nonetheless simulate magnetospheric processes that depend on reconnection? Since each global MHD code treats differently the dissipation process that produces magnetic reconnection (from simple, uncontrolled, numerical dissipation to algorithmically prescribed dissipation), one would expect the code outputs to differ widely if the microphysics of reconnection were crucial to determining the reconnection rate on a global scale.

Reconnection appears to be centrally important to magnetospheric convection and, as mentioned, to substorms. The question under consideration can therefore be restated in two parts. How well do MHD codes simulate magnetospheric convection? And how well do they simulate substorms? The question regarding magnetospheric convection was addressed by the "The First Magnetospheric Space Weather Metrics Challenge" in 1990–1991 (Hesse, 1991). Users of five different global MHD simulation codes were asked to reproduce the convection velocity across the polar cap – a direct gauge of magnetospheric convection – as measured by a polar-orbiting satellite. A number of profiles were required to be fitted. Figure 11.1 shows a particularly good fit between the observed velocities and those simulated by the two MHD codes used in this chapter. Most of the other codes did as well as the two shown here. Even the not-so-good fits are recognizable as fits to the data. The point is that, despite their different treatments of dissipation, the different codes produce outputs which do not differ widely and which come close to matching the actual measurements. The conclusion is that, at least in simulating the rate of global magnetic reconnection as manifested in magnetospheric convection, global MHD codes do pretty well even without a correct parameterization of the collisionless reconnection microphysics. The rate seems to depend instead on the macro-scale situation, rather as if each situation has an appropriate reconnection rate and the dissipation process, whatever it is, adjusts to provide the reconnection rate appropriate to the situation.

The question as it relates to substorms was addressed by the Geospace Environment Modeling (GEM) substorm campaign (Raeder and Maynard, 2001). Results from two global MHD simulations of the substorm chosen for the campaign were published in the January 2001 issue of the *Journal of Geophysical Research* (Raeder *et al.*, 2001a; Slinker *et al.*, 2001). The situation regarding the importance of the correct parameterization of collisionless reconnection is not as clear here as it is for magnetospheric convection, although studies independent of those involved in the GEM substorm campaign indicate that the actual microscopic physics might substantially influence reconnection rates (Birn and Hesse, 2001; Hesse *et al.*, 2001; Otto, 2001). This view is supported by the Raeder *et al.* (2001a) simulation. Nonetheless, all the substorm features and behavior that a global MHD simulation is capable of reproducing were reproduced in the two campaign simulations, but the timing, reconnection rates, and localization of reconnection were incorrect. As an indication of the importance of the treatment of magnetic reconnection, when simulating substorms, we note that one of the two codes produced a substorm with no explicit dissipation whereas the other code required a dissipation algorithm to produce a substorm. In view of the presently evolving capability of global MHD codes to simulate substorm phenomena, we will leave this aspect of global MHD simulations for some future treatment.

11.3 The solar wind at Earth

To understand what drives the magnetosphere it is useful first to look at the basic physical characteristics of the solar wind near Earth, as they determine much of the nature of the coupling and properties of the magnetosphere. The range of typical solar wind conditions is shown in Table 11.1, which is based on Feldman *et al.* (1977) and adopted from Wolf (1995).

Adopting typical solar wind values of 5 particles/cm^3 for the density, 400 km/s for the solar wind speed, and 5 nT for the magnetic field strength (values consistent with Table 11.1), we can evaluate the expected energy density of the solar wind. This can be broken down into three components, flow, magnetic, and thermal. The

Table 11.1. *Solar wind parameters*

Parameter	Mean	σ	median	5%–95% range limit	Observed min. and max. values
Number density (cm^{-3})	8.7	6.6	6.9	3–20	0.1–83
Velocity (km/s)	468	116	442	320–710	250–2000[a]
Ram pressure ρv^2 (nPa)	7	5.2	5.5	0.01–14.5	0.05[b]–28
Magnetic field (nT)	6.2	2.9	5.6	2.2–9.9	0[c]–85
Ion temp. (K)	1.2×10^5	0.9×10^5	1.0×10^5	0.1–3×10^5	10^4–3×10^5[d]
Electron temp. (K)	1.4×10^5	0.4×10^5	1.3×10^5	0.9–2×10^5	10^5–2×10^5[d]

[a] From Cliver *et al.* (1990). Measurements of high velocities in the Hapgood study were limited by instrumental effects.
[b] Taken from Smith, C. W. *et al.* (2001b).
[c] Indicates at least one interval with $B < 0.1$ nT.
[d] The upper limit is probably greater than this.

above values were adopted for the model results shown below. The flow energy density is estimated to be

$$\left(\frac{1}{2}\rho v^2\right)_{sw} \approx \frac{1}{2}m_{proton}nv_{sw}^2 \approx 7 \times 10^{-10} \left(\frac{n(cm^{-3})}{5}\right)\left(\frac{v(km/s)}{400}\right)^2 J\,m^{-3}$$
(11.1)

where n is the solar wind density in particles/cm^3 and v is the speed in km/s. The energy density of the solar wind's magnetic field is

$$\left\langle\frac{B^2}{2\mu_0}\right\rangle \approx 1.0 \times 10^{-11} \left(\frac{B(nT)}{5}\right)^2 J\,m^{-3} \approx 0.015 \left(\frac{1}{2}\rho v^2\right)_{sw},$$
(11.2)

where B is the magnetic field strength in nT, while the thermal energy density using values from Table 11.1 is

$$\left\langle\frac{3}{2}nk(T_e + T_i)\right\rangle \approx 2.5 \times 10^{-11} \left(\frac{n(cm^{-3})}{6}\right)\left(\frac{T_e(K)}{1.2 \times 10^5} + \frac{T_i(K)}{1.4 \times 10^5}\right) J\,m^{-3}$$

$$\approx 0.03 \left(\frac{1}{2}\rho v^2\right)_{sw}$$
(11.3)

where $T_{i,e}$ are the solar wind ion and electron temperatures. Taking the most probable values from Table 11.1, the above estimates show that the bulk of the energy in the solar wind at Earth is in the flow.

Depending upon whether the interplanetary magnetic field (IMF) is directed away from or towards the Earth, the observed interplanetary magnetic field near Earth tends to make an angle of $\sim 45°$ or $\sim 225°$ with the Sun–Earth direction.

The geo-effective north–south component of the IMF, which controls the rate of magnetic reconnection at the Earth, averages near zero and fluctuates on short time scales.

11.4 Magnetosheath modeling

In this section we will demonstrate the use of MHD simulations to give global quantitative information about the properties of the Earth's magnetosheath and will compare the results with analytical calculations where available. Recall from Chapter 10 that the supersonic super-Alfvénic solar wind interacts with the Earth's magnetic field to form a bow shock between the Earth and the Sun. The region of shocked solar wind sandwiched between the bow shock and the magnetopause is the magnetosheath (shown in Figs. 1.3 and 10.1). The distance between the bow shock and the nose of the magnetopause decreases as the Mach number of the solar wind increases, dropping from essentially infinity when the Mach number is unity to a minimum of approximately 20 percent of the radius of curvature of the magnetopause at the nose when the Mach number exceeds about 4 (Russell and Mulligan, 2002). The applicable solar wind Mach number in the MHD case is the geometric mean of the sonic and Alfvénic Mach numbers, called the magnetosonic Mach number, which in magnitude is usually 4 or greater. Thus, according to the rule just stated, the typical shock standoff distance should be about 20 percent of the radius of curvature of the magnetopause at the nose, which is typically about $15 R_E$, where R_E is the Earth's radius. The typical shock standoff distance, therefore, is about $3 R_E$. As the solar wind in the magnetosheath moves downstream from the nose, the width of the magnetosheath increases and the flow accelerates essentially back to the solar wind speed.

Spreiter *et al.* (1968) developed the first quantitative models of solar wind flow around the magnetosphere. By adapting a code designed to model supersonic aerodynamic flow around missile nose cones, Spreiter *et al.* were able to model the flow around an axisymmetric magnetopause. The location of the bow shock was determined as part of the model calculation by assuming a parabolic shape. The magnetosheath magnetic field was computed by assuming that the field lines were straight in the unshocked solar wind and frozen into the magnetosheath fluid. The approach was also successfully applied to the magnetospheres of Venus and Mars and even to the exterior heliopause (Spreiter *et al.*, 1966; Spreiter *et al.*, 1970; Spreiter and Stahara, 1980; Stahara *et al.*, 1989).

In Fig. 11.2, results for the two MHD simulation codes are given along with those for the Spreiter model. The figure shows contours of density in the equatorial plane (we will ignore the interior density contours in the simulations for now). One's first impression is that, whereas the Spreiter contours are smooth and the

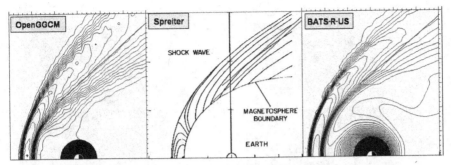

Fig. 11.2. The density contours for the two MHD simulation codes and the Spreiter analytical model, for comparison. The upstream solar wind conditions are the same for the two simulations (solar wind speed 400 km/s, density 5 protons/cm³, temperature 232 100 K, IMF 0 nT) and similar to those used in producing the Spreiter contours. The straight-line segments lie parallel to the ×2 density contour, which marks where the density in the magnetosheath is twice that in the solar wind.

boundaries sharp, the simulations have wiggly contours and diffuse boundaries. This, indeed, characterizes an important difference between analytical models and simulations. Wiggles and diffuseness in the simulations result from the finite grid resolution. Therefore, as the resolution increases with more powerful computers, so, too, will the wiggles reduce and the boundaries sharpen until, in principle, they are determined by real dissipation within the system rather than numerical dissipation within the code. Another marked difference is that the simulation contours are closed whereas after a certain point on the boundary the Spreiter contours are open. However, this is just another aspect of the diffuse boundaries of the simulation codes; the Spreiter contours are also closed but their return legs are hidden by being compressed into thin boundaries.

The simulations have the advantage of computing the shape of the magnetopause in a globally self-consistent way, whereas the Spreiter magnetopause is given by an analytical formula. Numerically, all three codes give about the same values for the contours, as can be seen by the straight-line segments in the simulation panels. These straight-line segments lie roughly parallel to the ×2 density contour, which marks a factor 2 greater density in the magnetosheath than in the solar wind. They are superimposed on the Spreiter figure, where they are seen to bracket the ×2 contour there also. Regarding the magnetosheath, therefore, the lessons here are that the simulations give global self-consistency but with irregularities determined by grid spacings and with diffuse boundaries determined by numerical dissipation, though ultimately by real dissipation. Outside regions where the gradients are set by numerical dissipation, the numerical values are reliable.

Magnetohydrodynamic simulations are particularly useful in displaying differences in global geometry resulting from different solar wind and IMF values.

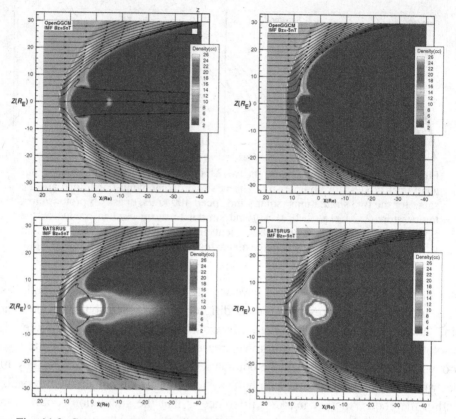

Fig. 11.3. Computed densities and flow streamlines in the noon–midnight plane obtained from from the OpenGGCM (upper panels) and the BATS-R-US (lower panels) global MHD codes, for the cases when the IMF B_z is 5 nT northward (left) or 5 nT southward (right). The black lines represent flow streamlines. See also the color-plate section.

Examples of MHD model flow solutions are shown in Fig. 11.3 (see also the color plate section). In each case, the solar wind speed was set to 400 km/s and the density to 5 particles/cm³; for the left-hand panels the IMF B_z was set to 5 nT and for the right-hand panels the IMF B_z was set to -5 nT. Figure 11.3 shows density and flow streamlines in the noon–midnight plane from the two global MHD models. The large densities seen in the BATS-R-US model for the interior magnetosphere are due to the density boundary conditions at the Earth (Zhang *et al.*, 2007). As in Fig. 11.2, the flow solutions resemble, to first order, many of the computed plots of Spreiter *et al.* (1968).

Figure 11.3 also shows that the codes agree on the following macro-scale properties.

(i) The dayside of the magnetosphere shrinks and the nightside expands when the IMF changes from northward to southward.

(ii) As a consequence, the magnetosphere as seen by the solar wind is blunter when the IMF is southward.

(iii) Therefore the density in the magnetosheath is higher over a wider north–south extent when the IMF is southward (the same holds in the east-west direction).

(iv) The dayside cusps move equatorward when the IMF changes from northward to southward, not only in distance because of dayside shrinkage but in angle relative to the Sun–Earth line.

The difference in size of the dayside and nightside magnetospheres reflects the transfer of magnetic flux from the dayside to the nightside when the IMF changes from northward to southward by the process of magnetospheric convection, as discussed in Sections 10.4 and 11.6.

Moreover, both codes agree that the solar wind does not flow directly into the tail through a "plasma mantle", as suggested in Fig. 1.3. The plasma–mantle mode of feeding solar wind into the plasma sheet, first proposed by Rosenbauer *et al.* (1975), has been widely held to be valid. However, MHD simulations consistently show that the mechanism does not work (Siscoe *et al.*, 2001), at least not in the straightforward way in which post-reconnection flow from the dayside intrudes into the vacuum of the tail through a slow-mode MHD expansion fan (Coroniti and Kennel, 1979; Sanchez *et al.*, 1990). Observations and simulations both show that a slow-mode expansion fan indeed exists but that it is not wide enough to overcome the radially diverging magnetosheath flow and so reach the plasma sheet (Siscoe *et al.*, 1994; Siscoe *et al.*, 1994, 2002a). Here MHD simulations have made a significant contribution to the way people think about plasma entry into the plasma sheet, which is a problem for which no solution has achieved community consensus at the time of writing.

One can learn more about basic magnetosheath properties using the codes by plotting key parameters along the stagnation streamline (at the axis of symmetry), as shown in Fig. 11.4 (the OpenGGCM code is used here because the upside of the wigglier contour lines in Fig. 11.2 is higher resolution). From the figure we learn the following.

(i) The shock standoff distance is, indeed, about $3R_E$ from the magnetopause, as aerodynamic texts lead one to expect, and it is bigger for the southward IMF case because here the magnetopause is blunter, as Fig. 11.3 shows, and so the radius of curvature at the nose is bigger.

(ii) The magnetopause is farthest from Earth in the northward IMF case and closest in the southward IMF case for the same IMF strength (the top two panels give a comparison);

(iii) There is a depletion layer in the density in the N IMF case (middle panel).

(iv) The density and magnetic field strength jump significantly across the bow shock.

(v) The scaling law for magnetosphere size as a function of the solar wind dynamic pressure holds (the bottom two panels give a comparison).

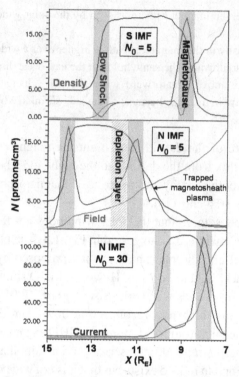

Fig. 11.4. Plots of number density, y component of the electric current, and magnetic field strength (the top, middle, and bottom curves at the left-hand side of each panel) along the stagnation streamline for (IMF, N_0) = (S, 5), (N, 5), and (N, 30), respectively, where N and S stand for northward and southward and N_0 stands for the upstream density (cm^{-3}). The solar wind speed is 400 km/s, the temperature 232 100 K in each case. The solid shadings identify the bow shock and the magnetopause and the hatched shading indicates the depletion layer. See also the color-plate section (OpenGGCM.)

The first point, concerning the *shock standoff distance,* needs no further comment, but the others need explanation. Consider the change in the location of the magnetopause seen between the top two panels: it depends only on whether the IMF is northward or southward. The explanation is that when the IMF is southward, magnetic reconnection takes magnetic flux away from the dayside and relocates it on the nightside and vice versa when the IMF is northward. In the southward case, the actual location of reconnection can be seen in the top panel of Fig. 11.4 at about $X = 9R_E$ in the magnetopause layer, where the field strength drops suddenly to zero. In the northward-IMF case, the location of reconnection is tailward of the cusps and so not evident in any parameter on the stagnation streamline in Fig. 11.4 (but it appears in Fig. 11.6).

There is an interesting *asymmetry* in the way that day-to-night flux transfer occurs in comparison with the night-to-day transfer. Day-to-night flux transfer occurs when reconnection at the nose causes magnetospheric magnetic flux to be carried by the solar wind flow from the dayside to the night side, where it builds up the tail. By contrast, night-to-day flux transfer occurs when reconnection behind the cusps results in tail-lobe field lines attaching to IMF field lines and being swept away by the solar wind, thereby unbuilding the tail, while the Earth-rooted, left-behind, part of the lobe field line, which is now free and fleeing with the solar wind, becomes attached to a magnetosheath field line that is draped over the dayside magnetopause. But if what we are describing is happening in the northern hemisphere, say, then the same thing is happening simultaneously in the southern hemisphere. Consequently the draped magnetosheath field line that has just been attached to Earth in the northern hemisphere has also become attached to Earth in the southern hemisphere. Hence a magnetosheath field line becomes a closed magnetospheric field line, and the magnetosheath plasma it carries is now inside the magnetosphere. The slow falloff in density inside the magnetopause seen in Fig. 11.4 in the northward IMF panel is the signature of the trapping of magnetosheath plasma inside the magnetosphere by this two-point reconnection process (the Song–Russell mechanism for forming the low-latitude boundary layer; Song and Russell, 1992). Thus, instead of moving magnetic flux from the night side to the dayside to reverse the flow of flux that occurs for a southward IMF, reconnection under a northward IMF simply ablates old flux from the night side and accretes new flux on the dayside. There is, to be sure, some night-to-day flow of flux but it is minor compared with the day-to-night flow that takes place under a southward IMF. The main N IMF process is the removal of flux from the nightside as it is carried downstream with the solar wind and the simultaneous plastering of flux from the incoming solar wind onto the dayside. Thus, for a northward IMF we have flux ablation and accretion, while for a southward IMF we have flux circulation.

The *plasma depletion layer* labeled in Fig. 11.4 for a northward IMF is a region of reduced density next to the magnetopause. It results from a buildup of magnetic pressure between the bow shock and the magnetopause (as can also be seen in the figure by the rise in field strength up to the magnetopause) and a consequent reduction in plasma pressure that keeps the total pressure roughly constant. Such a depletion layer was predicted in 1963, in the first paper in which the shape of the magnetopause was analytically calculated (Midgley and Davis, 1963) and has been subsequently modeled by analytical methods (Lees, 1964; Zwan and Wolf, 1976) and by MHD simulations (Siscoe *et al.*, 2002a). Wang *et al.* (2004) used the OpenGGCM to look in detail at the properties of the plasma depletion layer and concluded that it extends over a significant portion of the sub-solar magnetosheath, as one might infer from the field strength profile in the middle panel of Fig. 11.4. By contrast, in the top panel of Fig. 11.4, for the southward IMF case, the field strength shows no significant increase toward the magnetopause, the reason being that magnetic flux is removed at the magnetopause by reconnection. Thus for a southward IMF, there is no magnetic field pile-up and consequently no plasma depletion layer at the nose. Note also that there is no obvious plasma depletion layer in the bottom panel of Fig. 11.4 even though it depicts a northward IMF situation and the magnetic field strength grows toward the magnetopause. Here, however, the plasma pressure is so strong, because of the high density, that the magnetic pressure makes only a minor contribution to the total pressure. The situation approximates a pure gas-dynamic interaction. Despite these counter-examples, plasma depletion layers are a general phenomenon. They have been observed in the magnetosheaths of Earth (Paschmann *et al.*, 1978; Crooker *et al.*, 1979; and many others since then), Saturn (Violante *et al.*, 1995), the outer planets generally (Richardson, 2002), and Mars (Øieroset *et al.*, 2006) – but not Venus (Luhmann, 1995a) – and interplanetary coronal mass ejections (Farrugia *et al.*, 1997; Liu *et al.*, 2006).

Fourth in our list of the information that Fig. 11.4 provides concerns the *jumps* in density and field strength across the bow shock. The shock jump equations are linear for pure gas-dynamic flows but quartic for MHD flows (e.g. Siscoe, 1983); so, in general, MHD simulations offer a great advantage in determining the jump in solar wind and IMF parameters across the bow shock. It happens that the case displayed in Fig. 11.4, in which the magnetic field is perpendicular to the shock normal (the X-axis here), the jump equations are also simple in MHD. In the high-Mach-number limit, the equations predict that both density and field strength increase by about a factor 4 across the shock. Another easy case to solve is that in which the IMF is parallel to the shock normal. Then $\nabla \cdot \mathbf{B} = 0$ requires that there be no change in field strength.

The approximately fourfold jump in density for high Mach number can be checked in Fig. 11.4, since the base value for density in these plots is zero. Such a check cannot be made for the field strength, however, because here the base value is not zero. But one may go to the Community Coordinated Modeling Center (CCMC) website to readily verify that it, too, jumps by about a factor 4.

The relevant point is that the simulations give jump values everywhere along the shock regardless of the IMF's orientation relative to the shock normal. For example, Fig. 11.5 shows the field strength in the equatorial magnetosheath for the case in which the IMF lies along the Parker spiral (at 45° relative to the X-axis). Here the IMF is roughly perpendicular to the shock on the dawn (upper) side, parallel to it on the dusk (lower) side, and 45° to it at noon (left-hand side). As a result a marked dawn–dusk asymmetry in field strength occurs, in which on the dawn (upper) side there is little change in field strength between the IMF and the

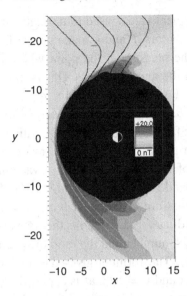

Fig. 11.5. Field strength and field lines in the equatorial magnetosheath for a Parker-spiral interplanetary magnetic field (BATS-R-US) seen from above the Earth's north pole with the Sun to the left. See also the color-plate section.

sheath (the high values here are in the magnetosphere) but on the dusk (lower) side, it jumps by about a factor 4. This asymmetry has been demonstrated in a statistical reconstruction of a cross section of the magnetosheath using a multi-year compilation of IMP-8 magnetometer data (White *et al.*, 1998). The final lesson that we wish to extract from Fig. 11.4 is to be drawn from a comparison of the middle and bottom panels. They differ only in that in the bottom panel the solar wind dynamic pressure ρv^2 is six times that in the middle panel. Thus, we may check the theoretical expectation that the distance between the Earth and the nose of the magnetopause, R_{MP} (the Chapman–Ferraro distance, Eq. 10.1) should scale as the one-sixth power of the dynamic pressure (Section 10.3). In the middle panel R_{MP} is about $11 R_E$ in the bottom panel and it is about $8 R_E$, which compares well with the theoretical value, $8.2 R_E$ that is obtained if one uses the one-sixth-power scaling rule to compress the magnetopause from $11 R_E$ under a factor 6 increase in dynamic pressure.

The value of R_{MP} can also be checked, for which we rewrite Eq. (10.1) as

$$R_{MP} = 10.8 R_E \left[\left(\frac{5}{\rho_{sw} \, (\text{cm}^{-3})} \right) \left(\frac{400}{v_{sw} \, (\text{km/s})} \right)^2 \left(\frac{0.885}{k} \right) \right]^{1/6}$$

$$\times \left[\left(\frac{30\,000}{B_0 \, (\text{nT})} \right) \left(\frac{f}{2.3} \right) \right]^{1/3} \tag{11.4}$$

where the numbers are typical values used to normalize the solar wind speed v_{sw} and the density ρ_{sw}. The factors f and k make up the factor ξ in Eq. (10.1) and are defined in terms of the pressure at the point where the velocity of the solar wind goes to zero against the magnetopause. This is the stagnation point, and the pressure there is called the stagnation pressure, p_{st}. It is also useful to define a stagnation field strength, B_{st}, that has a pressure equal to the stagnation pressure: $B_{st}^2/(2\mu_0) = p_{st}$. Then the factor f is defined as the ratio of B_{st} and the dipole field strength at the stagnation point. Its value is 2.3, as determined by an analytical solution to the problem of finding the shape of a vacuum magnetosphere under compression by an idealized solar wind (Mead, 1964). The factor k is the ratio of p_{st} and the dynamic pressure of the solar wind, ρv^2 (also known as the ram pressure). Its value in ordinary gas-dynamic theory is given by a complicated formula involving the Mach number and the ratio of specific heats γ (Landau and Lifshitz, 1959, Eq. 114.1). The formula gives the numerical value 0.885 in the limit of high Mach number and $\gamma = 5/3$, as is appropriate for an ideal monatomic gas. Thus, if we accept the theoretical values $f = 2.3$ and $k = 0.885$, Eq. (11.4) gives $R_{MP} = 10.8R_E$ for the solar wind values used to generate the middle panel in Fig. 11.4. The result agrees well with the position of the magnetopause seen there. The top panel uses the same solar wind values but has a smaller R_{MP} because of the day-to-night flux transport discussed above and to which we now return.

Figure 11.6 illustrates the features that we have been discussing, and the differences between them for northward and southward directions of the IMF.

11.5 Forces on the magnetosphere

The magnetopause in the top panel of Fig. 11.4 lies about $2R_E$ closer to Earth than it does in the middle panel, although the solar wind ram pressure is the same in both cases. The reason is that the mode of momentum coupling between the solar wind and the magnetosphere is different in the two cases. To understand the difference, think of momentum coupling as being the result of the solar wind's applying a force against the magnetopause. The force can be perpendicular to the magnetopause (a push) or tangential to it (a pull). The first option, in general aerodynamic parlance, is called a normal stress and the second a tangential stress (e.g. Prandtl and Tietjens, 1934; Shapiro, 1961). The case of a purely normal stress acting on the magnetopause was considered by the early workers in the field, who were interested in determining the shape of the magnetosphere and the distortion of the geomagnetic field that resulted from the electrical currents on the magnetopause. This is the so-called Chapman–Ferraro problem (see the review by Siscoe, 1988a). The Chapman–Ferraro problem turns out to be solvable analytically (as the limit

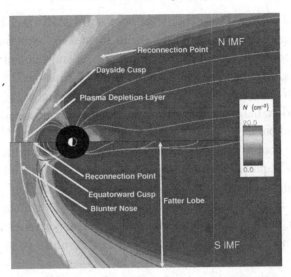

Fig. 11.6. The basic structural elements of the magnetosphere as depicted in global (OpenGGCM) MHD simulations can be seen here outlined in density contours and magnetic field lines. The figure identifies contrasts between the magnetosphere as shaped under northward (upper) and southward (lower) IMF conditions. The solar wind conditions are the same in both halves of the figure, except for the IMF direction. See also the color-plate section.

of a series) if one assumes a vacuum magnetosphere (no magnetotail, therefore) and the "Newtonian approximation" for normal stress; this sets the normal stress equal to the stagnation pressure times the square of the cosine of the angle between the normal to the magnetopause and the solar wind flow direction (cf. specular reflection). The solution thus obtained gives 2.3 for the factor f in Eq. (11.4) above, as already stated.

Regarding Fig. 11.4, in the case in which the IMF is northward (shown in the middle panel) the force exerted by the solar wind on the magnetopause approximates a normal stress. It is the sum of the plasma pressure force, which is always perpendicular to the boundary, and the magnetic field force, which is $\mathbf{J} \times \mathbf{B}$ where $\mathbf{J}$ is the electrical current flowing on the magnetopause and $\mathbf{B}$ is the magnetic field at the magnetopause. In this northward IMF case, $\mathbf{B}$ is parallel to the magnetopause, as Fig. 11.4 shows, so $\mathbf{J} \times \mathbf{B}$ is perpendicular to the magnetopause; that is, it constitutes a pure normal stress as well. Moreover, the magnetotail is diminished by ablation of the tail magnetic flux, which, as discussed earlier, applies to this case. Thus, the situation approximates well the normal-stress–no-tail problem treated by the early magnetosphere theorists and accounts for the agreement between the position of the magnetopause in the middle panel of Fig. 11.4 and the position predicted by Eq. (11.4) with $f = 2.3$.

The top panel in Fig. 11.4 depicts the southward IMF case, for which there is a substantial tangential stress on the magnetopause in addition to the normal stress. The tangential stress results from magnetic reconnection at the nose of the magnetopause, which gives a normal component of the magnetic field all over the dayside. Thus, in this case $\mathbf{J} \times \mathbf{B}$ has a tangential component at the magnetopause. As described earlier, the tangential stress moves magnetic flux from the dayside to the night side and builds up the tail. The situation, therefore, has both a tangential stress and a tail, doubly violating the ideal conditions under which an analytical solution to the Chapman–Ferraro problem is obtained. It would be fair to call the corresponding problem with a southward IMF the Dungey problem, since Dungey was the first person to describe it (Dungey, 1961). Unlike the Chapman–Ferraro problem, there is no analytic solution to the Dungey problem. Thus, MHD simulations are necessary to obtain a theory-based result to compare against the observed data.

The distinction between the normal-stress and tangential-stress modes of solar-wind–magnetosphere coupling is brought out by the different paths followed by their associated electrical current systems, as shown in Fig. 11.7. The upper half of the figure shows streamlines of the current system responsible for balancing the normal stress on the boundary (the Chapman–Ferraro current system). At upper left we have streamlines obtained with a global MHD simulation in which the IMF was set to zero and, for comparison, at upper right we have streamlines from an analytical calculation such as discussed above. One deduces that the streamlines of the Chapman–Ferraro current system form a nested set of ellipsoids. Although one cannot tell from the face-on perspective of the figure, the streamlines lie on the magnetopause. They do not enter the magnetosphere or the magnetosheath.

The lower half of Fig. 11.7 shows streamlines of the current system that exerts a tangential stress at the magnetopause. The image is from the output of an MHD simulation in which the IMF was set to 5 nT southward. Since no analytical solution exists in this case, there is no comparison panel. The name "region-1 current system" identifies it as the complete current circuit associated with the semi-permanent circumpolar rings (one in each hemisphere) of electrical currents that flow into and out of the ionosphere parallel to the magnetic field lines. The discoverers of these rings of field-aligned currents called them region-1 currents (Iijima and Potemra, 1976). There is another, lower-latitude, pair of field-aligned current rings, which they called region-2 currents. Region-2 currents, at present, are poorly simulated by MHD models and therefore we will discuss them no further despite their being very important for understanding the coupling of the ionosphere to the inner magnetosphere. We refer the reader to a review of this subject by Wolf (1983).

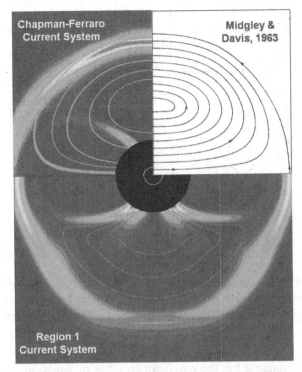

Fig. 11.7. Magnetohydrodynamic simulations (BATS-R-US) showing contours in the $x = 0$ plane (which contains the terminator) of current density and current streamlines of the Chapman–Ferraro current system (zero IMF) (upper half) and the region-1 current system (5 nT southward IMF) (lower half). Chapman–Ferraro current streamlines from the analytical calculation of Midgley and Davis (1963) are shown for comparison. Solar wind speed 400 km/s, density 5 cm^{-3}, temperature 232 100 K. See also the color-plate section.

One sees the magnetospheric part of the region-1 current system in Fig. 11.7 in the color contours (see the color-plate section) in the $x = 0$ terminator plane and in the streamlines. Again, although one cannot tell from the perspective of the figure, the streamlines of the region-1 current system close on the magnetopause, or beyond it in the magnetosheath, or on the bow shock (Siscoe and Siebert, 2006). Those in the figure close on the magnetopause. We will discuss their ionospheric closure below. For the present, we are interested in describing the difference in the way in which the solar wind momentum is coupled to the magnetosphere through the action of the two different current systems seen in the upper and lower halves of the figure.

Consider first the Chapman–Ferraro coupling for zero IMF (the upper half of Fig. 11.7). The force exerted by the solar wind on the magnetopause in this case is very nearly a pure normal stress applied by plasma pressure. The solar wind pushes

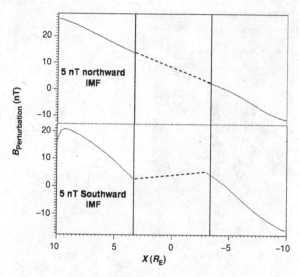

Fig. 11.8. Perturbation magnetic field along the X-axis for a northward IMF (upper panel) and southward IMF (lower panel). The broken lines connect computed values across the non-computable zone from $3.5R_E$. Solar wind speed 400 km/s, density 5 cm^{-3}, temperature 232 100 K (BATS-R-US).

against the magnetopause but the magnetopause has no mass, in comparison with the mass of the impacting solar wind. Neither has the magnetosphere, except through its connection to the Earth. So the force exerted by the solar wind on the magnetopause must be communicated to the Earth, a requirement achieved by the Chapman–Ferraro current system through the gradient in the magnetic field that it generates, B_{CF}. It has just the right gradient at the Earth to impart a force $M_E \nabla B_{CF}$ on the geomagnetic dipole (M_E is the geomagnetic dipole moment) that balances the total solar wind force on the magnetopause (Siscoe, 1966). This requirement is automatically satisfied in the analytical solutions found in the 1960s and in MHD simulations with a small northward IMF (to diminish the tail flux).

Figure 11.8 shows how the strength of the perturbation magnetic field generated by the force-coupling current systems varies as a function of distance along the x-axis from the nose of the magnetopause (left-hand margin), as obtained from MHD simulations. The gaps in the middle filled in by the broken lines occur because the BATS-R-US and OpenGGCM codes do not compute MHD variables inside a $3.5R_E$ Earth-centered sphere, for computational difficulties that arise because the propagation of waves becomes too fast close to Earth.

In the upper panel of Fig. 11.8 the IMF has been set to 5 nT northward, so the perturbation field is approximately that of the Chapman–Ferraro current system. The gradient across the Earth here is 1.7 nT/R_E, which may be compared with the value 1.5 nT/R_E obtained by the analytical calculations. Thus the MHD simulations agree well with the analytical calculations in this case. The force on the Earth is about 2×10^7 N directed away from the Sun. The effective momentum-capturing cross section of the magnetosphere, that is, the area over which the solar wind ram pressure if uniformly applied would produce a force of 2×10^7 N, is a circle of $11 R_E$ radius, which is intuitively reasonable since $11 R_E$ is the characteristic scale size of the magnetosphere given by Eq. (11.4).

The lower panel in Fig. 11.8 shows the corresponding result for the case in which the IMF is 5 nT southward (and, to repeat, for which there is no analytical solution for comparison). Obviously the situation is qualitatively different from the zero (or small northward) IMF case. Here the perturbation field, which is made up of contributions from the Chapman–Ferraro current system and the region-1 current system, has a sign reversal of its gradient across the Earth. A reversed gradient means that the force on the geomagnetic dipole is sunward. A naïve interpretation would be that the solar wind instead of pushing the Earth away from the Sun is sucking it toward the Sun, which makes no sense physically. The correct interpretation is that the region-1 current system delivers its force from the solar wind to the Earth not by a gradient across the dipole but by means of a $\mathbf{J} \times \mathbf{B}$ force in the ionosphere. This force is anti-sunward and is greater than the sunward force on the geomagnetic dipole so that the net force is indeed anti-sunward. To obtain the actual value of the total force that the solar wind exerts on the magnetosphere–Earth system in this case, one must integrate, over a surface that encloses the Earth (where all the force is applied), the total momentum stress tensor

$$\mathbf{S} = \left[\rho \mathbf{v} \mathbf{v} + \left(p + \frac{B^2}{2\mu_0} \right) \mathbf{I} - \frac{\mathbf{B} \mathbf{B}}{\mu_0} \right], \tag{11.5}$$

where ρ is the mass density, v is the velocity, p is the plasma pressure, $\mathbf{B}$ is the magnetic field, and $\mathbf{I}$ is the identity matrix. The total force acting on the mass inside a closed surface Σ is

$$\mathbf{F} = - \int_\Sigma \mathbf{S} \cdot \hat{\mathbf{n}} d\sigma \tag{11.6}$$

where $\hat{\mathbf{n}}$ is the outward pointing unit normal and $d\sigma$ is a differential surface element. Table 11.2 shows the result of carrying out the integration using the OpenGGCM and BATS-R-US codes at CCMC. Both codes give results roughly consistent with the above predictions based on a ram pressure acting on an area determined by the characteristic scale size of the magnetosphere; the results are within a factor 2 of each other. Nonetheless, the remaining differences are significant and indicate how much the numerical values can be affected by the particular coding scheme.

Using a different global MHD simulation code, Siscoe and Siebert (2006) carried out the surface integration for the case of zero IMF and for a strong southward IMF (30 nT). They found a force equal to 2.4×10^7 N in the first case (like the other

Table 11.2. *The x component of the force derived from Eq. (11.6). The computed forces are in newtons and are in the downstream direction*

Model	IMF $B_z = 5\,\text{nT}$	IMF $B_z = -5\,\text{nT}$
OpenGGCM	3.5×10^7	6.6×10^7
BATS-R-US	2.5×10^7	3.6×10^7

calculations) and 1.2×10^8 N in the second case. All the MHD models show that the net force increases when the IMF turns southward. The increase in force is attributable to the larger cross section against which the ram pressure can push and to an increase in tangential stress due to the coupling of the solar wind IMF and the magnetosphere via open field lines.

To recapitulate, the force exerted by the solar wind on the magnetosphere when the IMF is southward is about $5–10 \times 10^7$ Newtons. At least half comes from the tangential stress that the $\mathbf{J} \times \mathbf{B}$ force of the region-1 current system exerts against the solar wind at and beyond the magnetopause. Paradoxically, the $\mathbf{J} \times \mathbf{B}$ force exerted low down in the ionosphere by the region-1 current system is much larger (by about an order of magnitude) than the force that it simultaneously exerts high up, against the solar wind. The difference in the total $\mathbf{J} \times \mathbf{B}$ forces at the high- and low-altitude extremes of the region-1 current system arises because at high altitude the magnetic field is much weaker than at low altitude (as indicated in Fig. 11.7). This effect was first noted by Hill (1983), who likened it to a mechanical advantage. The amplification of the force was discussed further by Siscoe (2006b) and Vasyliūnas (2007), who explained that the apparent paradox (that the anti-sunward force that the region-1 current system exerts on the ionosphere at low altitude is many times greater than the sunward force it exerts on the solar wind at high altitude) does not violate Newton's third law (for every action there is an equal and opposite reaction) because the region-1 current system exerts a compensating (Newton-satisfying) sunward force on the geomagnetic dipole at the same time. The compensating force is indicated in Fig. 11.8 by the reversed gradient of the perturbation field.

The reversed gradient in Fig. 11.8 also explains the earthward displacement of the magnetopause seen in Figs. 11.4 and 11.6 for southward IMF cases. Such earthward displacement is known as magnetospheric erosion. It occurs because the region-1 current system generates a magnetic field that weakens the geomagnetic field on the dayside and strengthens it on the night side (hence the reversed gradient in Fig. 11.8). Under the condition of a weakened magnetospheric field, the magnetopause moves earthward until it reaches a point at which the magnetospheric magnetic field, though weakened, is nonetheless strong enough to balance the solar wind ram pressure. When the IMF is strong and southward the effect can be so pronounced that the magnetospheric field at the magnetopause is actually weaker than the dipole field at that distance (Siscoe *et al.*, 2002b). The magnetopause is correspondingly much closer to Earth than would be expected for a Chapman–Ferraro magnetopause.

11.6 Magnetospheric convection

As discussed in Chapter 10, solar-wind–magnetosphere coupling produces a circulation of plasma within the magnetosphere. Plasma in the high-latitude region connected to the outermost layers of the magnetosphere is driven anti-sunward by the tangential stress that we have been discussing, while plasma in the low latitudes connected to the inner magnetosphere generally flows sunward. But the plasma and magnetic flux are tied together, so we can describe circulation of plasma equally well as a circulation of magnetic flux. It is more useful, in fact, to speak in terms of the flow of magnetic flux because the total amount of magnetic flux in the magnetosphere is a known constant, which is not the case for plasma. The circulation of magnetic flux is called magnetospheric convection (Section 10.4.3).

There are two facts that now need to be brought together. First, since the flow of flux is induced by the tangential stress on the magnetopause, it is intimately connected with the region-1 current system. Second, since convection is the flow of magnetic flux (from day to night or vice versa), one can quantify it as a voltage. Here we are referring to the fact that a moving magnetic field generates a motional electric field $-\mathbf{v} \times \mathbf{B}$, and the width L of the channel on the magnetopause through which the flux flows then determines a voltage vBL. The voltage and the region-1 current are both transmitted along magnetic field lines from the solar wind coupling region to the ionosphere, where the ionospheric Ohm's law unites them through the equation for current conservation. The result is

$$\nabla \cdot (\Sigma \nabla \Phi) = -j_\| \sin i, \tag{11.7}$$

where Σ is the conductance tensor,

$$\Sigma \equiv \begin{pmatrix} \Sigma_p / \sin^2 i & \Sigma_H / \sin i \\ -\Sigma_H / \sin i & \Sigma_p \end{pmatrix}. \tag{11.8}$$

Here Σ_p and Σ_H are the Pedersen and Hall conductances (cf. Eq. 10.10 and the associated text), Φ is the voltage in the plane of the ionosphere, $\mathbf{j}$ is the region-1 current density flowing parallel to the magnetic field into and out of the ionosphere, and i is the magnetic dip angle. Our interest, in this chapter, is in how MHD codes represent magnetospheric convection. As mentioned earlier, at present the MHD

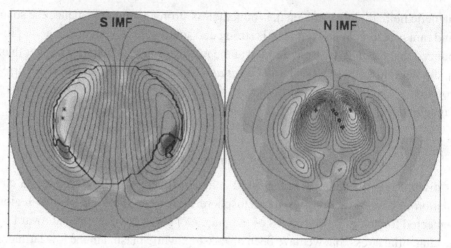

Fig. 11.9. Contours of electrical potential from the OpenGGCM model on northern-hemisphere polar plots (noon towards the top). The background images show parallel (region-1) current for 5 nT southward IMF (left) and 5 nT northward IMF (right). Other solar wind parameters are as before. The trans-polar voltages are 150 kV (left) and 26 kV (right). The range on the contour scales is the same in both figures, $\pm 1.6\,\mu A/m^2$. Red corresponds to the minus sign, blue to the plus sign. The latitude range extends from $50°$ (outer edge) to $90°$ (center). The black closed contour in the left-hand panel marks the border between open and closed field lines. Compare with the schematic diagram in Fig. 10.5, which shows the corresponding plasma flow (right) and electric field lines (left; the corresponding perspective in this figure is from the upper-left corner) for the case with southward IMF. See also the color-plate section.

parts of the BATS-R-US and OpenGGCM codes do not compute down to the ionosphere. They stop at a spherical surface at about 3.5 Earth radii and then extend the parallel currents along dipole magnetic field lines down to the ionosphere, where they determine the voltage via Eq. (11.7). The voltage is then projected back up to the inner boundary of the MHD code to impose the effect of the ionosphere on the global solution. An example is shown in Fig. 11.9.

In Fig. 11.9, the contour lines are both equipotentials and also flow lines of magnetospheric convection. The equivalence of the equipotentials and flow lines follows from their both being perpendicular to the electric field ($\mathbf{E} = -\nabla\Phi$ and $\mathbf{E} = -\mathbf{v} \times \mathbf{B}$, in standard notation). The left-hand panel of Fig. 11.9 shows a day-to-night flow of magnetic flux over the polar cap (enclosed by the black contour) and its return flow at lower latitudes. This is the situation that a southward IMF sets up through dayside magnetic reconnection, which leads to an anti-sunward tangential stress because of the resulting normal component of the magnetic field across the dayside magnetopause. The background color contours (see the color-plate section) show the location of the region-1 currents, which communicate the tangential stress to the ionosphere. In the ionosphere the tangential stress sets up the flow seen in the line contours. Region-1 current enters the ionosphere on the dawn side (parallel to the magnetic field, designated by blue) and leaves on the dusk side (antiparallel to the magnetic field, designated by red). Accordingly, the voltage is positive on the dawn side and negative on the dusk side. The voltage drop is 150 kV in this case. The dawn–dusk (left–right) asymmetry seen in the flow pattern and the color contours results from a finite Hall conductivity. The patterns are symmetric when the Hall conductivity is set to zero.

The right-hand panel in Fig. 11.9 shows the corresponding situation for a northward IMF that has lasted for a few hours following a period during which the magnetotail had been built up. One sees that the convection pattern has weakened and shrunk relative to the southward IMF case and, more significantly, that it has changed sign. The field-aligned current flows into the ionosphere on the dusk side (blue) and out on the dawn side (red). The voltage is correspondingly reversed: dusk is positive and dawn is negative for a sunward flow in the polar cap, which is opposite to the southward IMF case. Thus, the current system here is not the region-1 current system. It is called the NBZ current system, which stands for northward (IMF) B_z. One sees a vanishingly small residual of the former, region-1-type, flow pattern at lower latitude, but this is not where our interest lies. The weakness of the NBZ current and of its convection voltage reflects the discussion in Section 11.4, where we saw that in the northward IMF case magnetic reconnection takes place at high latitudes, tailward of the dayside cusps. An important consequence of this situation is that reconnection ablates magnetic flux from the night side and accretes incoming magnetosheath flux onto the dayside. Therefore, a night-to-day flow of magnetic flux is not needed to re-establish the day–night flux balance.

Nonetheless, as the NBZ flow pattern shows, some such cross-polar-cap night-to-day flow does occur, and for two reasons. First, global forces readjust the shape of the magnetosphere to accommodate the new day–night distribution of magnetic flux. Polar field lines in the ionosphere that formerly went into the tail now close on the dayside, so the field line moves sunward under global stresses that minimize the total magnetic energy. A second reason is not so obvious and perhaps more interesting, and this involves the Song and Russell (1992) mechanism for forming the low-latitude boundary layer (LLBL). In this scenario, the motion field that we see traced out in the ionosphere as the NBZ convection pattern results from the motion tailward along the inside of the magnetopause, under its own residual momentum and pressure gradient, of the magnetosheath plasma that has been entrained on closed magnetic field lines on the dayside. Thus a thick layer of magnetosheath-like plasma coats the inside of the magnetopause, forming the LLBL. In the readjustment scenario, the NBZ circulation is driven by the sunward motion of polar field lines. In the Song–Russell scenario, it is driven by plasma moving anti-sunward as the LLBL. In both cases, a complete NBZ circulation pattern results since the magnetic flux in the ionospheric cannot develop holes.

11.7 Energy flow in the magnetosphere

Up to now we have discussed the flow of mass (in the magnetosheath, mantle, and LLBL), of momentum (due to forces), and of magnetic flux (due to convection) through and around the magnetosphere. We now conclude with a brief discussion of the flow of energy (for a fuller treatment, see the review by Hill, 1983). To discuss the energy flow, we start with Poynting's theorem:

$$\frac{\partial}{\partial t}\left(\frac{B^2}{2\mu_0}\right) + \nabla \cdot \mathbf{S} = -\mathbf{j} \cdot \mathbf{E} = -\mathbf{v} \cdot (\mathbf{j} \times \mathbf{B}). \tag{11.9}$$

The first term in Eq. (11.9) is the rate of change of the electromagnetic energy density (the electric field component of which is negligibly small for non-relativistic plasmas; see Hill, 1983). The second term is the divergence of the electromagnetic energy flux (or Poynting vector $\mathbf{S} = \mathbf{E} \times \mathbf{B}/\mu_0$). On the right-hand side is the electromechanical energy conversion term, which is given first in general form then in MHD form, which follows from $\mathbf{E} = -\mathbf{v} \times \mathbf{B}$. In electrical engineering parlance, those places where the magnetic energy is increasing in time are referred to as dynamos and places where it is decreasing as loads. Thus, dynamos are regions where $\mathbf{v}$ opposes $\mathbf{j} \times \mathbf{B}$, i.e. where the $\mathbf{j} \times \mathbf{B}$ force acts to slow the flow, and conversely loads are regions where it acts to accelerate the flow. The former regions are places of field generation and the latter places of field dissipation.

For for a sufficiently long-term time average, or in the steady state, Eq. (11.9) becomes

$$\langle \mathbf{j} \cdot \mathbf{E} \rangle = -\langle \nabla \cdot (\mathbf{E} \times \mathbf{B})/\mu_0 \rangle \tag{11.10}$$

where the angle brackets $\langle \, \rangle$ denote a time-averaged quantity. The $\mathbf{j} \cdot \mathbf{E}$ term in Eq. (11.10) has been calculated from steady-state runs of both global MHD models and is shown in Fig. 11.10. Consider first the northward IMF case in the left-hand panels: the energy-conversion regions in the magnetosheath are easier to see in these panels. At the bow shock $\mathbf{j} \cdot \mathbf{E} < 0$, indicating that flow energy is being converted into magnetic field energy (the flow slows and the field strengthens across the shock). Inside the bow shock, close to the forward part of the magnetopause, again $\mathbf{j} \cdot \mathbf{E} < 0$, which shows that here the field is being compressed against the magnetopause, as seen in Fig. 11.4. This is where the plasma depletion layer develops. In the magnetosheath along the flanks of the magnetosphere $\mathbf{j} \cdot \mathbf{E} > 0$,

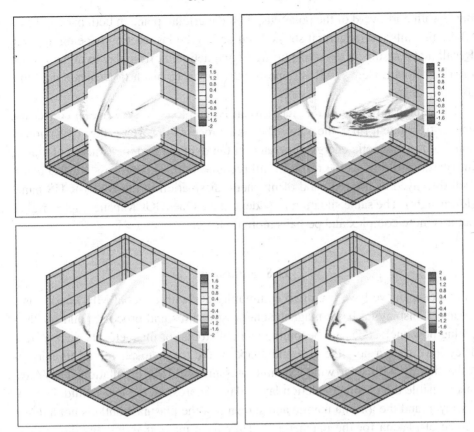

Fig. 11.10. Values of **j** · **E** (the scales are in pW/m^3) in the noon–midnight meridian (left) and equatorial plane (right), computed by the OPENGGCM (upper) and BATS-R-US (lower) MHD codes. In the left-hand panels $B_z = 5$ nT and in the right-hand panels $B_z = -6$ nT. See also the color-plate section.

showing that the magnetic field, which has been compressed at the nose, now decompresses and joins with the plasma pressure to accelerate the solar wind down the magnetosheath. Along the magnetopause, inside this acceleration region, viscous coupling is slowing the wind and stretching the magnetic field, making the magnetopause visible as a ribbon of blue (see the color-plate section). The patch of red visible in the meridian plane above the dayside cusp in both panels marks the location of magnetic reconnection for the northward IMF. It is red because reconnection converts magnetic energy into flow energy. Magnetic reconnection for a southward IMF is seen in Fig. 11.10 to be occurring at the magnetopause in the equatorial plane and in the meridian plane equatorward of the dayside cusp. The **j** × **B** force is accelerating the flow northward, tangentially to the magnetopause, hence the red color. The rate at which magnetic reconnection converts magnetic energy into kinetic energy is much greater here than for a northward IMF. The blue swath northward of the polar cusp in the meridian plane in both panels shows where the tailward tangential stress discussed in the last section is being applied. Recall that it results in the transporting of magnetic flux from day to night to form the magnetotail. Here $\mathbf{j} \times \mathbf{B}$ is opposing the magnetosheath flow, hence the blue, dynamo, color.

The equatorial plane of the magnetotail is seen to be very active in converting energy from one form to the other. The main process, however, is the conversion of some of the magnetic energy generated in the dynamo region, as just mentioned, into earthward flow as the result of tail reconnection. This process is particularly well displayed as the red band of magnetic dissipation in the BATS-R-US panel (lower right). The same situation is evident in the OpenGGCM panel (upper right), but it is more complex and perhaps more realistic.

11.8 Summary

In this chapter we have used two community-accessible global MHD codes as a means of displaying basic magnetospheric structures and processes and of illustrating how such codes are becoming valuable tools for magnetospheric research. They have shown how the size and shape of the magnetosphere and its magnetosheath vary with solar wind dynamic pressure and IMF orientation. They have made visible the plasma depletion layer, dayside erosion, the low-latitude boundary layer, and the plasma mantle and shown that the plasma mantle is not a direct source of plasma for the magnetotail. They have made manifest the dawn–dusk asymmetry in the magnetosheath when the IMF has a Parker-spiral orientation and have depicted the Chapman–Ferraro current system, the region-1 current system, and the patterns of direct and reverse magnetospheric convection in the ionosphere for southward and northward IMF, respectively. These codes have illustrated the mechanical leveraging of the region-1 current system through the reversed day-to-night gradient across the Earth of the magnetic field it generates. Furthermore, their use has shown that, whereas day-to-night flux transfer occurs principally by flux circulation under a southward IMF (i.e. by direct magnetospheric convection), night-to-day flux transfer occurs principally by flux ablation on the nightside and flux accretion on the dayside. They have enabled us to identify the regions of electromechanical energy conversion in the bow shock, magnetosheath, magnetopause, and magnetosphere under northward and southward IMFs. In general, they have confirmed previously held concepts regarding magnetospheric structure and dynamics (with the exception of the plasma mantle as a source for the plasma sheet in the tail) and have revealed new ones, e.g. that dayside erosion is caused by the region-1 current system; this can even cause the field at the dayside magnetopause to become weaker than the dipole value.

Global MHD codes are indeed useful and powerful tools. But, like any powerful tool, they can cause harm if used without caution. We have seen that two codes can yield different results for the same input conditions. This continues to be the case as the number of codes for comparison increases. Therefore, it is good practice to use more than one code in a research project, in order to be able to estimate the variance in the result. Also, skill comes with experience. Practising on known examples can give confidence when working on new but similar situations.

12

On the ionosphere and chromosphere

TIM J. FULLER-ROWELL AND CAROLUS J. SCHRIJVER

12.1 Introduction

In this chapter we review some basic physical processes in the upper atmospheres of planets, with a particular focus on Earth with its strong internal magnetic field. Less attention will be given to the upper atmospheres of planets with little or no internal magnetic field. The same basic physical processes operate in all planetary atmospheres. The main distinctions are in the species present (see the review by Bougher *et al.*, 2002), the rotation rate and gravitational force, and whether the planet has a significant intrinsic magnetic field. The vertical extent, absorption of radiation, interaction with the solar wind, and dynamical balances will clearly differ. Planets with a small magnetic field will also have quite different electrodynamic properties.

Earth's upper atmosphere can be categorized as a gravitationally bound, partially ionized, fluid. The fluid properties of the gas result from the frequent collisions between the atoms and molecules of the medium. Rather than having to accommodate the random nature of the forces exerted on individual gas particles the principles of kinetic theory can be invoked, so that the medium can be described by the bulk properties of the fluid such as pressure, density, temperature, and velocity. The vertical extent of the sensible atmosphere is usually defined by the altitude at which the fluid approximation is no longer valid, referred to as the exobase. Below the exobase, i.e. the topmost extent of an atmosphere, the distance and time between collisions is short compared with the scale sizes of interest in the dynamics and energetics of the fluid. For Earth this altitude is usually above 600 km, but it can be higher since it depends on the gas kinetic temperature and hence the degree of thermal expansion of the medium. It is more appropriate to state the vertical extent of the atmosphere in terms of the pressure level. Most physical models of Earth's upper atmosphere are limited to a top pressure level of about 10^{-7} Pa for this reason. At lower pressures, or greater heights, the mean free path or distance between collisions becomes very long, exceeding tens of kilometers, so the medium no longer behaves as a fluid but as a collection of non-interacting particles. The neutral atmosphere contrasts with space plasmas, which can be treated as fluids at much lower densities owing to the presence of magnetic and long-range Coulomb interactions. The effective "collisions" coming from the interaction of charged particles enable space plasmas to be treated as fluids in many parts of the heliosphere. In fact, it is the times and places when fluid approximations are no longer dominant that pose some of the most interesting science challenges in heliophysics.

Sections 12.1–12.7 review some physical processes from the perspective of the thermosphere–ionosphere standpoint. This discipline treats the concepts of electromagnetism from a somewhat different perspective than earlier chapters. Electric and magnetic fields are treated as a consequence of charges and their motion, and the correlation between electric and magnetic fields is assumed to be the result of correlation between their respective sources. Several assumptions are made in the ionosphere–thermosphere (IT) system that are specific to it. The first is that the source of the magnetic field comes largely from the currents flowing internal to the Earth and that their strength and direction change very slowly. The second is that the internal currents flowing in the IT system and in the magnetosphere make only a small perturbation (less than 1%) to the magnetic field strength at the altitudes of the IT system ($100 - 600 \, \text{km}$). This assumption makes it possible to use the magnetic field as the basis of a coordinate system for treating the ionospheric plasma. The third assumption is that electric fields can be expressed as the gradient of a potential. Electric fields either imposed from the magnetosphere at high latitudes or driven internally by neutral-wind dynamo action at mid and low latitudes play a fundamental role in redistributing ionospheric plasma.

In Section 12.8 we attempt to compare and contrast the IT system with a region of the solar atmosphere that can also be characterized as a gravitationally bound, partially ionized, plasma, namely the chromosphere. That section describes the physical processes from the solar standpoint while identifying some heliophysical processes common to the two disciplines. The differing terminology, particularly in the description of magnetic fields, makes this comparison a challenge. We hope that these contrasting perspectives of the two heliophysical domains will stimulate and motivate readers to deepen their understanding of the common physical processes.

12.2 Forces and flows in the neutral atmosphere

12.2.1 The gas law and hydrostatic balance

The frequent collisions of molecules in a gas close to thermal equilibrium enable the Maxwellian energy distribution of the individual particles to be characterized by the basic fluid properties of pressure p, temperature T, number density n, and mass density ρ, which are related by the perfect gas law

$$p = nkT = \frac{\rho RT}{M},\tag{12.1}$$

where k and R are the Boltzmann and universal gas constants, respectively, and M is the mean molecular mass. The fluid pressure in the atmosphere represents the weight of the column of gas above a point.

The neutral gas under the influence of the planet's gravitational force gives rise to the concept of hydrostatic balance, which states that the change in pressure with height, ∂p, is closely balanced by the weight of the fluid, $nmg\partial h$ (where m is the mean molecular mass in kilograms and h is the height), under the action of the planet's gravitational acceleration g. This is expressed mathematically as

$$\frac{\partial p}{\partial h} = -\rho g = \frac{-p}{H}.\tag{12.2}$$

This basic equation describes the exponential decrease in gas density with altitude, and results in the concept of the pressure scale height,

$$H = \frac{RT}{Mg}\tag{12.3}$$

which represents the altitude through which the gas pressure will decrease by a factor $1/e$. Earth's upper atmosphere extends for about a dozen scale heights above 100 km altitude, the scale heights changing from about 5 km to 50 km with increasing altitude as the temperature increases from about 180 K to over 1000 K (see Fig. 12.1), and the mean molecular mass decreases.

The assumption of hydrostatic balance implies that a vertical column of gas will respond to a heat source instantaneously. In practice, information about heating at a given altitude or pressure level is transferred to other regions by acoustic gravity waves, which have speeds of hundreds of meters per second, corresponding to a time scale for adjustment of typically 5 to 10 minutes. The treatment and understanding of dynamical time scales shorter than this period, or of spatial scales less than a scale height, has to include acoustic waves explicitly.

The quasi-equilibrium implied by hydrostatic balance does not exclude the possibility of vertical winds. The assumption simply demands that the rate of heating is such that the atmosphere adjusts at a comparable rate. The term "quasi-hydrostatic balance" is the more correct expression in the case where vertical winds are accommodated in the system. One component of the vertical wind is then defined as the rate of change in height of a pressure surface in the column of gas, $(\partial h/\partial t)_p$, the so-called barometric wind. Vertical winds in Earth's upper atmosphere of the order of 100 m/s can be accommodated within the quasi-hydrostatic assumption.

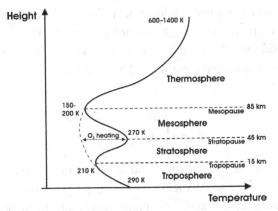

Fig. 12.1. Average vertical temperature profile through Earth's atmosphere. The general shape of the temperature profile is reasonably consistent, to the point where it can be used to define the four main neutral-atmosphere "layers", from the troposphere to the thermosphere. The temperature of the uppermost layer, the thermosphere, increases steeply with altitude due to the absorption of solar extreme ultraviolet (EUV) and far ultraviolet (FUV) radiation. The thermosphere and upper mesosphere are partially ionized by the same EUV radiation, which varies by a factor 3 over the solar cycle, and by auroral particle precipitation.

The assumption of hydrostatic balance has enabled the wide use of pressure as the vertical coordinate in atmospheric models. In fact, it is only fairly recently that this assumption has begun to be relaxed in models of the Earth's upper atmosphere, with explicit inclusion of a realistic adjustment process by acoustic waves. It is possible to extend the application of such models to the physical response at small length scales and on short time scales.

12.2.2 *Continuity equation*

The continuity equation captures one of the most fundamental and universal fluid concepts. The chemical production and loss of the major species in the planetary atmosphere is relatively slow, so that in the pressure coordinate system widely used in many atmospheric models the continuity equation can be expressed as

$$\frac{\partial \omega}{\partial p} = -\nabla \cdot \mathbf{v}, \tag{12.4}$$

where ω is the vertical wind dp/dt in the pressure coordinate system and the right-hand side represents the horizontal divergence of the neutral wind velocity $\mathbf{v}$ on a pressure surface. The fluid therefore appears incompressible in this natural coordinate system, where the horizontal divergence or convergence must be balanced by a vertical flow. This in fact describes the second component of the vertical wind, the so-called "divergence wind". A local heat source will cause the local column of gas to expand thermally (this is the barometric wind). The horizontal pressure gradients thus induced will drive a divergent wind that must be balanced by a vertical flow across the pressure surfaces. The total vertical wind v_z in this system can be expressed as

$$v_z = \left(\frac{\partial h}{\partial t}\right)_p - \frac{\omega}{\rho g}, \tag{12.5}$$

which is the sum of the barometric wind component and the divergence wind component $-\omega/(\rho g)$. Photochemical production and loss for some of the minor species is much more rapid than for the major species, so these physical processes must be considered in the continuity equations when studying the minor species individually.

12.2.3 Horizontally stratified fluid and the Coriolis effect

One common assumption when addressing the large-scale dynamics in atmospheric fluids is that the horizontal scale size is significantly greater than the vertical scale. This assumption implies that horizontal motions are constrained to follow the curvature of the planet. Combining this assumption with the rotation of a planetary body gives rise to the Coriolis force. The Coriolis effect is the apparent deflection of moving objects from a straight path when they are viewed from a rotating frame of reference. The apparent force is a consequence of the inertia of the fluid as it is constrained to move on a horizontal curved surface. For planets with a rotational direction the same as that of Earth, i.e. a prograde rotation, the Coriolis effect away from the geographic equator causes a parcel of fluid moving with respect to the planetary rotation velocity to be directed towards the right in the northern hemisphere and to the left in the south. For planets with opposite rotation the converse is true. The basic force per unit mass can be expressed as $-2\Omega \times v$, where Ω is the planet's angular velocity.

12.2.4 Ion drag

The basic forces in a non-ionized fluid are pressure gradients, the gravitational and Coriolis forces, and viscosity effects. Earth's upper atmosphere, however, is partially ionized and in the presence of the intrinsic internal magnetic field gives rise to an additional force in the atmosphere, known as ion drag. In the absence of electric fields and collisions with neutral particles, the partially ionized plasma in Earth's atmosphere is bound by the magnetic field. In this case, plasma can flow parallel to the magnetic field direction but is inhibited in flowing perpendicularly to the field.

In the upper thermosphere collisions between the ions and neutral particles (neutrals) are relatively infrequent; ions collide with a neutral about once per second at around 300 km. The collisions are, however, sufficient to cause a drag on the neutral flow. The basic force per unit mass can be expressed as $\nu_{ni}(\mathbf{v} - \mathbf{v}_i)$: the force is proportional to the difference between the neutral wind velocity $\mathbf{v}$ and the ion velocity $\mathbf{v}_i$, scaled by the neutral–ion collision frequency ν_{ni}. With the low ion fraction, neutrals collide with an ion very infrequently in the upper thermosphere, so it can take tens of minutes to accelerate the bulky neutral gas. At lower altitudes, although ions collide more frequently with neutrals the ion fraction is generally even less, so it can take an hour or more to accelerate the gas to appreciable velocities.

The simplicity of this formalism hides a great deal of complexity in both the ion motion and the drag force as functions of altitude through the atmospheric domain. The different collision frequencies between neutrals and ions and neutrals and electrons results in charge separation, current flow, and the generation of either polarization electric fields or field-aligned currents that demand a response from the magnetosphere. We expand upon the electromagnetic nature of the dynamo processes[†] and the response of the partially ionized plasma to externally imposed electric fields later. At this point it is sufficient to say that, in the altitude range of atmospheric gas, the current flow introduces only a small perturbation to the intrinsic internal magnetic field of the planet, typically by less than one percent. The variation in the magnetic field can therefore be assumed to be small for most applications to ion drag and neutral wind dynamics, in contrast with the situation at high altitudes in the region of the planetary magnetospheres.

12.2.5 Momentum equation

The horizontal acceleration, or the basic force on a unit mass of a parcel of neutral atmospheric fluid, is obtained on the basis of the principles behind the Navier–Stokes equations and can be expressed by the so-called "primitive" equation

$$\frac{d\mathbf{v}}{dt} = -\frac{\nabla p}{\rho} - 2\mathbf{\Omega} \times \mathbf{v} + \frac{\nabla(\mu(\nabla \cdot \mathbf{v}))}{\rho} - \nu_{ni}(\mathbf{v} - \mathbf{v}_i), \qquad (12.6)$$

[†] The term "dynamo process" has several different meanings in the heliophysical literature. At the most fundamental level, as used in this chapter, it refers to forces that can induce electric currents. In the language of the magnetosphere, it refers to processes that have a Lorentz force $\mathbf{j} \times \mathbf{B}$ acting against a flow. In astrophysical parlance, a "dynamo process" is a flow-field interaction that maintains magnetic energy at a finite level well beyond the resistive decay time scale.

where the terms on the right are the horizontal pressure gradient and the Coriolis, viscosity, and ion drag forces, respectively; Ω is the angular velocity and μ is the coefficient of molecular viscosity.

The basic equations of motion include nonlinearities of the system, which are embedded in the total derivative operator $d/dt = \partial/\partial t + \mathbf{v} \cdot \nabla$. At low velocities, the effect of transport by the wind field is relatively minor. The tenuous gas of the upper atmosphere, however, can support wind speeds far exceeding those typical of the lower atmosphere. The degree of nonlinearity of a dynamic system is often defined by the Rossby number $R_o = v/(fL)$, which is the ratio of the centripetal acceleration v^2/L and the Coriolis acceleration fv, where L is the scale length or the radius of curvature of a dynamical system and f is the Coriolis parameter. The magnitude of the Coriolis parameter $f = 2\Omega \sin\theta$ can be seen to depend on the angular velocity of the planet, Ω, and the latitude θ. A Rossby number[†] significantly greater than unity denotes a highly nonlinear system where transport or advection by the wind field is important. For example a Class 5 hurricane in Earth's troposphere has winds in excess of 156 mph or 70 m/s, with scale sizes of one to two hundred kilometers, indicating a highly nonlinear system with a Rossby number greater than 5. The hurricane-like vortex associated with the Great Red Spot in Jupiter's lower atmosphere has wind speeds in excess of 400 mph or 160 m/s and a scale size of 10,000 km. Surprisingly, although still in the range where nonlinearities are significant, this indicates a significantly smaller Rossby number (i.e. relatively more important effects related to the planet's rotation through the Coriolis term). In comparison, upper atmosphere winds on Earth are typically 160 m/s and can exceed 1000 m/s under the strong ion drag forcing imposed from the magnetosphere at high latitudes. The scale sizes tend to be larger so that typical Rossby numbers at high latitudes are similar to Earth's hurricanes. Other examples of the dynamical response in Earth's upper atmosphere will be considered later.

12.3 Neutral-gas mixing, fractionation, and global circulation

In the lower atmosphere of a planet, the fluid is very well mixed. For instance the relative proportion of molecular nitrogen and molecular oxygen in Earth's atmosphere is remarkably constant from the surface to about 110 km altitude. The atmosphere below this altitude is known as the homosphere and is constantly being mixed by turbulent wave eddies. It is only at altitudes above about 110 km

---

[†] The Rossby number can also be expressed as a ratio of time scales: $R'_o = P/\tau$, where P is the rotation period of the body and τ is the characteristic time scale for a flow to complete a full turn (also known as the turnover time). In astrophysical settings this is used to quantify, for example, the relative sensitivity of the overturning of convective motions to a star's rotation, which introduces a preferred orientation to the otherwise random convection. The Rossby number is thus a measure used to gauge dynamo efficiency (see Vol. III).

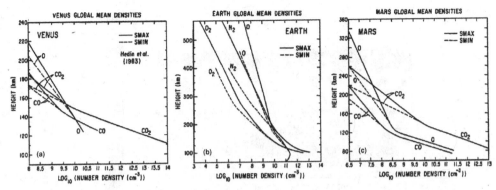

Fig. 12.2. Comparison of the global mean vertical profiles of the major species in the neutral upper atmospheres of (a) Venus, (b) Earth, and (c) Mars for low and high solar activity (from Bougher and Roble, 1991). The lines labeled SMIN and SMAX indicate solar minimum and maximum conditions. Note that the turbopause heights (where turbulent mixing and diffusive separation are comparable) are ~ 135, ~ 110, and ~ 125 km for Venus, Earth, and Mars, respectively.

that turbulent mixing gives way to molecular mixing processes, where each species begins to be distributed vertically under its own pressure scale height or hydrostatic balance; see Eq. (12.2). A heavy species such as carbon dioxide will decrease in concentration with height more rapidly than a lighter species, such as atomic oxygen (see Fig. 12.2). Each species i will have its own characteristic scale height H_i, where $H_i = RT/(M_i g)$ is the vertical distance over which a species decreases in partial pressure and number density by a fraction $1/e$. The upper atmosphere differs from the lower atmosphere in this respect, so that the mean mass of the fluid as well as other gas parameters such as the specific heat c_p, will change with altitude.

The vertical distribution of species is therefore affected by the balance between turbulent mixing and diffusive separation. Traditionally, the point of transition where both processes contribute equally has been termed the turbopause, which is typically assigned to an altitude of about 110 km for Earth's atmosphere. This altitude can vary with location and season, of course, depending on the strength of the gravity waves and other sources from the lower atmosphere responsible for the mixing and on the likelihood that the waves will break. Gravity-wave breaking is a complex field and will not be developed further here.

The vertical distribution of species also has a global seasonal or latitudinal structure due to large-scale advection. During the solstice period, the temperature difference between the summer and winter hemispheres introduces a pressure gradient force in the direction from the summer to the winter hemisphere. In the absence of drag, zonal winds would develop such that the Coriolis force from the zonal winds would balance the meridional pressure gradient, a condition known

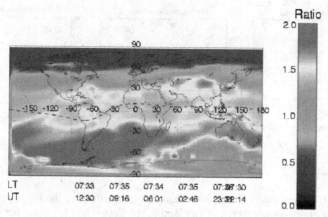

Fig. 12.3. Global distribution for 14 April 2002 of the column-integrated ratio of atomic oxygen and molecular nitrogen across the middle and upper thermosphere from observations of the GUVI instrument on the TIMED satellite (Paxton *et al.*, 1999). For this April season the global circulation is already more solstice-like, which produces low values in the northern-summer high latitudes and a maximum in the southern-winter hemisphere. The shift of the peak in the southern hemisphere from polar to winter mid latitudes is due to the competition between the solar-driven circulation and that driven by the high-latitude magnetospheric source of heat. See also the color-plate section. (Courtesy of L. Paxton.)

as geostrophic balance. In practice the zonal winds experience drag from collisions with ions or from viscosity, so that this pure geostrophic balance rarely occurs. The result of this imbalance is an inter-hemispheric circulation from summer to winter. Closure of this circulation drives an upwelling of material across pressure surfaces in the summer hemisphere and a downwelling in the winter hemisphere. The upwelling causes the heavier, molecular-rich, gas, which had diffusively separated out at lower altitudes, to be transported upwards, increasing the mean molecular mass in summer. In winter, downwelling reduces the mean mass (see Fig. 12.3). The mixing of the neutral composition by global circulation impacts the ionosphere. Ion loss rates are faster in the molecular-rich atmosphere, so the neutral composition structure gives rise to a seasonal ionospheric anomaly, where the dayside winter plasma densities are greater than in summer; this is the so-called winter anomaly.

The large-scale global seasonal circulation is analogous to a huge inter-hemispheric mixing cell, a global equivalent of the small-scale turbulent mixing cells in the lower atmosphere. The mixing effect is the same. The implication is that the atmosphere is better mixed at solstice. Through the Earth year there will be two peaks in mixing, in June and December, whereas at equinox the weaker global circulation allows the atmosphere to begin to separate out by molecular diffusion. The difference is subtle, but there are several consequences of this semi-annual variation. First, the globally averaged mean mass will vary semi-annually. Second, since the scale height or thickness of the atmosphere depends on the mean molecular mass, the atmosphere will be more compressed at solstice and more expanded at equinox. This semi-annual breathing of the upper atmosphere introduces a semi-annual variation in neutral density. Third, since the loss rate of the ionosphere is dependent on the neutral composition, a semi-annual variation in plasma density is introduced.

The relatively simple pattern of summer-to-winter circulation is augmented by the addition of the high-latitude magnetospheric or "geomagnetic" sources. Such a heat source tends to reinforce the solar-radiation-driven equatorward flow in the summer hemisphere and to compete with the poleward flow in winter, to add an additional complexity. During geomagnetic storms the high-latitude source can dominate the solar-driven circulation. In the giant planets the auroral energy always exceeds the solar input.

In the turbulent region, momentum and energy are also mixed by eddies. A parcel of gas that is displaced vertically by a turbulent eddy will undergo adiabatic heating or cooling, depending on whether the parcel has been displaced upward to a lower pressure level where it will expand and cool or downward to a region of higher pressure where it will be compressed and heated. In the absence of local heat sources, the equilibrium vertical temperature profile under the action of turbulence is the adiabatic lapse rate g/c_p. In the region of molecular diffusion, individual atoms and molecules rather than parcels of gas exchange locations, so the equilibrium profile migrates towards isothermal. The vertical transport of momentum by turbulent mixing or molecular processes is not affected by adiabatic heating and cooling so, in both cases, the viscosity of the medium tends to smear out gradients and, in the absence of momentum sources, the medium will approach a constant temperature.

12.4 Energy input and dissipation

12.4.1 Heating, ionization, and dissociation profiles

Much of the external sources of the heating, ionization, and dissociation of a planetary atmosphere come from the absorption of photons or particles impinging on the neutral atmosphere. The physics defining the altitude profile of the three processes is the same. For example, the rate of ionization q by solar radiation of intensity I at some layer in the atmosphere where the number density is $n(h)$ can be expressed as a product of four factors:

$$q = \eta \sigma_a n(h) I \qquad (12.7)$$

where σ_a is the absorption cross section and η is the ionizing efficiency; η could equally be the heating or dissociation efficiency. The intensity of the radiation

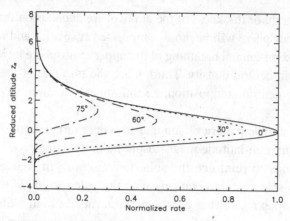

Fig. 12.4. Peak-normalized shape of the vertical projection of the classical Chapman profile for heating, ionization, and dissociation rates in a stratified hydrostatic atmosphere. The reduced height $z_H = (h - h_m)/H$ is obtained from the height h_m of the peak rate at zero zenith angle, and is expressed in density scale heights. The curves are labeled by the solar zenith angle.

gradually decreases along the path through the atmosphere, starting from initial intensity I_∞. The altitude deposition profile depends on the absorption coefficient and the atmospheric number density, which varies exponentially with height. Clearly the product of the intensity I of the radiation, which decreases as the source penetrates the atmosphere, and the atmospheric number density $n(h)$, which increases with increasing depth into the atmosphere, must reach a maximum at some altitude or, more correctly, at some pressure level. The level of penetration is referred to as the optical depth τ and is expressed mathematically as

$$\tau = \sigma_a n(h) H(h) \sec \chi, \tag{12.8}$$

where the product of the number density $n(h)$ at height h and the scale height $H(h)$ at that level represents the integrated content of the column of gas above that point and χ is the angle from the zenith at which the radiation penetrates a planar atmosphere.

The profile of the rate of heating, ionization, or dissociation from these processes takes the form of the classical Chapman profile (shown normalized and as a function of the reduced height in Fig. 12.4) and is given mathematically by

$$Ch = I_\infty \exp\left[-\sigma_a n(h) H(h) \sec \chi\right] \eta \sigma_a n(h). \tag{12.9}$$

The peak of the profile is at unit optical depth, which depends on the mass of the atmosphere traversed by an incoming energetic photon or particle. This corresponds to a fixed pressure level for a given angle of incidence. In pressure coordinates the depth of penetration into the atmosphere of a photon or particle therefore does not change with the gas temperature or the degree of thermal expansion. Even with the changing heating over the solar cycle or during a storm that might cause a thermal expansion of the atmospheric gas, that same radiation will still penetrate and produce heating or ionization at the same pressure level. The altitude associated with that pressure and the local number density would, of course, be different since they depend explicitly on the gas temperature.

12.4.2 Adiabatic warming and cooling processes

Adiabatic processes cause changes in the temperature of a fluid in response to the compression or expansion of a gas parcel. Solar heating initially imparts energy and provides the heat source for the upper atmosphere. Local or regional heating imposes horizontal pressure gradients that set the atmosphere in motion horizontally. The continuity of the global circulation is closed by upwelling in the region of divergence and downwelling in the region of convergence. The upwelling of a parcel of gas to a region of reduced pressure causes the parcel to expand and adiabatically cool whereas downwelling transports parcels of air to regions of higher pressure, causing compression and adiabatic heating.

12.4.3 Joule heating

As well as the momentum transfer involved in the neutral–plasma interactions there is also an energy exchange via frictional dissipation from the movement of the ions through the neutrals or the neutrals through the ions. This frictional dissipation, from the perspective of the neutral gas, is known as Joule heating. The effect can also be described as the dissipation of a current flowing through the resistive medium of the neutral gas. This concept will be elaborated upon later in the discussion on electrodynamics (Section 12.6).

12.4.4 Energy equation

The advective form of the internal energy equation can be expressed as

$$\frac{dc_p T}{dt} - \frac{\omega}{\rho} = Q \tag{12.10}$$

where $c_p T$ is the specific enthalpy or internal energy of the gas and the second term represents adiabatic heating or cooling; Q represents the various heat sources and sinks including solar and Joule heating, viscous dissipation, radiative cooling, and vertical heat conduction by both molecular and eddy transport. In the same way as for the momentum equation, the nonlinearities of the system are embedded in the total derivative.

12.5 Ionization fraction

The ionosphere is the ionized portion of the upper atmosphere that extends upward from about 60 km and at its upper levels merges with the plasmasphere at mid and low latitudes and with the magnetosphere at high latitudes. Unlike the neutral component, the large-scale ionized component does not always approach hydrostatic equilibrium. The production of ion and electron pairs from direct ionization or from dissociative ionization by energetic photons or particles follows the Chapman profile, but the resultant profile shape depends on the relative strengths of chemical loss and diffusion. If diffusion is the dominant process, which is true for the top-side ionosphere, the equilibrium profile tends towards the hydrostatic or diffusive equilibrium profile, with an exponential decrease with altitude. The scale height of the plasma is influenced by the Coulomb interaction between the ions and electrons, which does not allow the heavy (ions) and light (electrons) plasma species to separate out in the same way that neutrals do. This effect is called ambipolar diffusion and results in a plasma scale height, $2RT/(M_i g)$, for a given temperature which is twice that expected for the ion mass M_i.

By convention, the ionosphere is described by a series of layers or regions referred to by the letters D, E, F_1, and F_2. The history of the naming convention is described in Rishbeth and Garriott (1969). Typical dayside and nightside mid-latitude profiles are shown in Fig. 12.5. The reason for the appearance of a peak plasma density in the F region is that below this peak the diffusive profile is eroded away by chemical loss processes. The distinctions between layers arise from the nature of the loss process and the altitude profile of production. The dominant ion in the topmost F region is O^+, ionized primarily by solar coronal extreme ultraviolet emission. At progressively higher altitudes the dominant ions give way to He^+ and H^+. In the successively lower E and D regions, the dominant ions are O_2^+, NO^+, and N_2^+, which are close to chemical equilibrium. Molecular ions O_2^+ are ionized by coronal X-ray and transition region FUV photons and NO^+ is ionized primarily by chromospheric Lyman-α photons; O_2^+ and N_2^+ are ionized by hard X-ray photons.

There are two main loss processes for the charged-particle component, namely dissociation and radiative recombination, and these depend on the neutral medium in which the ions are embedded. The latter process typically proceeds through a multi-step path with charge exchange involving intermediate species. The spectrum of energetic photons in the extreme ultraviolet solar radiation produces overlapping Chapman profiles contributing to the production in the various layers. For instance the F1 layer appears during the day when production is strong, and the E-region can appear as a separate peak due to the production by Lyman-β photons. The ion continuity equation therefore must include photochemical production and loss processes and the transport of plasma must include both electrodynamic and collisional processes.

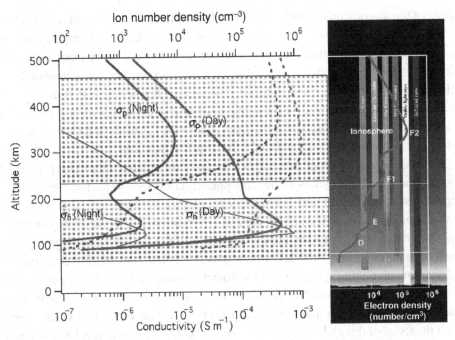

Fig. 12.5. Typical vertical profiles of the ion number density (broken line) and Pedersen and Hall conductivities for day and night conditions. The peak plasma densities in the F_2 layer are at around 300 km altitude (from Heelis, 2004). Also shown are the atmospheric penetration depths of various wavelengths of solar radiation from X-rays to infrared.

The height of the peak in the charged-particle density at the F region is initially controlled by the balance between the diffusive and chemical loss processes. The orientation of the magnetic field also has an impact. At mid latitudes on Earth the magnetic field is inclined to the horizontal, so that the meridional neutral atmosphere winds tend to push the plasma up and down along the direction of the magnetic field to a different altitude. The typical diurnal variation in F-region winds, poleward during the day and equatorward at night, imposes a diurnal variation on the height of the F region, so that it is lower by day and higher by night. Electrodynamic processes, particularly in the regions where the magnetic field is close to horizontal, also affect the vertical structure.

12.6 Electrodynamics

Ion drag and Joule heating, represented in the neutral momentum and thermal energy, respectively, are two manifestations of the electrodynamic interaction between the neutral and ionized components in Earth's upper atmosphere. The nature of the interaction depends on the height profile of the ionization fraction and on the mobility of the ions and electrons within the ambient medium pervaded by the magnetic field. Together they define the momentum and energy exchange and the current flow in the system. Note that, as mentioned earlier, the magnetic field below 600 km altitude in Earth's upper atmosphere changes by less than 1% as a result of local ionospheric or remote magnetospheric current systems. The usual assumption is that the magnetic field is invariant when considering the interaction between the ion and neutral medium in the upper atmosphere.

The electrodynamic properties can be conveniently separated into a high-latitude region, where the current flow in the ionosphere is connected to the magnetospheric current system, and a mid-and-low-latitude region, where the majority of the current flow and polarization electric fields are controlled internally by the thermosphere–ionosphere conductivity and dynamics. The exception is the leakage of high-latitude electrodynamics to low latitudes, which makes a small contribution even during quiet geomagnetic times; during storms, however, the so-called prompt penetration of the magnetospheric electric field can be a significant source of low-latitude electric field. The mechanism for this apparent prompt penetration of electric field is not the focus of this chapter. As will be shown, electric fields at low latitudes have a dominant impact on the redistribution of ionospheric plasma and the shaping of plasma structure.

This section provides a discussion of the physical principles underlying the electrodynamic properties of the high, mid, and equatorial ionosphere. Several existing texts discuss various aspects of ionospheric electrodynamics and some of the attendant phenomena; they include Heelis (2004), Richmond (1995), Kelley (1989), Schunk and Nagy (2000), Rishbeth and Garriott (1969), and references therein.

12.6.1 High-latitude electrodynamics

The magnetospheric electrodynamic sources are imposed upon the upper atmosphere by a current system that can flow along the highly conducting field lines linking the plasma processes in the magnetosphere with the ionosphere. The closure of currents originating in the magnetosphere requires the electrical and mechanical forces to move the charged particles in the atmosphere. An electric field $\mathbf{E}$ in the upper atmosphere provides a force $e\mathbf{E}$ on a particle with positive charge e (most positive ions in planetary atmospheres are singly charged). An electric field in any given frame can of course be removed by considering an alternative frame moving at a velocity with respect to the first. In this frame the electric force is replaced by an equivalent mechanical force. It is therefore important that the electrical and mechanical forces are considered in an integrated way. Two other significant forces on the plasma must be taken into account; these control the plasma motion and the current flow perpendicular to the magnetic field. The first is the Lorentz force $e\mathbf{v_i} \times \mathbf{B}$ on a charged particle moving with velocity $\mathbf{v_i}$ in a magnetic field $\mathbf{B}$; the Lorentz force causes a moving charge to gyrate around the magnetic field and, if the charge is made to accelerate further by an electric field or any other force, to move in a direction perpendicular to $\mathbf{B}$ and that force. The second force comes from momentum exchange due to collisions with the neutral gas. This force, $m_i \nu (\mathbf{v_i} - \mathbf{v})$, depends on the collision frequency ν between charged particles of mass m_i moving at velocity $\mathbf{v_i}$ and neutral-gas particles moving at velocity $\mathbf{v}$.

In the ionosphere, currents flowing perpendicularly to the magnetic field are produced by electric fields and neutral winds. Divergence in these perpendicular currents will either drive field-aligned currents or cause a charge build-up and the production of additional polarization electric fields, so that the divergence of the total current remains zero.

At altitudes above $\sim$160 km, the collisions of ions with the neutral gas are relatively infrequent, decreasing from tens of collisions per second to less than one above 300 km. In contrast, the ions gyrate about the magnetic field about two hundred times per second, so that the steady-state plasma drift perpendicular to $\mathbf{B}$ is a balance between the force from the electric field and the Lorentz force. The result is a steady drift velocity of the plasma perpendicular to the electric field and the magnetic field directions, with magnitude $\mathbf{E} \times \mathbf{B}/B^2$. At high latitudes the ions respond directly to the strong magnetospherically imposed electric fields, which cause ion drifts at many hundreds, if not thousands, of meters per second. Although collisions between ions and the neutral gas are relatively infrequent above $\sim$160 km, they are sufficient to accelerate the neutral component, i.e. the thermosphere, at high latitudes to many hundreds of meters per second over periods of tens of minutes or more. Figure 12.6 shows the ion drift and neutral wind during moderate geomagnetic activity ($K_p \sim 3$) poleward of 40° geographic latitude, in response to a fairly typical two-cell magnetospheric convection pattern (see Chapter 10). In the left-hand panel of Fig. 12.6, the vectors represent the plasma drift velocity in the upper thermosphere, where the ion drift velocity is close to $\mathbf{E} \times \mathbf{B}/B^2$, since the collisions with the thermosphere are relatively infrequent. The color contours (see the color-plate section) represent the plasma density close to 300 km altitude, near the F-region peak. The effect of transport on the plasma density is clearly visible with the formation of the sub-auroral trough and the advection of the dayside solar-produced plasma towards and across the polar region.

In the right-hand panel of Fig. 12.6 the neutral wind and temperature response over the same region is shown. Winds driven by solar heating alone would typically be anti-sunward and would reach $\sim$ 150 m/s at 300 km in the polar region. With

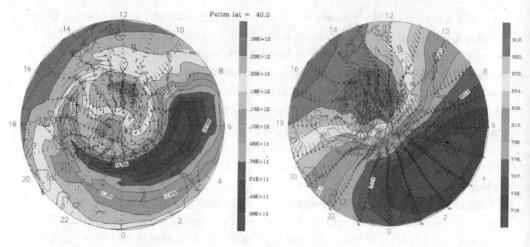

Fig. 12.6. (Left) Simulation of plasma drift and electron density at the altitude of the *F*-region peak in the upper thermosphere for moderate geomagnetic activity (for a global planetary K_p index ~ 3). The region poleward of 40° latitude is shown at 18 UT. (Right) Neutral wind and temperature over the same region at 300 km altitude in response to plasma drift. The maximum plasma drift velocity is ~ 700 m/s and the peak wind speed is ~ 380 m/s. (From Fuller-Rowell *et al.*, 2008.) See also the color-plate section.

the imposition of magnetospheric convection, even with the infrequent collisions at this altitude there is a sufficient momentum source to accelerate the medium to close to 400 m/s. The first reaction of the neutrals is to follow the ion convection, but other inertial and viscous forces act to introduce asymmetries in the circulation or cell structure. The clockwise cell excited in the dusk sector is particularly strong owing to an inertial resonance between the ion and neutral convection. The natural motion of the neutral gas is to form a clockwise vortex, owing to the action of the Coriolis force, so there is a natural tendency for this cell to develop and resonate. In contrast the dawn cell is less well formed, with weak or virtually non-existent sunward winds in the dawn-sector auroral oval in response to the plasma convection. This anticlockwise cell does not have a resonance with the ion convection since momentum is continually being transported out of the cell by the tendency for the vortex to diverge. The neutral temperature structure at this altitude, shown in the right-hand panel, has a fairly modest impact from Joule heating; it is controlled more by transport of the neutral gas by the elevated circulation. The warmer dayside gas tends to be advected poleward by the neutral circulation, in much the same way as the ionized component is transported poleward by the ion drift.

The fact that neutral winds are still accelerated in the upper thermosphere implies that the plasma drift cannot be purely $\mathbf{E} \times \mathbf{B}/B^2$: a small momentum exchange must be influencing the plasma velocities as well. This effect, however, is small and very difficult to detect in observations above 200 km altitude. In fact, the assumption is often made that a measurement of the plasma drift can be used to infer the electric field perpendicular to the magnetic field.

A similar computation in the lower thermosphere, where collisions between the neutrals and ions are more dominant, shows a quite different pattern. As the collision frequency increases in the denser lower thermosphere, the mobility of the ions is more profoundly affected and imposes large departures from the $\mathbf{E} \times \mathbf{B}/B^2$ drift. At low altitudes, $\sim 100\,\mathrm{km}$, the ions are forced to move with the neutral gas whether it is stationary or moving. The large-scale wind system at this altitude is driven by the tidal and planetary waves propagating from the lower-atmospheric terrestrial weather system, and the mass of the atmosphere is such that ion drag has little or no impact on the neutral dynamics. The narrow altitude range between 100 and 160 km altitude is responsible for most of the dissipation of electromagnetic energy from the magnetosphere. The neutral dynamics and conductivity in this boundary region between space and atmospheric plasma are critical.

If the neutral gas is stationary then the ion drift vector rotates from the $\mathbf{E} \times \mathbf{B}$ direction towards the electric field direction, and significantly reduces in magnitude, as the collision frequency increases. The degree of rotation depends on the ratio of the ion gyrofrequency eB/m (in SI units) and the ion–neutral collision frequency ν_{in} (this ratio quantifies the particle magnetization, see below). When the two are equal, which occurs at about 120 km altitude, the ion drift rotates by $45°$, halfway between the directions of $\mathbf{E}$ and $\mathbf{E} \times \mathbf{B}$, and is reduced in magnitude. With no electric field and infrequent collisions, the force from the collisions with the neutral gas moving with velocity $\mathbf{v}$ forces the plasma in a direction perpendicular to $\mathbf{B}$ and the neutral-wind direction $\mathbf{v}$. Again, with increasing collision frequency the drift direction rotates from this $\mathbf{v} \times \mathbf{B}$ direction towards $\mathbf{v}$. Figure 12.7 shows the ion drift magnitude and direction in response to either an electric field and no wind, or with wind alone and no electric fields (from Heelis, 2004).

Over the entire altitude range above 100 km altitude the electron collision frequency with the neutral gas remains low, so that electrons continue to drift in the $\mathbf{E} \times \mathbf{B}$ direction. It is only below 100 km that the electron mobility begins to be impacted by collisions with neutrals.

The relative mobilities of the ions and electrons give rise to a current flow

$$\mathbf{j} = n_i e(\mathbf{v}_i - \mathbf{v}_e), \tag{12.11}$$

where n_i is the plasma density and $\mathbf{v}_i$ and $\mathbf{v}_e$ represent the ion and electron velocities, respectively. The mobilities can be used to define the electrical conductivities of the medium in the Pedersen and Hall directions perpendicular to $\mathbf{B}$ (see Figure 12.5).

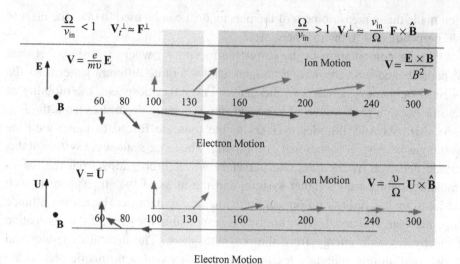

Fig. 12.7. Approximate values of the magnitude and direction of the ion and electron drifts produced by the direct action of neutral winds and electric fields. The approximate locations of the E- and F-regions are indicated by the narrower and wider shaded regions, respectively (from Heelis, 2004).

If we take the magnetic field to be aligned with the z-axis then the generalized Ohm's law $\mathbf{j} = \Sigma \cdot \mathbf{E}_0$ (where $\mathbf{E}_0$ is the total electric field, i.e. $\mathbf{E}_0 = \mathbf{E} + \mathbf{v} \times \mathbf{B}$) contains the conductivity tensor

$$\Sigma = \begin{pmatrix} \sigma_P & \sigma_H & 0 \\ -\sigma_H & \sigma_P & 0 \\ 0 & 0 & \sigma_\parallel \end{pmatrix}, \tag{12.12}$$

where the Pedersen, Hall, and parallel conductivities are given by

$$\sigma_P = n_e e^2 \left[\frac{1}{m_e \nu_{en}} \left(\frac{1}{1 + M_e^2} \right) + \frac{1}{m_i \nu_{in}} \left(\frac{1}{1 + M_i^2} \right) \right], \tag{12.13}$$

$$\sigma_H = n_e e^2 \left[\frac{1}{m_e \nu_{en}} \left(\frac{M_e}{1 + M_e^2} \right) - \frac{1}{m_i \nu_{in}} \left(\frac{M_i}{1 + M_i^2} \right) \right], \tag{12.14}$$

$$\sigma_\parallel = n_e e^2 \left(\frac{1}{m_e \nu_{en}} + \frac{1}{m_i \nu_{in}} \right) \tag{12.15}$$

(see e.g. Luhmann, 1995b, or Baumjohann and Treumann, 1997). Here $m_{e,i}$ are the electron and ion masses, $\nu_{en,in}$ are the electron–neutral and ion–neutral collision frequencies, and $M_{e,i} = \Omega_{e,i}/\nu_{e,i}$ are the electron and ion magnetizations. We note that $\Omega_{e,i} = |q_{e,i}| B / m_{e,i}$ are the electron and ion (with charge q_i) gyrofrequencies around the field of strength B. The effect of the collisions is to rotate the net current from the direction of $\mathbf{E}$ at high altitudes towards the negative $\mathbf{E} \times \mathbf{B}$ direction at low altitudes. The current and dissipation reach a peak at the altitude where the Pedersen and Hall conductivities are equal, around 125 km. For high-frequency currents, such as those that can occur in the solar chromosphere, the dissipation may increase markedly (see Section 12.8). Note that σ_P is generally dominated by the ion term.

The horizontal current flow $\mathbf{j}_h$, expressed as

$$\mathbf{j}_h = \Sigma \cdot (\mathbf{E} + \mathbf{v} \times \mathbf{B}), \tag{12.16}$$

contains the electric field $\mathbf{E}$ that is the net consequence of the magnetospheric and ionospheric dynamos and the mutual interaction between them. At high latitudes the magnetospheric dynamo usually dominates. In this electrodynamic context the equivalent expression for ion drag is $\mathbf{j} \times \mathbf{B}/\rho$. At high latitudes, where there is a direct connection between magnetospherically driven potentials and the upper atmosphere, the kinetic energy from the momentum forcing and the Joule heating from the dissipation of high-latitude currents imposes a load on the magnetosphere. The total electrodynamic energy,

$$\mathbf{j} \cdot (\mathbf{E} + \mathbf{v} \times \mathbf{B}) + \mathbf{v} \cdot (\mathbf{j} \times \mathbf{B}), \tag{12.17}$$

is the sum of the Joule heating (the first term) and the change in kinetic energy from ion drag (the second term) and is equal to $\mathbf{j} \cdot \mathbf{E}$.

These quantities can refer to a particular point in space or an integral along the magnetic field. The neutral-wind system is intimately involved in the energy deposited by the magnetospheric currents. Notice that, as the winds are accelerated in the direction of the ion drift, the load on the magnetosphere is reduced. If the winds match the ion drift then the dissipation drops to zero. Note also that the neutral wind system drives a current and the resulting electric field at high latitude is a combination or balance between the magnetospheric and thermospheric wind generators. Separating their relative contribution from a given electric field observation is impossible unless the height profiles of the conductivities and winds are known.

The electromagnetic energy flowing between the magnetosphere and upper atmosphere is also referred to as the Poynting flux; this is a measure of the field-aligned integral of the total Joule heating and kinetic energy from ion drag flowing into or out of the region. It can be quantified by observing the magnetic perturbations of the field-aligned currents flowing between the magnetosphere and ionosphere. The magnetic signature gives no information about the altitude profile of the dissipation, so the actual impact on the atmosphere, such as in a temperature or density increase, cannot be quantified from this information. The flow of energy is usually from the magnetosphere to the upper atmosphere, but there can be regions in the spatial distribution where the flow is reversed. For instance, when the momentum of the neutral atmosphere is transported to a region where the magnetospheric component of electromagnetic energy is small, the thermospheric dynamo can feed back energy to the magnetosphere. Another case is the classical flywheel effect, where winds have been accelerated to high velocities when the magnetospheric forcing is high but the external driver subsequently drops. The inertia of the thermospheric gas causes the winds to persist and again feed the magnetosphere with electromagnetic energy.

In addition to the Joule and kinetic energy contributions, a third component of energy exchange between the magnetosphere and upper atmosphere comes from the particle flux. Generally the particle energy flux is downward in the form of the precipitation of auroral electrons, although proton precipitation can be equally important in some spatial areas, particularly in the dusk sector. The particle precipitation, as well as depositing about 20% of the energy on average, also carries the field-aligned currents linking the regions. Upward currents are normally associated with electrons precipitating from the magnetosphere, downward currents are normally associated with upward flowing ionospheric electrons. Downward energetic-electron precipitation is also a source of ionization and hence conductivity. If the charges available are not sufficient to supply a current flow parallel to the magnetic field, potential drops can form that will accelerate the electrons to higher energies. Because of their greater mass, protons or positive ions play a smaller role in the energy exchange although upward flowing protons and O^+ ions from the ionosphere are a major source of ions in the magnetosphere. During a major storm event a significant fraction of the magnetospheric plasma consists of heavy ions from the ionosphere, which introduces a mass loading and influences the response of the magnetosphere to the solar wind dynamo. The upflow of heavy ions from the ionosphere to the magnetosphere can initially be driven by the thermal expansion but other forces, such as parallel electric fields, are required for the gravitational force to be to completely overcome, so that particles escape from the atmosphere.

On large spatial scales, exceeding about 100 km, the electric fields are expected to map efficiently between the regions. In the discussion so far the electrodynamic forcing has been assumed to be quasi-steady, whereby the magnetosphere and ionospheric dynamos are in balance. On large spatial scales and long time scales, i.e. tens of minutes, this is usually the case. However, on short time scales the two can be mismatched. Information about electric field shears or potential drops is carried between the magnetosphere and ionosphere by Alfvén waves, in much the same way that acoustic waves carry information about pressure changes in the neutral gas. The Alfvén waves are reflected back and forth to adjust potential structures and field-aligned currents. In so doing, single structures can form into multiple structures.

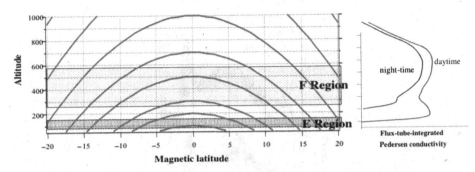

Fig. 12.8. Shape of the magnetic field at mid and low latitudes in Earth's atmosphere, which gives rise to its unique electrodynamic properties (from Heelis, 2004).

12.6.2 Mid- and low-latitude electrodynamics

The same electrodynamic processes occur at mid and low latitudes. The differences from the high-latitude case are, first, the fact that a connection with the magnetosphere is no longer dominant and, second, the particular configuration of the magnetic field, Fig. 12.8. At high latitudes the magnetic field connects ionospheric dynamo processes to those in the magnetosphere. At mid and low latitudes the electric fields arise largely from internal dynamo processes driven by the conversion of neutral-wind kinetic energy to electromagnetic energy and are typically an order of magnitude smaller (at just a few millivolts per meter) than the high-latitude fields. The energy involved is also much smaller. The importance of these small electric fields at low latitudes is no longer the Joule heating and momentum dissipation that they produce but rather their role in the redistribution of plasma.

During quiet times, the electric fields are driven by a combination of the E- and F-region dynamo processes. The net result at the magnetic equator is an upward plasma drift (due to the eastward electric field) during the day and a downward drift (due to the westward electric field) at night, with a pre-reversal enhancement (PRE) just after sunset. Figure 12.9 shows the diurnal variation in the vertical plasma drift at the magnetic equator during quiet geomagnetic conditions, from the Jicamarca incoherent-scatter radar. The upward plasma transport induced by electrodynamic drift on the dayside generates the equatorial ionization anomaly (EIA), or Appleton anomaly, which is the most dominant feature in the global ionosphere structure (see Fig. 12.10). This feature is a so-called anomaly because, although the ionizing effect of solar radiation maximizes under the overhead Sun, the peaks in plasma density are located at about $15°$ latitude on either side of the magnetic equator. The EIA forms as a result of the dayside eastward electric fields and horizontal magnetic field, which cause the equatorial plasma to $\mathbf{E} \times \mathbf{B}$-drift upward. The gravitational force subsequently causes the plasma to diffuse downward and

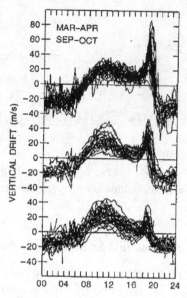

Fig. 12.9. Diurnal variation and day-to-day variability of the vertical plasma drift at Jicamarca, Peru, which is on the magnetic equator, at equinox and at low, mid, and high solar activity (from bottom to top). The drift is upward by day and downward by night and is punctuated by a strong post-sunset pre-reversal enhancement in the upward drift (from Fejer and Scherliess, 1997).

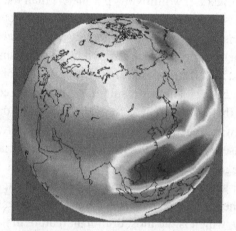

Fig. 12.10. Global structure of the *F*-region plasma density showing the equatorial ionization (or Appleton) anomaly at about 15° latitude on either side of the magnetic equator, from a numerical simulation using a coupled thermosphere, ionosphere, plasmasphere model with self-consistent mid- and low-latitude electrodynamics. See also the color-plate section.

poleward along the magnetic field to form the peaks on either side of the magnetic equator.

For a given wind system, and in the absence of magnetospheric penetration electric field, the ionospheric electric field $\mathbf{E}$ and current density $\mathbf{j}$ are determined by the dynamo equations as described above. The E-region dynamo is strong during the day; the wind system driving the currents arises from tides and planetary waves propagating upward from the lower atmosphere. The winds transport ions away from the local noon, since this is where the E-region plasma density and conductivity maximize. The $\mathbf{v}_i \times \mathbf{B}$ drift of the ions produces a charge accumulation at the terminators that is positive at dawn and negative at dusk. The fields map to the F region to produce the main diurnal variation of zonal electric field, as seen in Fig. 12.9, with an upward drift during the day and a downward drift at night at the magnetic equator. Mapped to the F region the meridional field produces a westward zonal drift during the day and an eastward drift at night.

Another significant difference at low latitudes, as compared with high latitudes, is that F-region dynamo processes are more important owing to the large volume of the flux-tube-integrated quantities, as a consequence of the magnetic field geometry. The E-region dynamo can explain the basic diurnal variation in the observed low-latitude $\mathbf{E} \times \mathbf{B}$ drifts, but it is the F-region dynamo that is responsible for the pre-reversal enhancement. On the dayside, wind-driven F-region current can map and close through the high-conductivity E region. After sunset a zonal wind at the magnetic equator, for instance, drives a vertical current that can no longer close through the E region. The current polarizes and, together with the local time gradient in the F-region flux-tube integrated conductivity, leads to the large increase in the zonal electric fields after sunset. A similar physical process can occur due to winds flowing across the dawn terminator, but the magnitude is smaller and the response more intermittent.

The strong PRE after sunset can also raise the ionosphere, giving rise to the Rayleigh–Taylor instability and producing conditions conducive to the generation of ionospheric irregularities. The latter are notoriously difficult to predict on a day-to-day basis. The physical process behind the appearance of irregularities is ion motion in the direction $\mathbf{g} \times \mathbf{B}$ induced by the gravitational acceleration $\mathbf{g}$. At low latitudes where the magnetic field is close to horizontal, the force leads to a zonal current. Divergence in this zonal current from the conductivity gradients will drive a current loop through the E region by day and electric polarization fields at night. Gravity-wave-driven perturbations on the bottom-side F region will cause low-density regions to $\mathbf{E} \times \mathbf{B}$-drift upward and high-density regions will drift downward (see Fig. 12.11). The perturbation will grow in time, and the growth rate is largest when the F region is raised to higher altitudes so that the ion–neutral collision frequency is low.

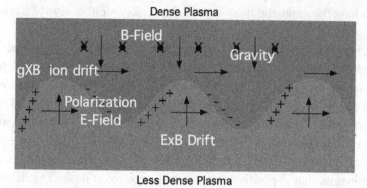

Fig. 12.11. Polarization fields (arrows pointing to the right) and associated $\mathbf{E} \times \mathbf{B}$ (arrows pointing upwards) drifts in the F region resulting from perturbations in the conductivity in the presence of the gravitationally driven ion current (from Heelis, 2004).

12.6.3 *System storm response*

Many of the basic physical principles coupling the ion and neutral dynamics and electrodynamics, and linking the ionosphere to the magnetosphere, have been presented above. Trying to evaluate the response of the system to extreme events, such as a geomagnetic storm, however, places a severe test on our understanding of the system.

During active storm conditions magnetospheric convection drives the high-latitude thermosphere–ionosphere system much more strongly. Plasma drift velocities and Joule heating rates increase dramatically. The neutral dynamics at high latitudes are more disturbed, with winds approaching 1 km/s. In these cases Joule heating becomes more prominent and can elevate the temperature at high latitudes by hundreds of kelvins in the upper thermosphere. This high-latitude Joule heating has global consequences: it is responsible for the launch of large-scale gravity wave or wind surges towards the equator and is the source of subsequent changes in the global circulation. The large-scale waves have typical wavelength 1000 km and phase propagation speeds ranging from 400 to 1000 m/s.

Changes in the neutral dynamics during storms provide the conduit for a series of changes in neutral composition, electrodynamics, and plasma density. The basic physical processes operating during a storm are the same at all phases of the solar cycle; however, the balance between solar-driven and storm-circulation processes, and the ensuing composition changes, will vary. Large storms can occur at any phase of the solar cycle but are more likely when the Sun is more magnetically active.

It is also important to remember that the high-latitude magnetospheric electric fields can also penetrate globally, giving the so-called prompt penetration electric field, and can have a direct and often dramatic influence on plasma restructuring at mid and low latitudes. A current challenge is to be able to separate and quantify the influence of the various physical processes in geospace during storms. Numerical models can be useful tools in this endeavor to unravel the physical processes.

Understanding the flare and storm response places a rigorous test on our understanding of the dynamics and electrodynamics of the system. Here we have introduced the basic physical processes; the detailed responses will be explored in Vol. II.

12.7 Outstanding issues and science questions

Our level of understanding of Earth's upper atmosphere is fairly mature; many of the controlling physical processes are understood at a reasonable level of sophistication. We are not at the level of the terrestrial weather, where the introduction of physically based data-assimilation techniques and a plethora of observations has brought the science to a level where a six-hour forecast of the system is just as accurate as current observations, an impressive position to have reached. Thermosphere–ionosphere science is not yet at that level, and some important characteristics of the system are still to be elucidated.

One outstanding challenge is to understand the relative contribution of the various physical processes. For instance, how important are non-hydrostatic as against hydrostatic processes in the dynamics and energetics of the atmosphere? We are in the enviable position, in comparison with some other heliophysical domains, of having a relatively high level of understanding. Figure 12.12 illustrates the plethora of dynamical, chemical, plasma, and electrodynamic processes present in the upper atmosphere of Earth, and its inherent complexity. (See also the color plate section.) This level of understanding poses the challenge of not only separating out the various physics processes in the response to, say, a geomagnetic storm or a solar flare but also of being able to bring our understanding to the quantitative level where a specification can lead to an accurate forecast of the system. Combining data with complex physical models, and the adoption of data-assimilation techniques, puts a much more severe constraint on our perceived understanding and it offers one of the new challenges for the future.

Another challenge is quantifying the impact of the lower atmosphere, i.e. the terrestrial weather, on the upper atmosphere. The energy content in the enthalpy and dynamics of the dense lower atmosphere far exceeds that of the upper atmosphere. Only a very small fraction of this huge source needs to propagate upward for it to have a dramatic influence on the upper atmosphere system. Such an energy and momentum transfer can be carried by any number of physical processes such as gravity waves, tides, or planetary waves. Our understanding of the nature of

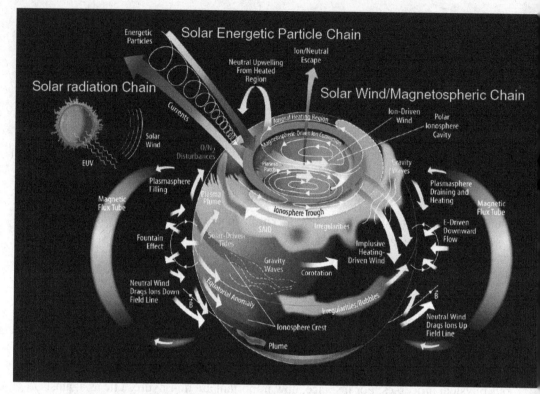

Fig. 12.12. Indication of the complexity of the ionosphere–thermosphere–mesosphere (ITM) system and the range of physical processes operating. This chapter has touched briefly on just a few of the physical profiles behind this complexity. Courtesy of J. Grebowsky (NASA GSFC). See also the color-plate section.

this coupling is at a very rudimentary level and is a significant challenge for the future. The differences between the apparently chaotic lower atmosphere and the strong external forcing of the upper atmosphere make the prospect of forecasting the system's behaviour extremely challenging.

A third challenging area is the interaction between space plasma and the atmosphere. The majority of the energy and momentum deposition of space plasma is constrained to a narrow altitude band in the lower thermosphere that is extremely inaccessible. In this 100 to 200 km altitude band, satellite lifetimes are very short so there is a strong reliance on remote sensing or the very occasional rocket probe. The way in which currents close through this critical region is very speculative, and the vertical structure and balance between momentum and energy dissipation is unknown. Yet it is this interface region with space that influences the connection with the magnetosphere and ultimately the drag on the solar wind imposed by the planet.

There are also several other areas still to be elucidated, such as the impact of space climate on global change, for example. Comparing and understanding how the common physical processes apply to the range of objects in the heliosphere such as the terrestrial and Jovian planets, or to a region of the solar atmosphere, is invaluable in achieving a deeper and more unified appreciation of heliophysics.

For further reading of the physical processes the reader should refer to Rishbeth and Garriott (1969), Holton (1972), Chapman and Cowling (1960), Bougher *et al.* (2002) and several of the other chapters in the AGU monograph *Atmospheres in the Solar System: Comparative Aeronomy* (2002), Heelis (2004), Schunk and Nagy (2000), Kelley (1989), and Forbes (1995).

12.8 Comparing the Sun's chromosphere and Earth's ionosphere

The layers from the solar surface up to a few thousand kilometers above it comprise a gravitationally bound partially-ionized plasma coupled marginally to a dominant neutral component, and this has many similarities to the planetary ionospheres that are the focus of this chapter. This short section can only touch on some processes in which the collisional coupling between charged and neutral particles that characterizes an ionosphere may also be important in a stellar atmosphere; references to publications will point the interested reader to the literature. Much of the description that follows is from the perspective of the solar discipline, in particular with regard to the physical properties of the magnetic field and in terminology. As stated before, an attempt is made, where possible, to show where common heliophysical processes are operating, but the differing terminologies sometimes make this task challenging.

12.8.1 Stratification, convection, rotation, diffusion

On average, the stratification of the atmosphere of the Sun follows the same principle as the stratification of Earth's atmosphere, of course: the action of gravity is balanced by a vertical pressure gradient. On a star without waves or a magnetic field, a chromosphere would formally exist in the sense that there would be a thin atmospheric layer above the photosphere that would be largely transparent except in a few very strong spectral lines. But the solar chromosphere is primarily thought of as an atmospheric domain above the photosphere in which temperatures range up to some $20\,000$ K, which may be compared to the ~ 6000 K of the photosphere below. The solar chromosphere exists only because of strong wave dynamics and electrodynamic processes that deposit heat into it and thus lift far larger amounts of material to the heights of the chromosphere (which lies some ten to twenty photospheric pressure scale heights above the solar "surface") than would exist there without these processes. The equivalent heat source in the Earth's thermosphere arises from the absorption of extreme ultraviolet radiation (EUV), which causes a similar increase in temperature from 180 K at the mesopause to more than 1000 K at the top of the thermosphere. The heating causes expansion of the atmosphere to greater heights in much the same way as in the chromosphere.

The dynamics of the magnetized chromospheric plasma is such that very substantial deviations from this average state occur. How these manifest themselves depends very much on the relative magnitude of the pressures associated with plasma and field, respectively, as expressed in the plasma β-value. This value depends very strongly on position and height, because of the intermittent nature of the solar magnetic field. Strong, kilogauss, field is clustered into so-called flux tubes caught in the downflows of the photospheric convection (see Sections 8.1 and 8.2 and Figs. 8.2 and 8.3). From there, this field expands with height until it forms a vaulted canopy at chromospheric heights; sufficiently near to the centers of the flux tubes, the plasma β lies below unity throughout the chromosphere. Beneath this vaulted canopy, over the interiors of the granular convection, we find a much weaker, much less organized field called the internetwork field; here, the plasma β varies from around unity to well above unity.

Like gravity waves from the Earth's lower atmosphere intruding into the lower thermosphere, solar convection overshoots into the high-β domains of the chromosphere. Well above the photosphere the expanding upflows are cooler than the descending downflows which, when imaged, yield a pattern referred to as "reversed granulation". Part of the cooling of the upflowing plasma in this case is due to expansion and part of it is radiative cooling (e.g. Cheung *et al.*, 2007). In addition to this relatively gradual overturning, intense sound waves steepen into shock waves with substantial undulations that cause the outer atmosphere at any one instant to deviate from the mean stratification.

Within magnetic, low-β, environments, the dynamic magnetic field frequently causes a variety of phenomena in which the chromospheric plasma is temporarily lifted up by Lorentz forces from its unattained equilibrium state, in fountains of plasma called by names such as spicules, sprays, or surges. Most of the time, the plasma in the low-β flux tubes is confined to flowing up and down along the field, because it is effectively in a frozen-in state (see Section 3.2.3.1). If we look at the average state (deviations from it are discussed in Vol. II), this results in a magneto(hydro)static stratification: whenever flows have negligible effects on the stratification, the force balance in Eq. (3.2) yields:

$$ -\nabla p + \rho \mathbf{g} + \frac{1}{4\pi}(\nabla \times \mathbf{B}) \times \mathbf{B} = 0. \tag{12.18} $$

As the Lorentz force has no component along the magnetic field, the stratification of the plasma within a strong-field flux tube as a function of height is the same regardless of the path of the flux tube through the atmosphere. For gradual changes in the temperature profile, the stratification adjusts by expanding or compressing the gas with height accordingly. In general the chromospheric field is so dynamic that both field and plasma are continually moving; the plasma in the tube remains frozen in, however, and, rather than executing a convective overturning, slides up and down along the field (with some differentiation between the ionized and neutral components, as we will discuss below).

The time scales on which the convecting plasma is mixed within the outer-atmospheric layers of the Sun are so short (from minutes to about a day, increasing with convective scale) compared with the rotation period, which is about 25 days, that Coriolis forces are not important. Only in the deep convective envelope are the turnover and rotation periods comparable, which is deemed an essential requirement for global dynamo action (see Vol. III). The relatively rapid recycling of chromospheric matter, and the very dynamic state in which it is kept, also means that chemical fractionation by diffusive gravitational settling is unimportant.

The chromosphere does appear to act as a fractionation filter for ions entering the corona and solar wind, however. This phenomenon, often referred to as the first-ionization potential (FIP) effect, reflects the fact that ions with FIPs below about 10 eV (such as Fe, Si, and Mg) are enhanced typically by a factor 3 over the photospheric abundances, while elements with higher ionization potentials (such as C, N, and O) are not. The fractionation by ionization energy points to a mechanism that operates within the partially ionized chromosphere, but no consensus exists as to its detailed nature. The mechanisms proposed include the differential heating of the ionized and neutral components (e.g. Schwadron *et al.*, 1999), flow-induced Lorentz forces and collisional coupling to the neutral gas (e.g. Arge and Mullan, 1998), and the differential action of relatively high-frequency Alfvén waves resulting in wave-pressure acceleration (e.g. Laming, 2004).

12.8.2 Heating, cooling, and ionization

One major difference between the solar domain from photosphere to chromosphere and a planetary ionosphere lies, of course, in the way that the plasma is heated. In quiescence, daytime ionospheres are heated primarily by UV and EUV radiation coming from the Sun, whereas the chromosphere is heated by non-radiative energy coming from within the Sun. Interestingly, the primary dissipation mechanism for the non-radiative energy going into the chromosphere remains unknown, and even the means by which that energy is transported into the chromosphere has not been unambiguously identified. In fact, conditions are likely to depend strongly on where one looks. Over areas of weak surface magnetic field, and relatively low in the atmosphere, the ubiquitous powerful acoustic waves turn into shock waves, because of the strong stratification of the atmosphere, and subsequently dissipate or radiate their energy. Where the field is stronger, or higher in the atmosphere, the magnetic field is more important; acoustic waves are modified into a variety of MHD waves, and even electric currents may play a significant role (see Chapter 8 for aspects of the heating of the chromosphere).

During solar flares, geomagnetic activity, and in the night-time ionosphere, ionization by energetic particle precipitation is important for both the chromosphere and the ionosphere. This is discussed in Vol. II.

A consequence of the highly dynamic nature of the chromosphere and its heating mechanism and of the external heating sources for ionospheres is that the ionization states often do not reflect the gas-kinetic temperatures. In acoustically heated parts of the low solar chromosphere, for example, the ionization state is a mean state that reflects the slowness of the recombination process relative to the variable heating by propagating shock waves, with ionization degrees at times orders of magnitude away from that expected for the gas-kinetic temperature (Leenaarts *et al.*, 2007).

12.8.3 *Partial ionization and electrical conductivity*

A primary similarity between ionospheric and chromospheric plasmas lies in the fact that the substantial neutral component introduces through collisions the possibility of differential motions between the charged and neutral particles. The collisional coupling can be expressed by a conductivity coefficient in a generalized Ohm's law (and, of course, in additional terms in the momentum and energy equations; see e.g. Draine, 1986). The anisotropy of the conductivity depends on the magnetic field strength and on the density and ionization degree of the plasma.

The Earth's ionosphere has a range of degrees of ionization, starting from the essentially neutral troposphere below, reaching an ionization fraction of about $10^{-4} - 10^{-3}$ at around 200 km, and exceeding a few percent by 1000 km (e.g. Schlegel, 2007). In the case of the chromosphere, the ionization fraction starts at about 10^{-4} around photospheric height, drops to 10^{-5} through the classical "temperature minimum" around 500 km in height, and then increases to a few percent at around 1500 km in height, continuing to near-complete ionization in the solar corona (e.g. Leenaarts *et al.*, 2007). Figure 12.13 gives the densities and ionization fractions for mean states characteristic of the ionosphere and chromosphere. Note that the neutral densities in the $D–F_2$ ionospheric regions are comparable with those in the chromosphere but the ion densities are at least 1000 times lower at any given neutral density, resulting in a much weaker ion–neutral coupling in the ionosphere than in the chromosphere.

Let us look back at Eqs. (12.13)–(12.15) and assess their meaning for both chromosphere and ionosphere. In the limit of a weak magnetic field or a high

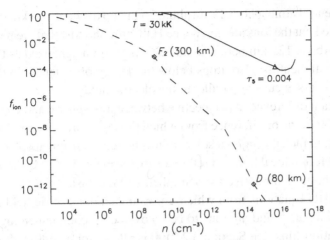

Fig. 12.13. Densities n (cm^{-3}) and ionization fractions f_{ion} for a characteristic dayside ionosphere (broken line) and the chromosphere (solid line). The diamonds mark the mean values for the ionospheric D and F_2 regions, centered on about 80 km and 300 km, respectively. The triangles denote respectively the base of the chromosphere (defined here as being at a continuum optical depth $\tau_5 = 0.004$) and the top of the chromosphere (where the temperature exceeds 30 000 K). The data are from Allen (1973).

collision frequency, the ion and electron magnetizations $M_{e,i} = \Omega_{e,i}/\nu_{e,i} \to 0$, and the conductivities $\sigma_P \to \sigma_\parallel$ and $\sigma_H \to 0$; hence currents are more readily aligned with the electric field, as expected. As the collision frequencies with the neutral population decrease, the current and field in the above-mentioned equations tend to align (as both $\sigma_{P,H} \to 0$); i.e. the field becomes force-free (as generally expected in the low-density solar corona outside flares and eruptions).

In the chromosphere of a solar active region we have that $M_e(500 < h < 2000\,\text{km}) = \mathcal{O}(100)$; M_e then decreases rapidly towards the photosphere to $M_e(h = 0) = \mathcal{O}(0.01)$ at the solar surface. Some studies find that the proton magnetization remains below unity throughout the chromosphere, up to the transition into the corona. (See e.g. Fontenla (2005); these findings depend on the atmospheric domain, of course, and on the models used. For example, Leake *et al.* (2005) suggested that the ion magnetization is, in fact, high throughout the chromosphere.) Consequently, the bulk of the active-region chromosphere has an anisotropic conductivity, with at least a factor 10 difference between the field-aligned and transverse components. Conduction in the corona is almost exclusively field-aligned (and thus essentially free of Lorentz forces), while photospheric conduction is nearly isotropic (e.g. Fontenla, 2005).

For the Earth's dayside ionosphere (e.g. Luhmann, 1995b; Baumjohann and Treumann, 1997), $M_{e,i} < 1$ below about 75 kilometers, so that the overall conductivity there is not strongly anisotropic – but it is also very low in general in the essentially neutral atmosphere. From about 75 km up to about 125 km the electrons are magnetized but the ions are not; hence Hall currents exist that are perpendicular to **E** and **B**. Above 125 km both ions and electrons are magnetized as the collision rate of ions with neutrals also drops below the ion gyrofrequency, so that the ions can carry a Pedersen current parallel to the electric field.

An important difference in perspective between ionospheric and chromospheric scientists is the frame of reference from which processes are described. In the chromosphere (as, in fact, throughout solar, heliospheric, and magnetospheric physics), there is no preferred local frame of reference for the magnetic field: plasma and field are in constant motion relative to each other, and the easiest way to describe this is through MHD equations from which, for example, the electric field and current have been eliminated (and are thus to be recovered by complementing them with auxiliary relationships, see Section 3.2.2); this allows for a continual, substantial, rate of change in the local magnetic field, $\partial \mathbf{B}/\partial t$.

From an ionospheric perspective, however, there is a natural reference frame, namely that at rest with respect to the planet underneath. This is a useful perspective because the Earth's magnetic field near the surface is essentially constant (i.e. the internal-dynamo time scale is very much longer than the ionosphere's characteristic time scales), while the externally induced changes of magnetic field are very much weaker than the background magnetic field. Consequently, for all but high-frequency processes we have $\partial \mathbf{B}/\partial t \approx 0$, so that a simple approximate equivalence results from Eqs. (3.4) and (3.19) and Ampère's law, Eq. (4.1):

$$\mathbf{j} \approx \sigma \mathbf{v} \times \mathbf{B}, \tag{12.19}$$

which is why thinking in terms of the electric current is generally useful for ionospheric physicists (in general, of course, one needs to allow for changes in the magnetic field associated with an electric field and for the anisotropy of the conductivity, as discussed in Section 12.6.1).

One consequence of this perspective is that ionospheric physicists often discuss processes by explicitly involving the electric field, whereas other space physicists rarely do. The dominant ionospheric electric field at high latitudes is driven by processes elsewhere in geospace deriving from the interaction of the solar wind and magnetosphere. At mid and low latitudes the electric fields are created by the action of the neutral winds driven by tidal forces colliding with the ions. In other space environments the necessity to include the electric field and current generally makes the reasoning more cumbersome. We have already touched upon these subjects in Sections 3.2.2 and 10.4.1. In essence, space plasma physicists use the local magnetic field as an elegant equivalent to a complex volume integral over all moving charges, whereas the ionospheric physicist considers electrical charges moving in a conducting layer. Observations of the perturbation of the magnetic field induced by the relatively weak ionospheric currents are a useful diagnostic of this current flow.

12.8.4 Magnetic fields, boundary layers, flows, and winds

In terms of the anisotropies of the plasma conductivity evidenced in Eq. (12.12), the ionospheric domain from 75 to 125 km and the domain from the top of the photosphere well into the magnetized parts of the chromosphere may be comparable. Other properties differentiate their behavior, however.

Below the solar photosphere the ionization fraction increases rapidly, as do the plasma density and the associated collision frequencies. The high collision frequencies result in a strong coupling between the ionized and neutral populations. As a result they flow together and carry the magnetic field with them: there appears to be relatively little transverse slippage between the magnetic field and the plasma flows.

In this environment, the general magnetization of the charged particles is low and the resulting conductivity tensor, Eq. (12.12), has diagonal components of comparable size with very small off-diagonal components. Hence the common MHD assumption that the conductivity is a scalar and thus that electric currents run in the direction of the total electric field is warranted to a very high degree of fidelity. Note that, despite the high collision rates, the conductivity remains high in the sense that the associated diffusion of the field through the plasma is ineffective for the large observable scales of the plasma: (sub-)photospheric plasma is advected in the turbulent flow without significant resistive diffusion. Consequently, when solar physicists talk about the "diffusion" of the photospheric magnetic field they mean the random-walk dispersal of advected field in the turbulent convective plasma motions, not the resistive diffusion to which a plasma physicist would be referring.

Immediately above the photosphere one still does not expect substantial effects from the differential motion of charged and neutral particles, despite the fact that the Hall, Pedersen, and parallel conductivities are significantly different. This is in large part a consequence of the organization of the magnetic field. The strong, nearly vertical, field that has most often been studied and that threads the chromosphere, corona, and heliosphere, is the field concentrated in the downflows between the convective granulation. Hence the convective motions near the surface primarily slide along the flux tubes, which consequently adjust their positions within the downflow lanes in response to pressure differences associated with aerodynamic drag. Exactly how important the Hall and Pedersen currents are to the ubiquitous weak field in the granular interiors remains to be seen; that field component has escaped observational capabilities until very recently, and its properties still defy detailed descriptions because much of its signal remains at or below noise levels even for the best observatories.

The Pedersen and Hall terms may be important, however, in three aspects of chromospheric processes. The first is during flux emergence from within the convective envelope. During this phase, magnetic field breaching the surface rapidly expands into the overlying atmosphere, while lifting neutral plasma up with it through collision-induced Lorentz forces. Numerical experiments with a strongly anisotropic conductivity suggest that they may cause a substantial weakening of the components of electric currents perpendicular to the direction of the magnetic field, effectively resulting in weakened $\mathbf{j} \times \mathbf{B}$ Lorentz forces compared to the scalar-conductivity case. This, in turn, could mean that less photospheric and chromospheric plasma is lifted into the high atmosphere during flux emergence (e.g. Arber *et al.*, 2007) and that the strong shearing motions in the coronal field seen in some simulations (e.g. Manchester, 2007) might be significantly different in simulations with anisotropic conductivity. Observations suggest that emerging magnetic flux often does not emerge as a simple arch but rather as an undulating field that transits the photospheric layer multiple times, in sea-serpent fashion. In such a geometry the magnetic field reaching into the chromosphere might drain its plasma load into the sub-photospheric dips in the field, and the trapped material there would not allow those field segments to rise. But they appear to do so anyway. It has been suggested that resistive effects are important in this process, either by allowing the material to leak out or by enabling rapid reconnection by means of which the mass-loaded segments can be separated from the rising loop segments (e.g. Pariat *et al.*, 2004). Just how important an enhanced or anisotropic resistivity is in the process of flux emergence remains to be established.

Second, the multi-fluid nature of the chromosphere (which includes heavier ions as well as protons, electrons, and neutrals) may be important for wave propagation and dissipation. This is readily argued for frequencies over 100 mHz (see e.g. Haerendel, 1992, and De Pontieu *et al.*, 2001), but the effects perhaps extend down to a few mHz (Goodman, 2000; Leake *et al.*, 2005). It is at present unclear whether these processes do indeed operate at important levels within the chromosphere.

Finally, the collisions between neutral and charged particles may be important within filaments and prominences. These are condensations of chromospheric material suspended in near-horizontal structures by the magnetic field, within an otherwise coronal environment in which only the gas pressure is available to stratify the plasma. Here, the neutral population may leak out of the filaments by diffusive percolation through the ion and electron populations (e.g. Gilbert *et al.*, 2007).

Although we can point to a body of literature relating to the fact of the partial ionization of the domain around the solar chromosphere, relatively little is known about its consequences. The primary reason for this is that numerical simulations of the solar atmosphere are in reality limited by the use of numerical conductivity that far exceeds the conductivity expected for the plasma environment. This is further compounded by the problem that gradients often build up in MHD simulations that become too steep for the codes to handle unless an artificial, often ad hoc, hyper-resistivity or anomalous resistivity is introduced. Such a resistivity can be regarded as a function of a high-order derivative in the magnetic field or as an ad hoc limiter of large electric currents (e.g. Caunt and Korpi, 2001; Yokoyama and Shibata, 1994). The bottom line is that we simply do not know how important the partial ionization of the chromosphere is for its energy budget and its response to dynamic magnetic fields.

In closing, we note that a fundamental difference between a chromosphere and a planetary ionosphere is that the latter sits above a neutral atmosphere with winds that flow across the essentially stationary magnetic field (see also the discussion in Section 10.4), whereas the photosphere below a stellar chromosphere is a good conductor in which the magnetic field is advected in the flows. Coronal and chromospheric electric currents may consequently connect to the stellar interior while photospheric flows simply drag the field along, so that neither process imparts observable currents in the chromospheric plasma by charged–neutral interactions. The ionospheric plasma, in contrast, responds to both electric fields induced from above and to the $\mathbf{v} \times \mathbf{B}$ force driven by the neutral winds below.

13

Comparative planetary environments

FRANCES BAGENAL

13.1 Introduction

The nature of the interaction between a planetary object and the surrounding
plasma depends on the properties of both the object and the plasma flow in which
it is embedded. A planet with a significant internal magnetic field forms a magnetosphere that extends the planet's influence beyond its surface or cloud tops.
A planetary object without a significant internal dynamo can interact with any
surrounding plasma via currents induced in an electrically conducting ionosphere.

All the solar system planets are embedded in the wind that streams radially
away from the Sun. The flow speed of the solar wind exceeds the speed of the
fastest wave mode that can propagate in the interplanetary plasma. The interaction
of the supersonic solar wind with a planetary magnetic field (either generated by
an internal dynamo or induced externally) produces a bow shock upstream of the
planet. Objects such as the Earth's Moon that have no appreciable atmosphere
and a low-conductivity surface have minimal electrodynamic interaction with the
surrounding plasma and just absorb the impinging solar wind with no upstream
shock. Interactions between planetary satellites and magnetospheric plasmas are
as varied as the moons themselves: Ganymede's significant dynamo produces
a mini-magnetosphere within the giant magnetosphere of Jupiter; the electrodynamic interactions of magnetospheric plasma flowing past the atmospheres of vol-
canically active Io (Jupiter) and Enceladus (Saturn) generate substantial currents
and supply more plasma to the system; moons without significant atmospheres
(e.g. Callisto at Jupiter) absorb the impinging plasma. The flow within magnetospheres tends to be subsonic, so that none of these varied interactions forms a
shock upstream of the moon. The types of plasma interactions are summarized in
Table 13.1.

The general principles of the structure and dynamics of planetary magnetospheres were presented in Chapter 10. The physical principles and observational

Table 13.1. *Types of interactions of planets with their embedding flow*

Plasma flow	No dynamo	Dynamo
Subsonic	Io (Jupiter) Enceladus (Saturn)	Ganymede (Jupiter)
Supersonic	Venus – atmosphere Moon – no atmosphere	Earth, Mercury – slow rotation Jupiter, Saturn – fast rotation Uranus, Neptune – oblique rotation

evidence for the dynamics of the Earth's magnetosphere (e.g. Cravens, 1997; Kivelson and Russell, 1995) are discussed in Vol. II. Here we will discuss the planets within our solar system, comparing their similarities and differences. A basic introduction is given in Van Allen and Bagenal (1998). Deeper studies of comparative magnetospheres range from the abstract to the specific (Siscoe, 1979; Vasyliūnas, 1988, 2004; Kivelson, 2007; Walker and Russell, 1995; Bagenal, 1992; Russell, 2004, 2006; Kivelson and Bagenal, 2007). In this chapter we take an intermediate path, with the goal of applying the general principles of Chapter 10 to specific planets but also providing a qualitative appreciation of the different characters of our local family of magnetospheres. We shall return to plasma interactions with non-magnetized objects in Sections 13.6.1 and 13.6.2.

13.1.1 Planetary magnetic fields

Spacecraft carrying magnetometers have flown to and characterized the magnetic fields of all the planets except Pluto. Tables 13.2 and 13.3 list the properties of each planet (the strength and direction of the planet's magnetic field and the rotation rate and direction of the planet's spin), the interplanetary medium (the strength and direction of the interplanetary magnetic field (IMF) and the speed, density, and temperature of the solar wind), as well as the characteristics of the magnetospheres observed to date.

While the theory of planetary dynamos has yet to reach the level of sophistication where it could predict with accuracy the presence (let alone the specific characteristics) of an internally generated magnetic field, it is generally understood that, for such a field to be present, planets need to have an interior that is sufficiently electrically conducting and that is convecting with sufficient vigor (see Chapter 3). The iron cores are potential dynamo regions of terrestrial planets. The high pressures inside the giant planets Jupiter and Saturn put the hydrogen into a phase where it has the electrical conductivity of liquid metal. Inside Uranus and Neptune the pressures are too weak to make hydrogen metallic and it is postulated

Table 13.2. *Properties of the solar wind and scales of planetary magnetospheres*

	Mercury	Venus	Earth	Mars	Jupiter	Saturn	Uranus	Neptune	Pluto
Distance a_p (AU)[a]	0.39	0.72	1[b]	1.52	5.2	9.5	19	30	40
Solar wind dens. (cm^{-3})	53	14	7	3	0.2	0.07	0.02	0.006	0.003
IMF strength[c] (nT)	41	14	8	5	1	0.6	0.3	0.2	0.1
IMF azimuth angle[c]	23°	38°	45°	57°	80°	84°	87°	88°	88°
Radius R_p (km)	2439	6051	6373	3394	71 400	60 268	25 600	24 765	1170±33
Sidereal spin period (d)	58.6	−243	0.9973	1.026	0.41	0.44	−0.72	0.67	−6.39
Magnetic moment/M_E[d]	$(3-6) \times 10^{-4}$	$<10^{-5}$	1[d]	$<10^{-5}$	20 000	600	50	25	?
Surface field[e] B_0 (nT)	200–400	—	30 600	—	430 000	21 400	22 800	14 200	?
R_{MP}[f] (R_p)	$1.6R_M$	—	$10R_E$	—	$42R_J$	$19R_S$	$25R_U$	$24R_N$	?
Observed size of magnetosphere	$1.5R_M$	—	$(8-12)R_E$	—	$(50-100)R_J$	$(16-22)R_S$	$18R_U$	$(23-26)R_N$	?

[a] 1 AU = 1.5×10^8 km.

[b] The density of the solar wind fluctuates by about a factor 5 around typical values $\rho_{sw} \sim 7 (cm^{-3})/a_p^2$.

[c] Mean values. The azimuth angle is $\tan^{-1}(B_\phi/B_r)$. The radial component of the IMF, B_r, decreases as $1/a_p^2$ while the transverse component, B_ϕ, increases with distance (Gosling, 2007).

[d] $M_E = 7.9 \times 10^{25}$ gauss cm^3 = 7.9×10^{15} tesla m^3.

[e] The magnitude of an eccentric dipole: for Earth and the outer planets from Connerney (1993); for Mercury from Connerney and Ness (1988); upper limits for Mars and for Venus (strictly speaking $M_V < 10^{-5} M_E$) from Russell (1993).

[f] R_{MP} is calculated using $R_{MP} = \xi (B_0^2/2\mu_0\rho v_{sw}^2)^{1/6}$ for typical solar wind conditions of ρ_{sw} given above and $v_{sw} \sim 400$ km/s and ξ an empirical factor of ~ 1.4 to match Earth observations (Walker and Russell, 1995).

Table 13.3. *Planetary magnetic fields*

	Ganymede	Mercury	Earth	Jupiter	Saturn	Uranus	Neptune
$B_{\mathrm{dip,eq}}$ (nT)[a]	719	200–400	30 600	430 000	21 400	22 80	14 200
$B_{\mathrm{max}}/B_{\mathrm{min}}$[b]	2	2	2.8	4.5	4.6	12	9
Dipole tilt[c]	$-4°$	$\sim 10°$	$11.2°$	$-9.4°$	$-0.0°$	$-59°$	$-47°$
Dipole offset[d]	—	—	0.076	0.119	0.038	0.352	0.485
Obliquity[e]	$0°$	$0°$	$23.5°$	$3.1°$	$26.7°$	$97.9°$	$29.6°$
$\delta\phi_{\mathrm{sw}}$[f]	$90°$	$90°$	$67°–114°$	$87°–93°$	$64°–117°$	$8°–172°$	$60°–120°$

[a] Surface field at dipole equator. Values derived from modeling the magnetic field as an eccentric dipole (with magnitude, tilt, and offset). Values for Mercury from Connerney and Ness (1988), for Earth and outer planets from Connerney (1993).
[b] Ratio of the maximum surface field to the minimum (equal to 2, for a centered dipole field). This ratio tends to increase with the planet's oblateness.
[c] Angle between the magnetic and rotation axes. Positive values correspond to a magnetic field directed north at the equator. The magnetic poles of the Earth's field are currently located at 83°N and 65°S latitudes and moving about 10° per century (Natural Resources Canada, Australian Antarctic Division).
[d] Values (in planetary radii, R_{p}) from eccentric dipole models of Connerney (1993).
[e] The inclination of a planet's spin equator to the ecliptic plane.
[f] Range in the angle between the radial direction from the Sun and the planet's rotation axis over an orbital period. In Ganymede's case, the angle is between the corotational flow and the moon's spin axis.

that their dynamos must be generated in regions of liquid water where, as in Earth's ocean, small concentrations of ions provide sufficient conductivity.

Given the disparity in scale between the giant and terrestrial planets (e.g. the volume of Jupiter is 1400 times that of the Earth) it is perhaps not surprising that the four terrestrial planets have far weaker magnetic fields generated in their interiors than the giant planets (Russell, 1993; Connerney, 1993; Stevenson, 2003). Extensive geophysical measurements have revealed substantial information about the distribution of density, temperature, and flows inside the Earth. Moreover, the remanent magnetization of surface rocks tells us how the Earth's field has changed over geological time. These geophysical data are powerful constraints on the geodynamo (Glatzmaier, 2002). For other planetary objects the presence or absence of a magnetic field is an important constraint on their interiors.

The apparent lack of an active dynamo inside Venus puts interesting constraints on the thermal evolution of that planet (Stevenson *et al.*, 1983; Schubert *et al.*, 1988). A common misconception is that it is the slowness of the rotation of Venus that prevents a dynamo. In fact, very little rotation is needed for a dynamo and all objects in the solar system have sufficient rotation (Stevenson, 2003). So, the question becomes "Why is Venus' core not convecting"? One possibility is that Venus' core temperature is too high for a solid iron core to condense (the differentiation of solid iron from an outer liquid sulfur–iron alloy drives Earth's dynamo). The lack of plate tectonics at Venus may be limiting the cooling of the planet's upper layers, further suppressing internal convection. Why planetary neighbors that are almost twins should have suffered such different internal histories is a major mystery of planetary geophysics (Smrekar *et al.*, 2007).

Measurements of its remanent crustal magnetism suggest that Mars has had an active dynamo and experienced changes in polarity over geological time scales (Acuna *et al.*, 2001; Connerney *et al.*, 2004) but stopped generating an internal field some four billion years ago, the dominant explanation being a transition from convection to conduction in cooling the core (Stevenson, 2001).

Having radii of $\sim 40\%$ of the Earth's radius, Mercury and Ganymede were originally expected to have cooled off, shutting down any internal dynamo. But spacecraft fly-bys showed each object to have a significant magnetic field. Thermal models of the particularly large iron core ($>70\%$ of the radius) of Mercury suggest that at least an outer region is likely to be liquid and possibly convecting (Stevenson *et al.*, 1983; Schubert *et al.*, 1988). However, the observed field is much weaker than standard dynamo theory would predict (Stevenson, 2003). Tidal heating in Ganymede's geological past may have kept the giant moon warm, but maintaining a dynamo in the smaller iron core ($\sim 30\%$ of the radius) may have needed enhanced amounts of sulfur, which suppresses the freezing point, and/or additional radiogenic heating (Stevenson, 2003).

13.1.2 *Planetary magnetospheres*

Figure 1.3 presents a schematic of the Earth's magnetosphere showing the bow shock and magnetopause boundaries as well as the major regions. In Chapter 10 (Eq. 10.1) we derived a characteristic scale for the sub-solar distance of the magnetopause, R_{MP}, by assuming that the pressure of the planet's magnetic field, assumed to be dipolar, balances the ram pressure of the solar wind. Table 13.2 shows that this is a reasonable approximation to the observed magnetospheric scale except in the case of Jupiter, where substantial plasma pressure inside expands the magnetosphere. Figure 13.1 illustrates the huge range in scale of the planetary magnetospheres. The magnetospheres of the giant planets encompass most of their extensive moon systems, including the four Galilean moons of Jupiter as well as Titan and Triton. Earth's Moon, however, resides almost entirely outside the magnetosphere, spending less than 5% of its orbit crossing the magnetotail.

The magnetospheres of Mercury, Earth, and Jupiter form a "small, medium, large" triad (Fig. 2.7): Earth tends to be considered as the standard of comparison for other magnetospheres. It is natural that our home planet's magnetosphere is

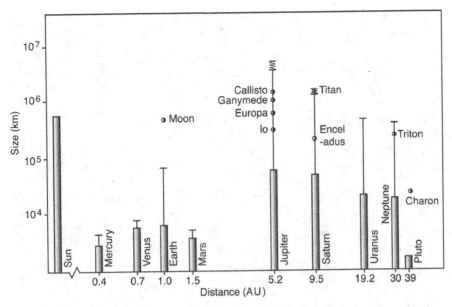

Fig. 13.1. A logarithmic plot of size of object vs. distance from the Sun, for the planets (solid bars), their magnetospheres (thin bars), and the orbital radii of their primary moons. The ranges in size of the magnetospheres of Jupiter and Saturn are shown by the zigzag lines.

better explored and its vicissitudes studied in detail, but it is also important to test our understanding of the magnetospheric principles derived at Earth by applying these concepts to other planets. Figure 2.7 illustrates the vast range in scales: each magnetosphere fits into the volume of the next-larger planet. The expanding solar wind of the heliosphere, as it moves through the interstellar medium, has similarities to a magnetosphere (e.g. it has a heliopause and bow shock). The passage of Voyager 2 through the termination shock at 94 AU gave a scale for the heliosphere that dwarfs the few-AU scale of Jupiter's magnetosphere.

Table 13.3 gives the magnetic moment of each planet and the surface value of the field at the equator on the assumption that each planetary field is dipolar. In reality, when we look closer at a planetary magnetic field we see greater complexity. The standard technique is to describe the internal magnetic field as a sum of multipoles or spherical harmonics (e.g. Walker and Russell, 1995; Connerney, 1993; Merrill *et al.*, 1996), the higher orders being functions that drop off increasingly rapidly with distance so that one needs to get very close to the planet to see any effects of these high-order multipoles. The amplitude of each multipole is derived by fitting magnetic field observations obtained by magnetometers on spacecraft flying past the planet (e.g. Connerney, 1981). The extensive coverage afforded by low-orbiting spacecraft at Earth provides an International Geomagnetic Reference Field with

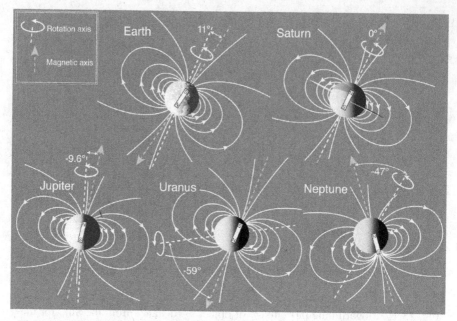

Fig. 13.2. The tilt angles between the spin and magnetic axes are shown for the five main magnetized planets. Considering the horizontal direction of the diagram as parallel to the ecliptic plane and the vertical direction as the ecliptic normal, then the spin axis is shown for conditions of maximum angle from the ecliptic normal (i.e. at solstice). Each planet's magnetic field can be approximated as a dipole, where the orientation and any offset from the center of the planet is illustrated by a bar magnet located at the center of the dipole.

196 harmonic coefficients (IGRF). The level of spacecraft coverage at most other planets limits the description of a planet's magnetic field to little more than a dipole tilted with respect to the spin axis and sometimes offset from the center of the planet. The values in Table 13.3 for the ratio of the minimum and maximum surface magnetic fields, which would have the value 2 for a centered dipole, illustrate the net importance of the non-dipolar components for some planets. At Mercury the observations are too limited to constrain the dipole tilt or offset (Connerney and Ness, 1988). The high values of the maximum/minimum ratios at Uranus and Neptune are symptomatic of highly irregular magnetic fields which can each be crudely characterized as a highly tilted dipole significantly offset from the center of the planet, as illustrated in Figure 13.2. The apparently close alignment of the magnetic field of Saturn with its rotation axis continues to be a puzzle because various of its magnetospheric phenomena exhibit spin modulation (to be discussed further below), which is not expected of an axisymmetric magnetic field.

Finally, when we discuss the dynamics of magnetospheres it will be clear that an important factor is the orientation of the planet's magnetic field relative to the interplanetary magnetic field (Chapter 9). The obliquity is the angle of the planet's spin axis relative to the ecliptic plane normal. As a planet orbits the Sun, if it has a large obliquity it will experience not only large seasonal changes but also a wide range in angles between the upstream solar wind (and embedded IMF) and the planet's magnetic field. Moreover, the large tilt of Uranus' and Neptune's magnetic fields with respect to their spin axes means that these magnetospheres also see a modulation of this solar wind angle over their spin period (i.e. a planetary day). While the solar wind remains flowing within a few degrees of radially from the Sun, the IMF forms a spiral of increasingly tangential field. At Earth the average spiral angle is 45°, at Jupiter it averages 80°, and at farther planets the field is basically tangential to the planet's orbit. The polarity changes several times during the ~25 day solar rotation (more frequently during solar maximum). Most important for magnetospheric dynamics is the variation in the north–south component of the IMF, which fluctuates about the ecliptic plane.

13.1.3 Plasma sources

The plasma found in a planetary magnetosphere could have a variety of sources (see Section 9.5): it could have leaked across the magnetopause from the solar wind, it may have escaped the planet's gravity and flowed out of the ionosphere, or it may be the result of the ionization of neutral material coming from satellites or rings embedded in the magnetosphere. The study of the origin of plasma populations and their evolution as they move through the magnetosphere is a detective story that becomes more complex the deeper one delves (see the review by Moore and Horwitz, 2007).

The clearest indicator of which sources are responsible for a particular planet's magnetospheric plasma is the chemical composition of the latter (Table 13.4). For example, the O^+ ions in the Earth's magnetosphere must surely have come from the ionosphere and the sulfur and oxygen ions at Jupiter have an obvious origin in Io's volcanic gases. But the source of protons is not so clear – protons could be either ionospheric, particularly for the hydrogen-dominated gas giants, or from the solar wind. One might consider that a useful source diagnostic would be the abundance of helium ions. Emanating from the hot (millions of kelvins, a few $100\,eV$) solar corona, helium in the solar wind is fully ionized as He^{++} ions, and comprises ~ 5% of the number density. Ionospheric plasma is much cooler (thousands of kelvins, $< 0.1\,eV$), so that ionospheric helium ions are mostly singly ionized. Thus, a measurement of the abundance ratios He^{++}/H^+ and He^+/H^+ would clearly distinguish the relative importance of these sources. Unfortunately, measuring the composition to such a level of detail is difficult for the bulk of the plasma, with energies in the range $1\,eV$ to $1\,keV$ (e.g. Young, 1997, 1998). Measurement of

Table 13.4. *Plasma characteristics of planetary magnetospheres*

	Ganymede	Mercury	Earth	Jupiter	Saturn	Uranus	Neptune
Max. plasma dens. (cm^{-3})	~400	~1	~4000	~3000	~100	3	2
Neutral density (cm^{-3})				~50	~1000		
Major ion species	O$^+$, H$^+$	H$^+$	O$^+$, H$^+$	O^{n+}, S^{n+}	O$^+$, H$_2$O$^+$, H$^+$	H$^+$	N$^+$, H$^+$
Minor ion species		O$^+$, Na$^+$[a]		H$^+$, H$_3^{+}$[b]			
Dominant source	Ganymede	solar wind	ionosphere[c]	Io	Enceladus	atmosphere	Triton
Neutral source[d] (kg/s)	10			600–2600	2–300		
Plasma source[e] (kg/s)	5	~1	5	300–900	10–15 [g]	0.02	0.2
Plasma source[f] (ions/s)	10^{26}	10^{26}	2 × 10^{26}	> 10^{28}	(3–5) × 10^{26} [g]	10^{25}	10^{25}
Lifetime[h]	minutes	minutes	hour–days	20–80 days	30–50 days	1–30 days	~1 day

[a] Mercury's tenuous atmosphere is a likely source of heavy ions.

[b] An ionospheric source that may be comparable by number to the primary, iogenic source.

[c] Ionospheric plasma dominates the inner magnetosphere; solar wind sources are significant in the outer regions.

[d] Net loss of neutrals from satellite and ring sources (Jupiter: Delamere *et al.*, 2004. Saturn: Hansen *et al.*, 2006; Waite *et al.*, 2006; Tokar *et al.*, 2006; Jurac and Richardson, 2005).

[e] Net production of plasma density (does not include charge exchange processes).

[f] Assumes that 15% of the impinging solar wind flux enters the magnetopause.

[g] Assumes a 5% net ionization rate of neutrals (Delamere *et al.*, 2007).

[h] Typical residence time in the magnetosphere. Plasma stays inside the plasmasphere for days but is convected through the outer magnetosphere in hours.

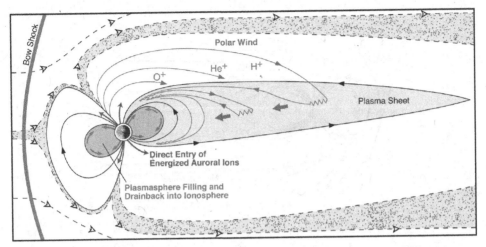

Fig. 13.3. Sources of plasma for the Earth's magnetosphere (after Chappell, 1988). The shaded and dotted area illustrates the boundary layer through which solar wind plasma enters the magnetosphere.

composition is more feasible at higher energies but then one needs to consider whether the process that has accelerated the ions within the magnetosphere since they left the source region is mass or charge dependent.

The temperature of a plasma can also be an indicator of its origin. Plasma in the ionosphere has characteristic temperatures of $< 0.1\,\mathrm{eV}$; the ionization of neutral gases produces ions with energies associated with the location flow speed while plasma that has leaked in from the solar wind tends to have energies of a few keV. But, again, we need to consider carefully how a parcel of plasma may have heated or cooled as it moved through the magnetosphere to the location at which it is measured. Figure 13.3 illustrates various ways in which ionospheric plasma enters the Earth's magnetosphere and evolves by different processes. As we explore other magnetospheres we should expect similar levels of complexity.

Table 13.4 summarizes the main plasma characteristics of the six planetary magnetospheres. To a first approximation one can say that sources from removal of material from the satellites dominate the magnetospheres of Jupiter, Saturn, and Neptune, ionospheric sources being secondary. Uranus having fewer, smaller, satellites, its weak ionospheric source probably gives the main contribution. With only the most tenuous of exospheres, Mercury's magnetosphere contains mostly solar wind material, but energetic particle and photon bombardment of the surface may be a significant source of O^+, Na^+, K^+, Mg^+, etc. (Slavin, 2004). At Earth the net sources from the solar wind and ionosphere are probably comparable, though the most recent studies suggest that the ionospheric contribution seems to be dominant (e.g. Moore and Horwitz, 2007).

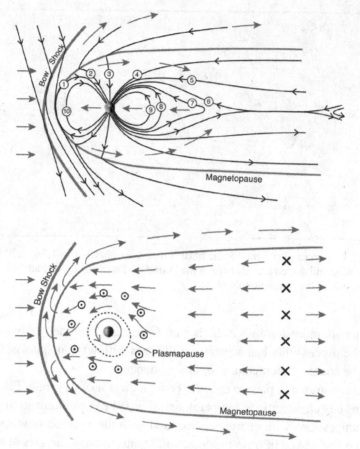

Fig. 13.4. Magnetospheric dynamics associated with the Dungey cycle, which is driven by the solar wind. (Upper panel) The view in the noon–midnight meridian plane. The numbers show the time sequence for a flux tube being reconnected at the dayside magnetopause and convected through the magnetosphere. (Lower panel) The view in the equatorial plane. After Dungey (1961).

13.1.4 Plasma dynamics

In Chapter 10 we developed a general theory of how magnetospheric plasma motions are driven by coupling either to the solar wind or to the rotation of the planet. Figure 13.4 shows how reconnection of the planet's magnetic field with the interplanetary field (often involving flux-transfer events; see Section 6.5.4) harnesses the momentum of the solar wind and drives the circulation of plasma within the magnetosphere; this circulation is sometimes called the Dungey cycle (Dungey, 1961). Figure 13.5 illustrates the main alternative dynamical process whereby the magnetospheric plasma is coupled to the angular momentum of the spinning planet. We will now apply these ideas to the specific planetary magnetospheres. Table 13.6

Table 13.5. *Energetic-particle characteristics in planetary magnetospheres*

	Earth	Jupiter	Saturn	Uranus	Neptune
Phase space density[a]	20 000	200 000	60 000	800	800
Plasma β[b]	< 1	> 1	> 1	~ 0.1	~ 0.2
Ring current[c] ΔB (nT)	10–23	200	10	< 1	< 0.1
Auroral power (watts)	10^{10}	10^{14}	10^{11}	10^{11}	< 10^8

[a] The phase space density of energetic particles (in this case 100 MeV/gauss ions) is measured in units of $c^2 (cm^2\,sr\,MeV^3)^{-1}$ and is listed near its maximum value.
[b] The ratio of the thermal energy density and the magnetic energy density of a plasma, $\beta = nkT/(\mu_0^{-1}B^2)$. These values are typical for the body of the magnetosphere. Higher values are often found in the tail plasma sheet and, in the case of the Earth, at times of enhanced ring current.
[c] The magnetic field produced at the surface of the planet due to the ring current of energetic particles in the planet's magnetosphere.

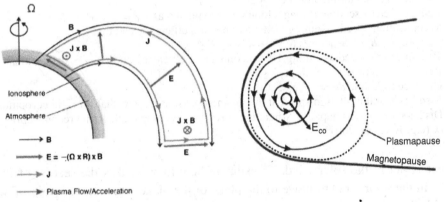

Fig. 13.5. Dynamics of a magnetosphere dominated by rotation, viewed from the side (left) and in the equatorial plane (right). Compare with Fig. 10.6.

lists various dynamical parameters of the different planetary magnetospheres that quantify the relative importance of rotational as against solar wind influences in each case.

First, let us quantify the spatial and temporal scales over which the Dungey cycle would operate at each planet. Let us further suppose that for some fraction of the time there is a component of the IMF that is opposite to the direction of the planetary magnetic field at the magnetopause (e.g. a negative B_z for Earth and a positive B_z for Jupiter and Saturn; we ignore the complexities of Uranus and Neptune for the moment). Such a configuration allows the reconnection of planetary and interplanetary fields at the dayside magnetopause (see step 1 of Figure 13.4). Now we have one end of the flux tube attached to the planet and the

Table 13.6. *Estimated dynamical characteristics of planetary magnetospheres*

	Mercury	Earth	Jupiter	Saturn	Uranus	Neptune
$R_{MP}{}^a$ (km)	4000	6.5×10^4	6×10^6	1×10^6	6×10^5	6×10^5
v_{sw} speed^b	370	390	420	430	450	460
$t_{N-T}{}^c$	10 s	3 min	4 hr	45 min	20 min	20 min
$R_T{}^d$ (R_p)	3	20	170	40	50	50
$R_T{}^e$ (km)	8000	1.3×10^5	1.2×10^7	2.3×10^6	1.3×10^6	1.2×10^6
$v_{rec,1}{}^e$	40	22	16	16	16	16
$v_{rec,2}{}^f$	37	39	42	43	45	46
$t_{rec}{}^g$	3 min	1 hr	80 hr	15 hr	8 hr	7 hr
$d_X{}^h$ (R_p)	30	200	1700	400	500	500
$v_{co}/v_{rec,2}{}^i$	4×10^{-5}	0.04	8	1.3	0.4	0.4
$d_{pp}{}^j$ (R_p)	0.03	6.7	350	95	70	70

[a] Sub-solar magnetopause radius (see Section 9.1).
[b] $v_{sw} = 387(a_p/a_E)^{0.05}$ km/s, from Belcher *et al.* (1993).
[c] Solar wind nose–terminator time: $t_{N-T} \equiv R_{MP}/v_{sw}$.
[d] Radius of cross section of magnetotail, approximated as $R_T = 2R_{MP}$.
[e] Reconnection speed assuming 20% reconnection efficiency and that
$v_{rec,1} \sim 0.2 v_{sw} B_{sw}/ B_{MP}$ km/s (e.g. Kivelson, 2007);
[f] Reconnection speed assuming 10% reconnection efficiency and $v_{rec,2} \sim 0.1 v_{sw}$ km/s.
[g] Reconnection time $t_{rec} = R_T/v_{rec,2}$ s.
[h] Distance to X-line $d_X = v_{sw} t_{rec}$.
[i] Assumes that the rotation speed at the magnetopause is $\sim 30\%$ that for rigid corotation.
[j] Distance to the plasmapause, where the corotation is comparable to the reconnection
flow (e.g. Kivelson 2007).

other is out in the solar wind. To estimate how long it takes the section of flux
tube in the solar wind to move to the plane of the planet's terminator (step 3), we
divide the sub-solar magnetopause distance R_{MP} by the local solar wind speed. For
Table 13.6 we used an empirical fit to Voyager data that includes a modest increase
in the solar wind speed with distance from the Sun, but the basic results would not
be very different if a constant value for the solar wind (say ~ 400 km/s) were used.
One immediately sees the effect of the vast scale of the giant magnetospheres of
the outer planets: the nose-terminator time scale is a mere 10 seconds at Mercury,
3 minutes at Earth, and as much as 4 hours at Jupiter.

The next step is to calculate how long the open flux tube would take to convect
to the equator or central plane of the magnetotail (from step 3 to step 6 in Figure 13.4). For simplicity, the radius of each magnetotail has been approximated
as twice the sub-solar standoff distance (i.e. $2R_{MP}$). This probably underestimates
the cross-sectional radius of real magnetotails. We need to divide this distance by
a convective speed to estimate a minimum convective time scale. The traditional
approach to calculating the speed of circulation in the magnetosphere driven by

solar wind, coupling was to calculate the electric field associated with an object moving with the planet relative to the solar wind, $\mathbf{E}_{sw} = -\mathbf{v}_{sw} \times \mathbf{B}_{IMF}$, assume that some fraction (say, 20%) of this electric field permeates the whole magnetosphere (i.e. the convective electric field $\mathbf{E}_{con} \approx 0.2\mathbf{E}_{sw}$), and then estimate how magnetospheric plasma would drift in this convection electric field and the local planetary magnetic field ($\mathbf{v}_{con} = \mathbf{E}_{con} \times \mathbf{B}_{planet}$) (e.g. Cravens, 1997; Bagenal, 1992). However, this approach begs the question of how the electric field permeates the magnetosphere and in which reference frame one should calculate the electric field. An alternative approach that avoids such a conundrum was presented in Chapter 10 (elaborated further in Southwood and Kivelson, 2007). Here, to obtain a rough upper estimate for a reconnection-driven convection speed we have just taken 10% of the solar wind speed (roughly 40 km/s at all planets). Again, the large scales of the giant planet magnetospheres mean that even with generous values for the convection speed one obtains long time scales for flux tubes to convect to the equator from the upper and lower magnetopause boundaries. At Jupiter this time scale is 80 hours, equivalent to eight full rotation periods. The time scales for steps 3–6 of the Dungey cycle for the other giant planets are much less, but they are still several hours and comparable with the planetary rotation rate. By contrast, this convection time scale is just an hour at Earth and a few minutes at Mercury.

The Dungey-cycle time scale mentioned above can also be used to estimate the length of the magnetotail, by multiplying the reconnection time scale and the solar wind speed. More accurately, it gives us the distance down the tail to the X-line, where further reconnection closes the open magnetic flux (hence conserving, on average, the total magnetic flux emanating from the planet). The re-closed magnetic flux tube then convects sunward (steps 7–10 in Figure 13.4) to begin the Dungey cycle again at the dayside magnetopause. Table 13.6 shows that values for this X-line (often called, for obscure reasons, the distant Earth neutral line). This X-line distance is about $20R_{MP}$ if one takes the reconnection-driven convective speed v_{con} to be 10% of v_{sw} and the tail radius to be $2R_{MP}$. Lower estimates of v_{con} give larger distances to the tail X-line. In practice, we know that the Earth's tail extends for several thousand R_E while Jupiter's magnetotail was encountered by Voyager 2 as it approached Saturn at a distance greater than $9000R_J$ or 4 AU downstream of Jupiter. The estimates of distances to magnetotail X-lines derived from simple Dungey cycle principles shown in Table 13.6 illustrate the vast scales of the magnetospheres of the outer planets and the huge distances that flux tubes reconnecting (re-closing) in the tail would need to travel back to the planet if these magnetospheres were driven by Earth-like processes.

Next we need to compare the relative importance of the reconnection-driving Dungey cycle and the effects of the planet's rotating magnetic field. Again, the traditional approach has been to consider electric fields and to compare the convection electric field $\mathbf{E}_{sw}$ with the corotational electric field, $\mathbf{E}_{cor} = -(\mathbf{\Omega} \times \mathbf{R}) \times \mathbf{B}_{planet}$ (e.g. Cravens, 1997; Bagenal, 1992). In Chapter 10 we compared the electric potentials across the polar cap associated with the two types of flow (Eq. 10.18). We compared the corotation speed $v_{cor} = |\mathbf{\Omega} \times \mathbf{R}|$ with our upper estimate of the convection flows driven by reconnection, v_{con}. The very low values in Table 13.6 of v_{cor}/v_{con} for Mercury and Earth confirm that the dynamics of these magnetospheres are dominated by coupling to the solar wind while it is clearly the case that rotation dominates Jupiter and Saturn. Uranus and Neptune, once again, are not simple cases with speed ratios of order unity that would suggest the comparable importance of rotation and solar-wind-driven circulation.

In a general sense, close to the planet where the magnetic field is strong and rotation speeds are low one expects strong coupling to the planet's rotation. At larger distances from the planet, one expects decreasing corotation and an increasing influence of the solar wind. Finally, we can estimate the size R_{pp} of a region (called the plasmapause at Earth) within which rotation flows dominate solar-wind-driven flows. The values for R_{pp} in the bottom row of Table 13.6 further illustrate how the planets' magnetospheres span the range between the extremes of Jupiter (where $R_{pp} \gg 1$ and rotation dominates throughout) and Mercury (where $R_{PP} \ll 1$ means that there is no region of corotating plasma in the tiny magnetosphere).

13.1.5 *Energetic particles*

At all magnetospheres there are substantial populations of particles with energies far greater than at their original sources. In Table 13.5 some of the properties of nonthermal particles in different magnetospheres are given for comparison. Figure 13.3 shows some of the processes whereby ions and electrons in the ionospheric plasma at Earth are accelerated; in the Earth's auroral regions this occurs by intense local electric and magnetic fields. In the polar regions heated ionospheric oxygen, helium, and hydrogen ions escape the planet as a polar wind that flows away from the planet on the nightside. The lighter ions extend farther down the tail before drifting towards the plasma sheet that sits at the nightside magnetic equator. In the plasma sheet ions are scattered and accelerated by the local electric and magnetic fields. Plasma in the plasma sheet is also accelerated as it convects from the tail towards the planet in the second half of the Dungey cycle (Fig. 13.4). As particles reach energies of tens of keV they experience significant drifts due to magnetic field gradients (e.g. Cravens, 1997). The ions and electrons drift in opposite directions, producing a ring of electrical current that circles the planet. Further energy is transferred to the convecting energetic particles by low-frequency oscillations of the Earth's magnetic field, producing the radiation-belt particles at $\sim$ tens of MeV energies. The sources and losses of these energetic particles depend strongly on geomagnetic activity.

In the rotation-dominated magnetospheres of Jupiter and Saturn the plasma is accelerated as it moves outward. The details of the acceleration mechanism(s) are far from understood but it is likely that the source of energy is the rotation of the planet, coupled by to the plasma by its strong magnetic field. The interaction of magnetospheric ions with neutral atoms and molecules in the extended satellite atmospheres involves charge-exchange reactions whereby a corotating ion becomes neutralized. The momentum of the neutralized particle is well above the planet's escape speed, so the particles flee the system as energetic neutral atoms. These escaping neutral atoms have been imaged as Jupiter's giant neutral-sodium cloud (Mendillo *et al.*, 1990; Thomas *et al.*, 2004) and detected in situ at both Jupiter and Saturn. A small fraction of these escaping neutrals become re-ionized by solar photons in the outer magnetosphere or neighboring solar wind. The large rotational energies farther out mean that these new ions pick up substantial gyro-energy (perhaps MeV) on ionization. As these fresh energetic ions move inwards into stronger magnetic fields they gain further energy through conservation of the first adiabatic invariant. Such processes are the source of the high fluxes of energetic particles in the inner magnetosphere that bombard the moons and make exploration with spacecraft that carry sensitive electronics so challenging. Most of the inward-moving energetic particles are absorbed by satellites or their neutral clouds. Some particles, however, make their way (over time scales of years) to the inner radiation belts at Jupiter which produce intense synchrotron emission at decimetric wavelengths (see the review by Bolton *et al.*, 2004). The smaller physical scale and shorter time scales of the Saturn system result in less net acceleration and weaker fluxes of energetic particles. Absorption by the majestic ring system further prevents the build-up of comparable fluxes close to the planet, so that there are no synchrotron-emitting belts at Saturn. Significant populations of energetic particles were detected at Uranus and Neptune but the fluxes were much lower than at Jupiter and Saturn. It could be that the shorter residence times in these smaller magnetospheres limit the amount of acceleration or it may be much harder for particles to be stably trapped in such non-dipolar fields.

13.2 Jupiter

Jupiter is a planet of superlatives: the most massive planet in the solar system, which rotates the fastest, has the strongest magnetic field, and has the most massive satellite system of any planet. These unique properties lead to volcanos on Io and a population of energetic plasma trapped in the magnetic field that provides a physical link between the satellites, particularly Io, and the planet Jupiter. For those seeking further details, the jovian magnetosphere is reviewed in seven chapters of *Jupiter: The Planet, Satellites and Magnetosphere* covering topics of plasma interactions with the satellites (Bagenal *et al.*, 2004).

Clear indications that Jupiter traps electrons in its magnetic field were apparent as soon as astronomers turned radio receivers to the sky. Early radio measurements showed that Jupiter has a strong magnetic field tilted about 10° from the spin axis, that energetic (MeV) electrons were trapped at the equator close to the planet, and that Io must be interacting with the surrounding plasma and triggering bursts of emission. The magnetometers and particle detectors on Pioneer 10 (1973) and Pioneer 11 (1974) revealed the vastness of Jupiter's magnetosphere and made in situ measurements on energetic ions and electrons. The Voyager 1 fly-by in 1979 revealed Io's prodigious volcanic activity, thus explaining why this innermost Galilean moon plays such a strong role. Additional data came from subsequent traversals by the Ulysses (1992) and Cassini (2000) spacecraft, but it was the 34 orbits of Galileo (1995–2003) around Jupiter that mapped out magnetospheric structures and monitored their temporal variability. As at Earth, magnetospheric activity is projected onto the planet's atmosphere via auroral emissions; this has been observed from X-rays to radio wavelengths with ground- and space-based telescopes. Jupiter has the advantage for us over the rest of the outer planets of not just being very large but also being much closer, allowing high-quality measurements to be made from Earth.

The magnetosphere of Jupiter extends well beyond the orbits of the Galilean satellite system (Fig. 13.1), and it is these moons that provide much of the plasma (Table 13.4) and some interesting magnetospheric phenomena. In particular, Io loses about 1 tonne per second of atmospheric material (mostly SO_2 and dissociation products), which, when ionized to sulfur and oxygen ions, becomes trapped in Jupiter's magnetic field. Coupling to Jupiter causes the magnetospheric plasma to corotate with the planet. Strong centrifugal forces confine the plasma towards the equator. Thus, the densest plasma forms a torus around Jupiter at the orbit of Io (see the review by Thomas *et al.*, 2004).

Compared with the local plasma, which is corotating with Jupiter at 74 km/s, the neutral atoms are moving slowly, close to Io's orbital speed of 17 km/s. When a neutral atom becomes ionized (via electron impact) it experiences an electric field, resulting in a gyromotion of 57 km/s. Thus, new S^+ and O^+ ions gain 540 eV and 270 eV in gyro-energy. The new "pick-up" ion is also accelerated up to the speed of the surrounding plasma. The necessary momentum comes from the torus plasma, which is in turn coupled, via field-aligned currents, to Jupiter – the jovian flywheel being the ultimate source of momentum and energy for most processes in the magnetosphere. About one-third to one-half of the neutral atoms are ionized to produce additional fresh plasma while the rest are lost via reactions in which a neutral atom exchanges an electron with a torus ion. On becoming neutralized the particle is no longer confined by the magnetic field and flies off as an energetic neutral atom. This charge-exchange process adds gyro-energy to the ions and extracts momentum from the surrounding plasma, but it does not add more plasma to the system.

The Io plasma torus has total mass ~ 2 megatonnes, which would be replenished by a source of ~ 1 tonne/s in ~ 23 days. Multiplying by a typical energy ($T_i \approx 60$ eV, $T_e \approx 5$ eV) we obtain $\sim 6 \times 10^{17}$ J for the total thermal energy of the torus. The observed UV power is about 1.5 TW, emitted via more than 50 ion spectral lines, most of which are in the EUV. This emission would drain all the energy of the torus electrons in ~ 7 hours. Ion pickup replenishes energy, and Coulomb collisions feed the energy from ions to electrons, but not at a sufficient rate to maintain the observed emissions. A source of additional energy, perhaps mediated via plasma waves, seems to be supplying hot electrons and a comparable amount of energy as ion pickup.

Voyager, Galileo, and, particularly, Cassini observations of UV emissions from the torus show temporal variability (by about a factor 2) in torus properties (Steffl et al., 2004, 2006). Models of the physical chemistry of the torus match the observed properties in regard to the production of neutral O and S atoms, a radial transport time, and a source of hot electrons (Delamere and Bagenal, 2003). Furthermore, the variation in torus emissions observed over several months by Cassini reflect the observed changes in the output of Io's volcanic plumes (Delamere et al., 2004).

13.2.1 Plasma transport

The earliest theoretical studies concluded that the magnetosphere of Jupiter is "all plasmasphere" with little influence of solar-wind-driven convection (Brice and Ioannidis, 1970). Indeed, rotation dominates the plasma flows observed in the jovian magnetosphere out to distances $\sim 70 R_J$ (Frank et al., 2002; Krupp et al., 2001, 2004). Yet, the presence of sulfur and oxygen ions in the middle magnetosphere, far from Io, indicates that plasma is transported outwards, in directions transverse to the magnetic field.

Rotation-dominated magnetospheres can be thought of as a giant centrifuge with outward radial transport being strongly favored over inward transport. Radial transport of the Iogenic plasma is thought to occur through a process of flux-tube interchange, a diffusive process analogous to the Rayleigh–Taylor instability of fluid dynamics. Flux tubes laden with denser, cooler, plasma move outwards and relatively empty flux tubes containing hotter plasma from the outer magnetosphere move inwards. The 20–80 day time scale (equivalent to 50–200 rotations) for the replacement of the torus indicates surprisingly slow radial transport that maintains a relatively strong radial density gradient. Numerical modeling suggests that radial shear in the azimuthal flow (i.e. increasing lag behind corotation with increasing distance) stabilizes the interchange motion and drives the characteristic size of interchanging flux tubes to small scales (Pontius *et al.*, 1998; Wu *et al.*, 2007).

The net radial transport is thought to be slowest near Io's orbit (~ 15 m/s) and to speed up farther out (~ 50 m/s beyond $10R_J$). Plasma from the Io torus spreads out from Jupiter as a $\sim 5R_J$-thick plasma sheet throughout the magnetosphere. While the flow direction remains primarily rotational, both a lag behind corotation and local time asymmetries increase steadily with distance from the planet. Bursts of flow down the magnetotail are observed and also, on the dawn flanks, occasional strong bursts of super-rotation (Krupp *et al.*, 2004). Below we return to these deviations from co-rotation and discuss how they relate to auroral structures.

13.2.2 *Field structure*

As the equatorial plasma rotates rapidly it exerts a radial (centrifugal) stress on the flux tubes. Additional stress is provided by the radial pressure gradient of the plasma, inflating the magnetic field (see Fig. 13.6). The net result is a stretching of the initially dipolar field lines away from the planet, in a configuration that implies an azimuthal current in the near-equatorial disk (Fig. 13.6(a)). The lower two panels of Figure 13.6 show magnetic field lines derived from models that include the internally generated field plus the effects of currents on the magnetopause and in the plasma sheet. Figure 13.6(d) shows magnetic field lines projected onto the equatorial plane and illustrates how the field lines also bend or "curl" in the azimuthal direction, which means that there are also radial currents in the equatorial plasma sheet (Fig. 13.6(b)). Alternatively one can think of sub-corotating plasma pulling the magnetic field away from radial. At Jupiter, the field is more or less azimuthally symmetric out to about $50R_J$ but Fig. 13.6(d) shows that strong local time asymmetries develop in the outer magnetosphere (Khurana, 2001, 2005).

An important consequence of a strong internal plasma source and an equatorial plasma sheet is that the magnetosphere becomes more compressible. A simple pressure balance between the ram pressure of the solar wind and the magnetic pressure of a dipole produces a weak variation in the terrestrial dayside magnetopause distance R_{MP} for a solar wind density ρ and speed v_{sw} such that $R_{MP} \propto (\rho v_{sw}^2)^{-1/6}$. Measurements of the magnetopause locations at Jupiter indicate a much stronger variation, $R_{MP} \propto (\rho v_{sw}^2)^{-1/3}$. Consequently, a factor 10 variation in ram pressure at Earth changes the magnetopause distance by only 70% while at Jupiter the tenfold variations in solar wind pressure often observed at 5 AU cause the dayside magnetopause to move between $\sim 100R_J$ and $\sim 50R_J$. This greater compressibility of the jovian magnetosphere is due to a significant contribution of the plasma pressure in

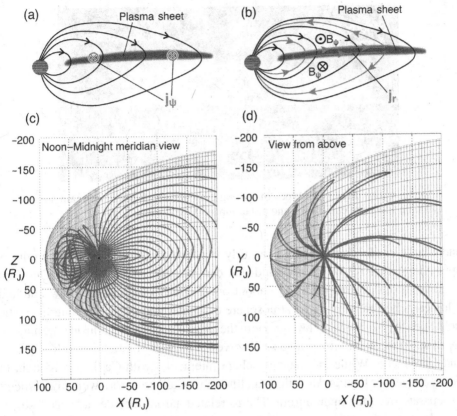

Fig. 13.6. Magnetic field configuration and current systems in Jupiter's magneto-sphere. The upper panels show the (a) azimuthal and (b) radial current systems. The lower panels show the magnetic field configuration (c) in the noon–midnight meridian plane and (d) in the equatorial plane; they were derived from in situ magnetic field measurements (Khurana and Schwarzl, 2005). Compare with the schematic representation in Fig. 10.6 discussed in Section 10.4.4.

the equatorial plasma sheet as well as a substantial system of azimuthal currents that weaken the radial gradient of the magnetic field compared to that of a dipole (illustrated in Fig. 10.2).

13.2.3 Aurora at Jupiter

Just as at Earth, the auroral emissions at Jupiter are important indicators of mag-netospheric processes. With limited spacecraft coverage of these magnetospheres, auroral activity is a projection of magnetospheric processes, communicated via precipitating energetic particles, onto the atmosphere; thus it allows us to study global processes not yet accessed by spacecraft. Figure 13.7 illustrates the three main types of aurora at Jupiter (see the reviews by Bhardwaj and Gladstone, 2000,

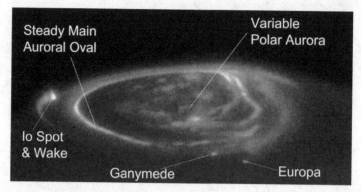

Fig. 13.7. The three main types of auroral emissions at Jupiter: the main aurora, satellite footprint emissions, and polar aurora (Clarke *et al.*, 2004).

and Clarke *et al.*, 2004). There is a fairly steady main auroral oval that produces approximately 10^{14} W globally and that can exceed $1\,\mathrm{W\,m^{-2}}$ locally. This oval is quite narrow, corresponding to about one degree in latitude or a few hundred kilometers horizontally in the atmosphere of Jupiter and mapping along magnetic field lines to $(20\text{–}30)R_J$ at the equator in the magnetosphere, well inside the magnetopause. Auroral emissions are also observed at the feet of flux tubes at Io, Europa, and Ganymede. While the magnetosphere interaction with Callisto is thought to be much weaker than for the other satellites, any Callisto aurora would be difficult to separate from the main aurora. The Io-related aurora includes a "wake" signature that extends half-way around Jupiter. The third type of jovian aurora is the highly variable polar aurora, which occurs at higher latitudes than the main aurora, corresponding to greater magnetospheric distances.

The fact that the shape of the jovian main auroral oval is constant and fixed, in magnetic coordinates (including an indication of a persistent magnetic anomaly in the northern hemisphere), tells us that the auroral emissions correspond to a persistent magnetospheric process that causes a more or less constant bombardment of electrons onto Jupiter's atmosphere. Unlike the terrestrial auroral oval, the jovian oval has no relation to the boundary between open and closed field lines of the polar cap; it maps to regions well within the magnetosphere. It is difficult to map the magnetic field lines accurately because of the strong equatorial currents, which are variable and imprecisely determined. But it has become clear that the main aurora is the signature of Jupiter's attempt to spin up its magnetosphere or, more accurately, Jupiter's failure to spin up its magnetosphere fully.

Figure 13.6(b) shows the simple current system proposed by Hill (1979). As the Iogenic plasma moves outwards, the conservation of angular momentum would suggest that the plasma should lose angular speed. In a magnetized plasma, however, electrical currents easily flow along magnetic fields and couple the magnetospheric plasma to Jupiter's flywheel. Hill (1979) argued that at some point the load on the ionosphere increases to the point where the coupling between the ionosphere and corotating atmosphere – manifested as the ionospheric conductivity – is not sufficient to carry the necessary current, causing the plasma to lag behind corotation. Using a simple dipole magnetic field, Hill (1979) obtained an expression for the critical distance for corotation lag that depended on the mass production and transport from Io and the (poorly determined) ionospheric conductivity. Matching his simple model to the Voyager observations of McNutt *et al.* (1979), Hill (1980) found he could model the observed profiles of azimuthal flow with a source giving 2–5 tonne/s and an ionospheric conductivity equal to 0.1 mho. Over the past five years Jupiter's main aurora has become an active area of study. Researchers have considered the effects of the non-dipolar nature of the magnetic field, the narrowness of the auroral emissions, realistic mass-loading rates, the non-linear feedback of ionospheric conductivity responding to electron precipitation, and the development of electrostatic potential drops in the region of low density between the ionosphere and torus (Cowley *et al.*, 2002, 2003a,b; Nichols and Cowley, 2004, 2005). The understanding of plasma processes developed in the terrestrial magnetosphere is being applied to the different regimes at Jupiter and will ultimately be tested when the Juno spacecraft goes into a close polar orbit (planned for 2016).

The auroral emissions poleward of the main auroral oval (see Fig. 13.7) are highly variable; they are modulated by the solar wind and controlled in local time, being usually dark on the dawn side and brighter on the dusk side (see the reviews by Grodent *et al.*, 2003a; Clarke *et al.*, 2004). The region of magnetic field lines that is open to the solar wind in the polar cap is thought to be very small ($< 10°$). Thus, most polar auroral activity reflects activity in the outer magnetosphere, occurring on closed magnetic field lines. Polar auroral activity has been associated with polar cusps (Pallier and Prange, 2004; Bunce *et al.*, 2004), as well as tail plasma sheet reconnection and the ejection of plasmoids down the magnetotail (Grodent *et al.*, 2003b). Spectral observations of auroral X-ray flares suggest that energetic ions are bombarding the polar atmosphere and may be the signature of the plasma sheet return (downward) current (Waite *et al.*, 1994) or accelerated solar wind ions (Gladstone *et al.*, 2002).

13.2.4 Outer magnetosphere dynamics

A major interest in studying the aurora is to explore how the various emissions are related to the dynamics of the outer magnetosphere; see Kivelson and Southwood (2005) and the reviews by Khurana *et al.* (2004); and Krupp *et al.* (2004). The innermost region, which we will call the Hill region, comprises the equatorial plasma disk where rotation dominates the flow. At a distance of about $20R_J$ the lag of plasma in the equatorial plasma sheet behind strict corotation drives upward currents, and the associated electron bombardment of the atmosphere causes the main aurora.

The middle magnetosphere is a compressible region (sometimes called the "cushion" or Vasyliūnas region, after his seminal article (Vasyliūnas, 1983) in which the dynamics of the outer magnetosphere was first addressed in a substantial fashion). On the dayside of the magnetosphere the ram pressure of the solar wind compresses the magnetosphere. Inward motion on the dawn side reduces the load on the ionosphere, producing a correspondingly dark region in the dawn polar aurora (Fig. 13.7). On the dusk side the plasma expands outwards and strong currents try to keep the magnetospheric plasma corotating. These strong currents produce the active dusk polar aurora. Kivelson and Southwood (2005) argued that the rapid expansion of flux tubes in the afternoon to dusk sector means that the second adiabatic invariant is not conserved, which results in the heating and thickening of the plasma sheet. As the plasma rotates around onto the nightside it is no longer confined by magnetopause currents, moves farther from the planet, and stretches the magnetic field with it. At some point either the coupling to the planet breaks down completely (e.g. because the Alfvén travel time between the equator and the poles becomes a substantial fraction of a rotational period) or the field becomes so radially extended that an X-point develops and a blob of plasma detaches and escapes down the magnetotail, as suggested by Vasyliūnas (1983). Kivelson and Southwood (2005) pointed out that the stretched equatorial magnetic field becomes so weak that the gyroradii of the heavy ions become comparable to the scales of local gradients. It is possible that the plasma diffuses across the magnetic field and "drizzles" down the magnetotail. If the process were entirely diffusive then the magnetic flux would remain connected to Jupiter. The flux tubes would become unloaded and presumably dipolarize as they swung around to the dayside. This is in contrast with the concept of a "planetary wind" (Brice and Ioannidis, 1970; Hill *et al.*, 1974) where a super-Alfvénic plasma wind (in pre-Voyager days assumed to come from the planet) blows the magnetic field open and carries flux down the tail (analogously to the solar wind). As the Voyager spacecraft exited the dawn magnetosphere at distances of about $150R_J$, strong tailward bursts of kiloelectronvolt Iogenic heavy ions were detected, which Krimigis *et al.* (1981) called a magnetospheric wind.

The volume of the magnetosphere that is open to the solar wind is completely unknown. Cowley *et al.* (2003a) postulated that there is a Dungey cycle (similar to that of Earth), driven by dayside reconnection, that carries flux over the poles. Cowley *et al.* argued that the return flow (after tail reconnection) proceeds around to dayside, flanking the dawn magnetopause. There is no evidence as yet of such a solar-wind-induced convection pattern, nor do we know how much polar flux is open to the solar wind. Furthermore, that the time scale for open flux tubes to complete a full Dungey cycle is hundreds of hours or tens of rotation periods raises the issue of the topology of the flux tubes that remain connected to the planet at one end while the other is carried down the tail and towards the equator.

13.2.5 Jovian magnetotail

Pursuing evidence for Vasyliūnas' argument that plasmoids are ejected down the jovian magnetotail, Grodent *et al.* (2003b) found evidence of spots of auroral emission poleward of the main aurora connected to the nightside magnetosphere that flashed with an approximately 10 minute duration. Such events were rare, recurring only about once per 1–2 days. These flashes seemed to occur in the pre-midnight sector, and Grodent *et al.* (2003b) estimated that they are coupled to a region of the magnetotail that was about $5R_J$ to $50R_J$ across and located further than $100R_J$ down the tail. Studies of in situ measurements by Russell *et al.* (2000) and Woch *et al.* (2002) led to the conclusion that plasmoids on the order of $\sim 25R_J$ in scale were being ejected every 4 hours to 3 days, with a predominance for the post-midnight sector and distances of $70R_J$–$120R_J$. Could such plasmoids account for most of the plasma loss down the magnetotail? Bagenal (2007) approximated a plasmoid as a disk of plasma sheet $2R_J$ thick having diameter $25R_J$ and density of 0.01 cm^{-3}, so that each plasmoid has a mass of about 500 tonnes. Ejecting one such plasmoid per day is equivalent to losing 0.006 tonne/s. Increasing the frequency to once per hour raises the loss rate to 0.15 tonne/s. Thus, on the one hand even with optimistic numbers the loss of plasma from the magnetosphere due to such plasmoid ejections cannot match the canonical plasma production rate, 0.5 tonne/s. On the other hand, a steady flow of plasma of density 0.01 cm^{-3}, in a conduit that is $5R_J$ thick and $100R_J$ wide, moving at a speed of 200 km/s would provide a loss of 0.5 tonne/s. Such numbers suggest that a quasi-steady loss rate is feasible. The question of the mechanism remains unanswered. Bagenal (2007) proposed three options: a diffusive "drizzle" across weak, highly stretched, magnetotail fields, a quasi-steady reconnection of small plasmoids, below the scale detectable via auroral emissions, or a continuous but perhaps gusty magnetospheric wind.

In the spring of 2007 the New Horizons spacecraft flew past Jupiter, getting a gravitational boost on its way to Pluto, and made an unprecedented passage down the core of the jovian magnetotail, exiting on the northern dusk flank. For over three months, while covering a distance of $2000R_J$, the spacecraft measured a combination of iogenic ions and ionospheric plasma (indicated by H^+ and H_3^+ ions) flowing down the tail (McComas *et al.*, 2007a; McNutt *et al.*, 2007). The fluxes of both thermal and energetic particles were highly variable on time scales of minutes to days. The tailward fluxes of internally generated plasma led McComas and

Bagenal (2007) to argue that perhaps Jupiter does not have a complete Dungey cycle but that the large time scale for any reconnection flow (see Table 13.6) suggests that magnetic flux that is opened near the sub-solar magnetopause re-closes on the magnetopause before it has traveled down the tail. They suggested that the magnetotail comprises a pipe of internally generated plasma that disconnects from the planetary field and flows away from Jupiter in intermittent surges or bubbles, with no planetward Dungey return flow.

13.3 Saturn

Before the Cassini mission it was tempting to dismiss the magnetosphere of Saturn as merely a smaller, less exciting, version of the jovian magnetosphere. Cassini measurements of the particles and fields in Saturn's neighborhood have shown processes similar to those at Jupiter (e.g. satellite sources, ion pickup, flux tube interchange, corotation, etc) but they have also revealed substantial intriguing differences. The magnetosphere of Saturn is strongly dominated by neutral atoms and molecules. The number-density ratio of neutrals to ions is 15 : 1 in the Enceladus torus compared with 1 : 50 in the Io torus. In contrast with Jupiter's steady main aurora, Saturn's auroral emissions are strongly modulated by the solar wind. While one might expect the alignment of Saturn's magnetic axis with the planet's spin axis to produce an azimuthally symmetric magnetosphere, observations show an intriguing rotational modulation. But, more mysteriously, the rotational modulation is only observed in a limited region of the magnetosphere. The magnetosphere of Saturn is shown in Fig. 13.8. Below we provide a brief summary of current ideas about these topics, which are under active research as the Cassini spacecraft continues to orbit Saturn.

13.3.1 Plasma sources

One of the great discoveries of the Cassini mission to Saturn has been the active volcanism of the small icy moon Enceladus. While Enceladus is a mere one-seventh the size of Io, this small moon suffers tidal heating that drives the eruption of geysers from the south polar region. The geyser plumes, extending over 500 km from the surface, seem to be mostly ice particles with water vapor and minor quantities of molecular nitrogen, methane, and carbon dioxide (Porco *et al.*, 2006; Hansen *et al.*, 2006; Waite *et al.*, 2006). Enceladus' geysers eject water molecules at about one-third the rate of Io's neutral production (Hansen *et al.*, 2006) but few of the products become ionized. The nearly three orders of magnitude difference in the ion–neutral density ratios of the two magnetospheres can be been explained in terms of a much lower energy input into the Saturn system (Delamere *et al.*,

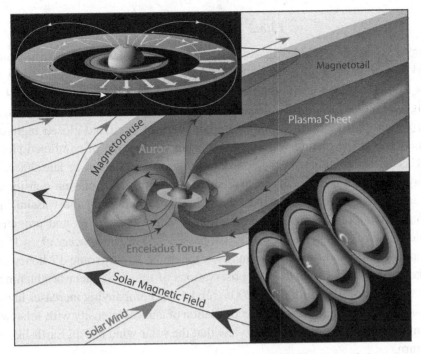

Fig. 13.8. (Center) Three-dimensional schematic representation of the magnetosphere of Saturn. (Top left) The asymmetric plasma disk; the arrows on the disk show the density and speed of the flow. The thin loops show the magnetic field. Gurnett (2007) proposed that the observed density variations are caused by a pattern of asymmetric radial outflows. (Bottom right) Hubble Space Telescope observations of Saturn's auroral emissions on 24, 26, and 28 January 2004 (Clarke, 2005).

2007). At Saturn the plasma flowing past Enceladus (at an orbital distance of ∼ four saturnian radii) has a slower speed than the plasma flow past Io (at ∼ six jovian radii). A factor 2 difference in relative motion (i.e. 26 km/s at Enceladus as against 57 km/s at Io) means that new ions pick up a factor 4 less energy. With less pickup energy the ions deliver less energy to the electrons. At low electron temperatures the ionization rates plummet and, correspondingly, plasma production drops. In fact, Delamere *et al.* (2007) showed that without an additional source of hot electrons (similar to that in the Io plasma torus) the Enceladus plasma torus would not be sustained.

The weaker plasma source at Saturn results in weaker centrifugal stresses and weaker magnetospheric currents. Thus the field structure at Saturn is similar to that shown in Fig. 13.6 for Jupiter but with less pronounced distortion from dipolar. The plasma pressure is also much reduced, so that Saturn's magnetosphere is less compressible that Jupiter's and shows a less dramatic response to changes in solar wind dynamic pressure.

13.3.2 Aurora at Saturn

Figure 13.8 shows Hubble Space Telescope (HST) images of Saturn's aurora (Clarke *et al.*, 2005). In contrast with Jupiter's large main auroral oval, which maps to regions deep inside the magnetosphere, Saturn's small auroral oval and strong variations in auroral intensity with solar wind conditions indicates that Saturn's aurora, like Earth's, marks the boundary of open and closed regions of magnetic flux. The picture was clarified during a campaign of combined Hubble and Cassini observations as the spacecraft approached Saturn in late 2000. For 22 days Cassini's instruments measured the magnetic field, plasma density, and plasma velocity in the solar wind while Hubble cameras and the Cassini radio antennas monitored Saturn's auroral activity. Nature cooperated and provided a couple of interplanetary shock waves that passed the Cassini spacecraft on 15 and 25 January 2001 and then hit the magnetosphere of Saturn some 17 hours later. Clarke *et al.* (2005) reported HST observations of the subsequent brightening of auroral emission, and Kurth *et al.* (2005) reported accompanying increases in radio emission. Crary *et al.* (2005) show a correlation of auroral intensity with solar-wind dynamical pressure, supporting the view that the solar wind has an Earth-like role at Saturn.

But further study showed that it was compression of the magnetopause by the solar wind that correlates with auroral intensity rather than reconnection of the solar and planetary magnetic fields. Crary *et al.* (2005) pointed out that, at Saturn's orbit, the solar magnetic field is essentially tangential so that the solar and planetary fields are largely orthogonal to each other: far from optimal conditions for magnetic reconnection. The magnetospheric processes driving Saturn's aurora should be better understood after Cassini moves to higher magnetic latitudes. In the mean time, the difficulties in measuring Saturn's rotation rate have wreaked havoc with our simple ideas of magnetospheric dynamics.

13.3.3 Planetary rotation at Saturn

So how do we establish how fast the interior of a gas planet is spinning? The usual trick is to measure the periodicity of radio emissions modulated by the planet's internal magnetic field. In this method it is assumed that the magnetic field is tilted and that the dynamo region where the field is generated spins at a rate representative of the bulk of the planet. Recent Cassini data indicate that apparent changes in Saturn's spin could in fact be caused by processes external to the planet. This raises new questions about how we measure and understand the rotation of the large gas planets. Saturn at first dumbfounded planetary theorists who study dynamo models by being observed to have a highly symmetric internal magnetic field. A field that is symmetric about the rotation axis violates a basic theorem of magnetic dynamos (Cowling, 1933). The second puzzle came with the detection of a systematic rotational modulation of the radio emission similar to a flashing strobe, which should not occur for a symmetric magnetic field. Meanwhile, radio measurements have revealed that Saturn's day appears to have become about 6 to 8 minutes longer – it is now roughly 10 hours and 47 minutes – since the 1980s when measured by the Voyager missions (Kurth *et al.*, 2007). Furthermore, the spin rate seems to keep changing and may be modulated by the solar wind speed (Zarka *et al.*, 2007).

A fundamental issue is whether the magnetospheric observations, including the radio emissions, do actually require the magnetic field emanating from the interior of Saturn to be asymmetric. Nearly 30 years ago, Stevenson suggested that strong shear motions in an electrically conducting shell surrounding the dynamo might impose symmetry around the rotational axis (Stevenson, 1981). That the rotational modulation of magnetospheric phenomena seems to be fairly constant with radial distance, that dynamic changes occur in the external plasma structures around Saturn, and that there is an apparent modulation by the solar wind speed indicate that an external explanation for Saturn's apparently erratic spin rate seems far more plausible than perturbations in the massive interior of the planet. Yet, localized magnetic anomalies (i.e. high-order multipoles) at high latitudes remain possible and may be affecting the currents that couple the magnetosphere to the planet (Southwood and Kivelson, 2007).

13.3.4 Magnetospheric dynamics

Gurnett *et al.* (2007) showed how Saturn's radio emission, the magnetic field measured in the magnetosphere, and the density of the plasma trapped in the magnetic field are all modulated with the same drifting period. They argued that the process that transports plasma radially outwards could be stronger on one side of Saturn than the other, as illustrated in the top left of Fig. 13.8. Gurnett *et al.* (2007) suggested that this circulation pattern also produces higher plasma densities in the region of stronger outflow and proposed that plasma production stresses the electrodynamic coupling between the magnetosphere and the planet, causing the pattern of weaker or stronger outward flow to slowly slip in phase relative to Saturn's internal rotation. What causes the proposed asymmetric convection pattern? In the 1980s, researchers tried to explain variations in the Io plasma torus (Dessler *et al.*, 1981) by invoking a convection pattern that rotated with the planet; however, evidence of such a flow pattern in the jovian magnetosphere remains elusive. Alternatively, a system of neutral winds in Saturn's atmosphere could drag the ionosphere around, which would stir up the magnetosphere electrodynamically and provide a source of hot electrons. Could small variations in the high-energy electron population in the Enceladus torus, similar to those in the Io torus, be causing the dramatic changes in plasma density observed by Cassini? If so, large-scale convection patterns in the magnetosphere may not be necessary, just minor modulations in the electrical currents that flow along the magnetic field between the equatorial plasma disk and the planet's ionosphere, bringing small fluxes of ionizing high-energy electrons to the torus. Delamere and Bagenal (2008) showed that a modulation in the small hot-electron population could produce the factor-2 variation in plasma density observed by Cassini.

Undoubtedly, the issue of Saturn's rotation rate and its coupling to the magnetosphere will be a vital area of exploration over the next few years. Similarly, it will be important to investigate whether material is ejected down the tail in the manner and to the extent of the jovian system. Only a few plasmoids have been detected to date at Saturn but this may be a result of limited coverage by the Cassini spacecraft. The substantial polar cap, marked by the aurora, and the influence of the solar wind on the auroral intensity indicate that the Dungey reconnection cycle plays a substantial role at Saturn. The extent and mechanism whereby any return, planetward, flow operates in the magnetotail awaits further exploration.

13.4 Uranus and Neptune

The Voyager fly-bys of Uranus (1986) and Neptune (1989) revealed what have to be described as highly irregular magnetospheres. The non-dipolar magnetic fields and the large angle between the magnetic and rotation axes not only pose interesting problems for dynamo theorists but also challenge the ideas of magnetospheric dynamics. Unfortunately, little study has been made of these odd magnetospheres for the past 15 years and there is little hope of further exploration in the foreseeable future. Thus, there is not much to add to the comparative reviews of their fields by Connerney (1993) and of their magnetospheres by Bagenal (1992). Here we provide a brief précis of these reviews to which the reader should turn for original references.

Tables 13.2 and 13.3 as well as Fig. 13.1 show Uranus and Neptune to have substantial magnetospheres that envelope most of their satellites. Figure 13.2 gives a sense of the irregularity of their magnetic fields, approximated as large tilts and offsets. Table 13.6 tells us that from just the solar wind and planetary parameters we should expect both rotation and solar wind coupling to affect the dynamics of these magnetospheres (though the weak IMF of the outer heliosphere suggests that reconnection will be much weaker than at planets closer to the Sun). Next, we take the orientations of these planets' magnetic fields shown in Fig. 13.2 and consider how these configurations, which rotate about the planet's spin axis every

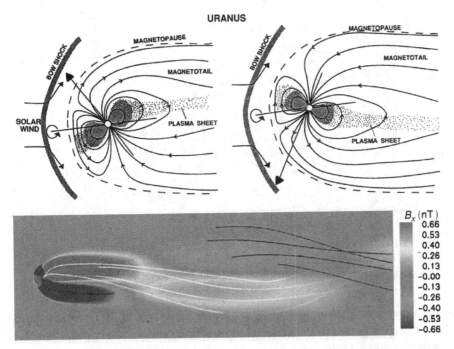

Fig. 13.9. The magnetosphere of Uranus at solstice (the time of the Voyager 2 flyby). The upper left and right panels show the configuration at different phases of the planet's 18-hour spin period (Bagenal, 1992). The lower panel shows a numerical simulation of the helical magnetotail (Tóth *et al.*, 2004). See also the color-plate section.

16–17 hours, might affect the solar wind coupling process illustrated in Figure 13.4. For Uranus around solstice (the Voyager era of the mid 1980s), when the spin axis is pointed roughly towards the Sun, the large tilt of the magnetic axis will result in a magnetosphere that to first approximation resembles that of the Earth but revolves every 17 hours. The finite propagation (at the Alfvén speed) of this rotational modulation down the magnetotail produces a helical plasma sheet and braided lobes of oppositely directed magnetic field (Fig. 13.9). At Neptune, the planet's obliquity being similar to Earth and Saturn one might have expected the fairly simple configurations of either of those planet's magnetospheres. But the large tilt angle discovered by Voyager results in a configuration that changes dramatically (the tail current sheet changes from a plane to a cylinder) over the 16 hour rotation period (Fig. 13.10).

The large range of the "solar wind angle" (see the last row of Table 13.3) indicates that substantial changes in orientation of the planet's spin with respect to the radial direction of the solar wind occur over the (long) orbital periods of these planets. Thus, one has the interesting challenge of imagining how the

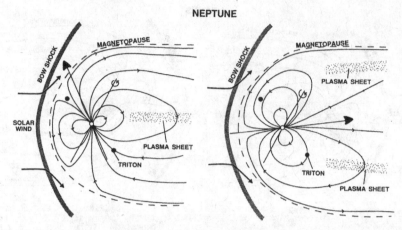

Fig. 13.10. The magnetosphere of Neptune in the configuration corresponding to the time of the Voyager 2 fly-by (Bagenal, 1992). Over the 19-hour spin period the magnetospheric plasma sheet in the tail changes from roughly planar to cylindrical.

magnetosphere of Uranus was behaving during equinox in 2007, when the spin axis was perpendicular to the solar wind direction (and parallel or anti-parallel to the IMF direction). Unfortunately we are unlikely to have any measurements to test the output of our imaginations. Such speculations are not wasted, however, since it is quite possible that such configurations – and many others – could have occurred in earlier epochs of Earth's history (as modeled by Zieger *et al.*, 2004) or may now be occuring in any of the giant planets detected in other solar systems.

13.5 Mercury and Ganymede

The smallest objects with internal dynamos are Mercury and Ganymede. These mini-magnetospheres were recently reviewed by Kivelson (2007). The small inner-most planet and the solar system's largest moon are about the same size and both are believed to have iron cores. Approximately dipolar magnetic fields have been detected; these hold off the surrounding plasma flow to make small but distinct magnetospheres. Just two brief fly-bys by Mariner 10 in the early 1970s gave a glimpse of Mercury's magnetosphere (see the review by Slavin, 2004). These early observations revealed a magnetosphere that, while small, seemed to have most of the main properties observed at Earth (Fig. 2.7), including trapped energetic-particle populations, mini-substorms, and particle injections from the magnetotail, which seem roughly consistent with simple magnetospheric scaling laws. The anticipated arrival of the MESSENGER spacecraft in 2011 and the future launch of the Bebi Colombo mission have provoked further thought about this largely forgotten little magnetosphere and we shall soon see if the details match up to expectations.

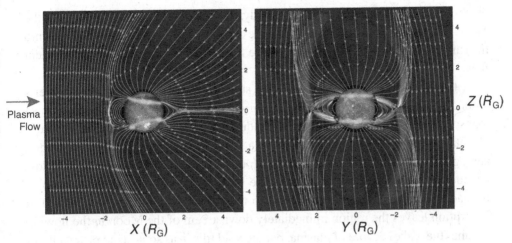

Fig. 13.11. Numerical model of the magnetosphere of Ganymede, with the satellite and the location of the auroral emissions superimposed (based on Jia *et al.*, 2008). (Left) The view looking at the anti-Jupiter side of Ganymede. (Right) The view looking in the direction of the plasma flow at the upstream side (orbital trailing side) of Ganymede, with Jupiter to the left. The shaded areas show the regions of currents parallel to the magnetic field. See also the color-plate section.

Ganymede's magnetosphere sits deep within the magnetosphere of Jupiter (for the background and discussion of Galileo observations see Kivelson *et al.*, 2004). Unlike the supersonic flows of the solar wind, the magnetospheric plasma impinging on Ganymede is subsonic and sub-Alfvénic. There is no upstream bow shock, therefore, and the flowing magnetospheric plasma convects Jupiter's magnetic field, which is roughly anti-parallel to that of Ganymede, towards the upstream magnetopause. The net result is a unique magnetospheric configuration with a region near the equator of magnetic flux that closes on the moon and with polar magnetic flux that connects the moon to Jupiter's north and south ionospheres (Fig. 13.11). A Dungey-style reconnection cycle seems to operate: upstream reconnection opens previously closed flux, convects flux tubes over Ganymede's pole, and re-closes the flux downstream. Computer simulations are helpful in visualizing the process (Jia *et al.*, 2008) but lack of information about the conductivities of Ganymede's tenuous patchy atmosphere and icy surface limit our understanding of the circuit of electrical currents that couple the magnetosphere to the moon.

13.6 Objects without dynamos

Having discussed the seven objects that have internally generated magnetic fields, we return to the objects without dynamos. As summarized in Table 13.1, the nature of the interaction between such bodies and the plasma in which they are embedded depends on the Mach number of the surrounding flow but is determined principally by the electrical conductivity of the body. If conducting paths exist across the planet's interior or ionosphere then electric currents flow through the body and into the surrounding plasma, where they create forces that slow and divert the incident flow.

In the case of an object sitting in the supersonic solar wind, the flow diverts around a region that is similar to a planetary magnetosphere. Mars and Venus have ionospheres that provide the required conducting paths. The barrier that separates planetary plasma from solar wind plasma is referred to as an ionopause (and is analogous to a magnetopause). Earth's Moon, with no ionosphere and a very low conductivity surface, does not deflect the bulk of the solar wind incident on it. Instead, the solar wind runs directly into the surface, where it is absorbed. The absorption leaves the region immediately downstream of the Moon in the flowing plasma (the wake) devoid of plasma, but the void fills in as solar wind plasma flows towards the center of the wake.

When the flow impinging on an object is subsonic, no upstream shock forms. But the flow will be absorbed or diverted depending on whether electrical currents flow within the object or within its ionosphere and into the surrounding plasma. Objects interacting with subsonic flow are exemplified by Io; similar processes occur, albeit to a lesser extent, at Enceladus, Titan, Triton, Europa, and several satellites embedded in the giant planet magnetospheres.

13.6.1 Venus and Mars

The magnetic structure surrounding Mars and Venus is similar to that around magnetized objects, because the interaction causes the magnetic field of the solar wind to drape around the planet. The draped field stretches out downstream (away from the sun), forming a magnetotail (Fig. 13.12). The symmetry of the magnetic configuration within such a tail is governed by the orientation of the magnetic field in the incident solar wind, and that orientation changes with time. For example, if the interplanetary magnetic field (IMF) is oriented northward then the symmetry plane of the tail is in the the east–west direction, and the northern lobe field points away from the sun while the southern lobe field points towards the sun. A southward-oriented IMF would reverse these polarities, and other orientations would produce rotations of the tail's plane of symmetry.

The solar wind brings in magnetic flux tubes that pile up at high altitudes at the dayside ionopause where, depending on the solar wind's dynamic pressure, they may either remain for extended times, thus producing a magnetic barrier that diverts the incident solar wind, or penetrate to low altitudes in localized bundles. Such localized bundles of magnetic flux are often highly twisted structures stretched out along the direction of the magnetic field. Such structures, referred to as flux

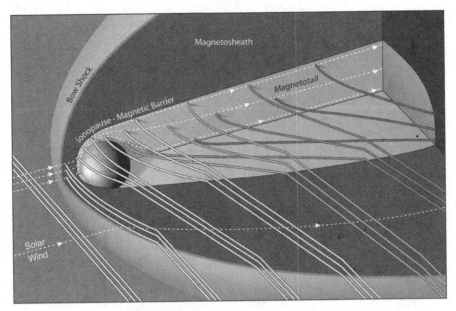

Fig. 13.12. The draping of tubes of solar magnetic flux around a conducting ionosphere such as that of Venus. The flux tubes are slowed down and sink into the wake to form a tail (after Saunders and Russell, 1986).

ropes, are discussed in Chapter 6. These flux ropes may be dragged deep into the atmosphere, possibly carrying away significant amounts of atmosphere.

While Mars' remarkably strong remanent magnetism extends its influence > 1000 km from the surface (Brain *et al.*, 2003), the overall interaction of the solar wind with Mars is more atmospheric (Nagy *et al.*, 2004) than magnetospheric. Mars interacts with the solar wind principally through currents that link to the ionosphere, but there are portions of the surface over which local magnetic fields block the access of the solar wind to low altitudes (Fig. 13.13). It has been suggested that "mini-magnetospheres" extending up to 1000 km form above the regions of intense crustal magnetization in the southern hemisphere; these mini-magnetospheres protect portions of the atmosphere from direct interaction with the solar wind. As a result, the crustal magnetization may have modified the evolution of the atmosphere and may still modify energy deposition into the upper atmosphere.

Several processes involved in the solar wind interaction could have contributed to atmospheric losses at Venus and Mars (Fig. 13.13). The outer neutral atmospheres of Venus and Mars extend out into the solar wind where neutral atoms are photoionized and carried away by the solar wind. Newly ionized ions pick up substantial energy and correspondingly large gyroradii. These energetic ions bombard the upper atmosphere, causing heating and ionization. At times of particularly

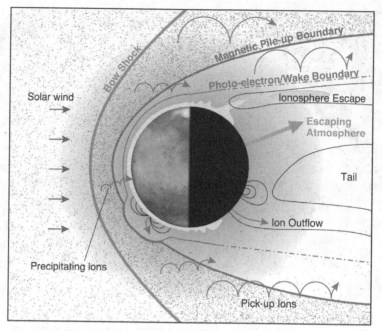

Fig. 13.13. Interaction of the solar wind with the atmosphere, ionosphere, and magnetized crust of Mars. The several processes whereby the planet may have lost much of its atmosphere are shown.

high solar wind pressure the ionosphere can be stripped away in the solar wind. Fresh ionization in the upstream solar wind also generates plasma waves. The solar wind convects the plasma waves towards the planet and into the upper layers of the ionosphere; it is possibly funneled by localized magnetic fields, in the case of Mars, that heat the ions and drive ion outflows, in a similar way to processes in the polar regions at Earth. Quantitative analyses of these different processes, both currently occurring and in the past, are active areas of research and the scientific targets of future missions to Mars.

13.6.2 Io

The discovery of Io's broad influences on the jovian system predated spacecraft explorations. Bigg (1964) discovered Io's controlling influence over Jupiter's decametric radio emissions. Brown and Chaffee (1974) observed sodium emission from Io, which Trafton *et al.* (1974) soon demonstrated to come from extended neutral clouds and not Io itself. Soon thereafter, Kupo *et al.* (1976) detected emissions from sulfur ions, which Brown (1976) recognized as coming from a dense plasma. With the prediction of volcanism by Peale *et al.* (1979) just before its discovery by Voyager 1 (Morabito *et al.*, 1979), a consistent picture of Io's role began to emerge.

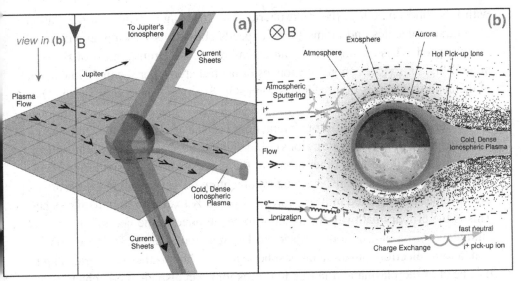

Fig. 13.14. Two views of the interaction between Io and the plasma torus. (a) A three-dimensional view showing the current sheets that couple Io and the surrounding plasma to Jupiter's ionosphere. (b) Cross section of the interaction looking down on the north pole of Io, in the plane of Io's equator, when Io is located between the Sun and Jupiter (orbital phase 180°, local noon in magnetospheric coordinates).

Voyager 1's discovery of Jupiter's aurora and extreme UV emission from the torus (Broadfoot *et al.*, 1979), along with its in situ measurements of the magnetosphere, extended our awareness of Io's effect on the larger system.

The ensuing 25 years of observation by interplanetary missions, Earth-orbiting observatories, and ground-based telescopes has deepened our understanding of Io's influences (see the reviews by Thomas *et al.*, 2004, and Schneider and Bagenal, 2007). Highlights include Galileo's many close fly-bys of Io, with detailed fields-and-particle measurements of Io's interaction with the magnetosphere, and Cassini's months-long UV observation of the torus. Progress from Earth-based studies include the Hubble Space Telescope's sensitive UV observations of the footprint aurora and of Io's atmospheric emissions and ground-based observations of new atomic and molecular species in Io's atmosphere and the plasma torus.

Over the age of the solar system, the tonne/s loss of Iogenic material to the magnetosphere accumulates to a net decrease in radius of about 2 km. While this loss is significant, Io is not in danger of running out of SO_2 in the life time of the solar system. It is plausible, however, that other volatile species such as H_2O were originally present on Io but were completely lost early in its history through processes now depleting Io of SO_2.

Figure 13.14 presents a sketch of the interaction of Io with the surrounding plasma that illustrates some of the processes. Inelastic collisions of torus ions with Io's atmosphere heat the atmospheric gases, causing a significant population of neutral molecules and atoms to gain speeds above Io's 2.6 km/s gravitational escape speed. These neutrals form an extensive corona circling most of the way around Jupiter. Io loses about 1–3 tonnes of neutral atoms per second. How much of the neutral escape is in molecular form (SO_2, SO, or S_2) as against atomic O or S is not known.

The various ion–electron–atom interactions each have a key effect on the magnetosphere. Most importantly torus ions collide with neutral atoms in the atmosphere, which in turn collide with other atoms in the process known as sputtering. Typically, one torus ion can transfer enough momentum for several atmospheric atoms or molecules to be ejected into Io's corona or possibly to escape from Io altogether. This is the primary pathway for material to be supplied to the neutral clouds and ultimately to the plasma torus. A second key reaction is electron impact ionization: a torus electron ionizes an atmospheric atom, which is then accelerated up to the speed of the plasma and leaves Io. Torus ions can also charge-exchange with atmospheric neutrals, which results in a fresh ion and a high-speed neutral. Elastic collisions between ions and atoms can also eject material at speeds between those resulting from sputtering and charge exchange. Finally, electron-impact dissociation breaks down molecules into their component atoms.

Figure 13.14 shows that the strong magnetic field of Jupiter affects the interaction in such a way that the flow around Io resembles fluid flow around a cylinder. (Note that a strong intrinsic magnetic field at Io has been ruled out by Galileo fly-bys over the poles.) Io's motion through the plasma creates an electrical current. While its surface or interior may be modestly conducting, the current is more likely to be carried in other conducting materials surrounding Io, such as its ionosphere and the plasma produced by ionization of its neutral corona. Currents induced across Io are closed by currents that flow along field lines between Io and Jupiter's polar ionosphere in both hemispheres. Observations by the Voyager 1 and Galileo spacecraft indicate that the net current in each circuit is about three million amps. The relative contributions from the conduction current through Io's ionosphere and the current generated by ion pickup in the surrounding plasma remains an issue of debate that awaits more sophisticated models (e.g. see the review by Saur *et al.*, 2004).

A major question regarding Jupiter's magnetosphere is whether most mass loading happens in the near-Io interaction or in the broad neutral clouds far from Io. There is no doubt that substantial pickup occurs near Io, simply owing to the exposure of the upper atmosphere to pickup by the magnetosphere. Pickup near Io is also supported by evidence of fresh pick-up ions of molecules (SO_2^+, SO^+, S_2^+, H_2S^+) near Io with dissociation lifetimes of just a few hours. But a closer look shows that the bulk of the Iogenic source comes from the ionization of atomic sulfur and oxygen farther from Io. Galileo measurements of the plasma fluxes downstream of Io suggest that the plasma source from the ionization of material in the immediate vicinity (within $\sim 5R_{\mathrm{Io}}$) of Io is less than 300 kg/s, which is $\sim 15\%$ of the canonical net tonnes-per-second Iogenic source. The remainder must come from ionization of the extended clouds. It is not clear whether the observations were made during a typical situation, nor it is well established how much the net source and relative contributions of local and distant processes vary with Io's volcanic activity.

While most impacting plasma is diverted to Io's flanks, some is locked to field lines that are carried through Io itself. This $\sim 10\%$ of upstream plasma is rapidly decelerated and moves slowly (~ 3–7 km/s) over the poles. Most particles are absorbed by the moon or its tenuous polar atmosphere, so that the almost-stagnant polar flux tubes are evacuated of plasma. Downstream of Io, the Galileo instruments detected a small trickle of the cold dense ionospheric plasma that had been stripped away. This cold dense "tail" had a dramatic signature ($>$ ten times the background density) but the nearly stagnant flow (~ 1 km/s) means that the net flux of this cold ionospheric material is at most a few percent of the Iogenic source and quickly couples to the surrounding torus plasma (Delamere *et al.*, 2003).

The strong electrodynamic interaction generates Alfvén waves that propagate away from Io along the magnetic field (reviewed by Saur *et al.*, 2004). Other MHD modes that propagate perpendicularly to the field dissipate within a short distance. The intense auroral emission in Jupiter's atmosphere at each "foot" of the flux tube connected to Io tells us that electrons are accelerated somewhere between Io and the atmosphere. The strong correlation of decametric radio emissions with Io's location also tells us that electrons stream away from Jupiter along the Io flux tube and field lines downstream of Io. But how much of the Alfvén wave energy propagates through the torus and reaches Jupiter is not known. Magnetohydrodynamic models suggest that much of the wave energy is reflected at the sharp latitudinal gradients of density in the torus. Furthermore, how the Alfvén wave evolves as it moves through the very low density region between the torus and Jupiter's ionosphere is far from understood. Early ideas suggested that multiple bounces of the Alfvén wave between ionospheres of opposite hemispheres could explain the repetitive bursts of radio emission. More recent studies suggest that the process is more complex, however. Ergun *et al.* (2006) suggested that a resonance is set up whereby Alfvén waves reaching Jupiter's ionosphere accelerate electrons responsible for the short bursts of radio emission (S-bursts). As flux tubes are carried downstream of Io a steady-state current system is set up (Su *et al.*, 2003, 2006). In upward current regions, a few R_{J} above the ionosphere, potential drops develop that accelerate electrons into the ionosphere to produce the wake aurora (Fig. 13.7).

13.7 Outstanding questions

The tables presented in this chapter quantify the characteristics of the seven magnetospheres of our solar system. The schematics give a glimpse of the diversity of their natures. While magnetospheres must share the same underlying basic physical processes, it is the application to very different conditions at the different planets that makes the study of planetary magnetospheres so interesting and tests our understanding. Below are major outstanding questions in planetary magnetospheres.

- How do magnetic dynamos work in the wide range of planetary objects? Why do tiny Mercury and Ganymede have magnetic fields while Earth's sister planet Venus does not? What do the irregular magnetic fields of Uranus and Neptune tell us about their interiors?
- At Saturn, what causes the spin-periodic variability in radio emissions, magnetic field, and plasma properties? What causes the apparent fluctuation in the periodicity?
- How is plasma heated as it moves radially outward in rotation-dominated magnetospheres?
- How is material lost down the magnetotails of Jupiter and Saturn?
- What causes the $\sim$ three-day periodicity in particle fluxes in the magnetosphere of Jupiter?
- Do Jupiter and/or Saturn have return, planetward, Dungey flows in the magnetotails? If not, how do flux tubes opened by dayside reconnection close and conserve magnetic flux?
- What processes lead to the decoupling of the middle magnetosphere of Jupiter from the planet's rotating ionosphere and cause the narrow auroral oval? What role do parallel potential drops play?
- What processes relate the solar wind variability to the apparent changes in Saturn's main aurora and the polar aurora at Jupiter?
- How do electrical currents couple the magnetospheres of Ganymede and Mercury to these planets with very tenuous atmospheres?
- How are particles accelerated and trapped in the mini-magnetospheres of Ganymede and Mercury?
- What processes have been responsible for removing atmospheric gases (particularly water) over the geological history of Mars and Venus?
- What processes are involved in the interactions of Io and Enceladus with their surrounding plasmas? What causes the similarities and differences between the two systems?

Appendix I Authors and editors

Frances Bagenal
LASP – University of Colorado
392 UCB
Duane Room 153
Boulder, CO 80309-0392

Thomas J. Bogdan
NOAA Space Weather Prediction Center
DSRC 2C109
325 Broadway
Boulder, CO 80305

Terry G. Forbes
Space Science Center
University of New Hampshire
Morse Hall
39 College Road
Durham, NH 03824-3525

Tim Fuller-Rowell
CIRES University of Colorado and
NOAA Space Weather Prediction Center
Boulder, CO

Viggo H. Hansteen
Institute of Theoretical Astrophysics
University of Oslo
PO Box 1029, Blindern
Oslo NO-0315
NORWAY

Dana W. Longcope
Physics Department
Montana State University

EP 260D
Bozeman, MT 59717-3840

Mark Moldwin
Department of Earth & Space
Sciences and Institute of Geophysics
and Planetary Physics, UCLA
405 Hilgard Avenue
Los Angeles, CA 90095-1567

Matthias Rempel
NCAR HAO
PO Box 3000
Boulder, CO 80307

Carolus J. Schrijver (editor)
Solar and Astrophysics Laboratory
Lockheed Martin Advanced
Technical Center
3251 Hanover Street, Building 252
Palo Alto, CA 94304-1191

George L. Siscoe (editor)
Boston University
725 Commonwealth Avenue
Boston, MA 02215

Charles W. Smith
Dept of Physics and Astronomy
Space Science Center
Institute for Earth, Oceans, and Space
University of New Hampshire
207 Morse Hall
Durham, NH 03824

Frank R. Toffoletto
Physics and Astronomy Department
Rice University
MS108
6100 Main Street
Houston, TX 77005-1892

Vytenis M. Vasyliūnas
MPI Sonnensystemforschung
Max-Planck-Strasse 2
Katlenburg-Lindau 37191
GERMANY

List of illustrations

List of tables

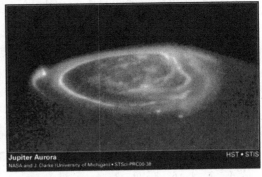

Plate 1 (Fig. 2.9). Aurorae occur near the magnetic polar caps of planets. They result from the precipitation of energetic particles, usually electrons, along the conduits formed by the dipolar magnetic field lines. The magnetic field lines that emanate from the polar caps often change their connectivity between the planetary field anchored at the opposite antipode and the magnetic fields carried by the solar wind. Rotation and the orientation of the solar wind magnetic field often determine when topological changes in connectivity will occur. Reconnection and current sheets are responsible for these changes and are also implicated in the acceleration of particles that gives rise to the aurora in the ionosphere and thermosphere. The figure gives for comparison two similar but distinct types of aurora around (left) Saturn and (right) Jupiter (see Chapter 13; the association of the features in Jupiter's aurora with its magnetic field and satellites is shown in Fig. 13.7). (Courtesy of NASA.)

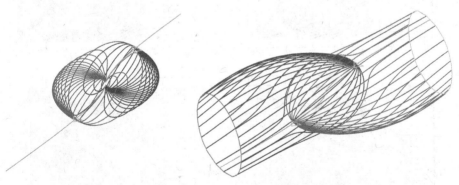

Plate 2 (Fig. 4.4). Perspective views, from near the equator, of fan surfaces from the two null points for the field shown in the lower panel of Fig. 4.3. The left-hand panel shows the inner portions of the fan surface from A and B. The broken field lines labeled Σ_i^A and Σ_i^B in the lower panel of Fig. 4.3 are included in these. The right-hand panel shows the outer portions. Here the surfaces are truncated at the faces of a viewing box. Each fan surface originates from a triangle.

*Plates 1-16 are available for download in colour from
www.cambridge.org/9780521110617

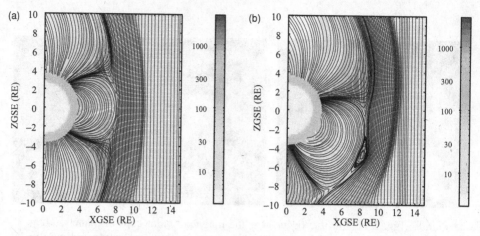

Plate 3 (Fig. 6.10). Illustration of the role that dipole tilt plays in the formation of flux-transfer events. (a) No dipole tilt; (b) a dipole tilt of 34°. Multiple X-lines lead to the formation of flux-transfer events. The color scale represents the plasma pressure. (From Raeder, 2006.)

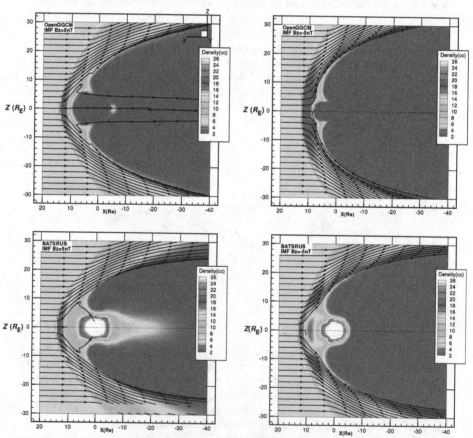

Plate 4 (Fig. 11.3). Computed densities and flow streamlines in the noon–midnight plane obtained from from the OpenGGCM (upper panels) and the BATS-R-US (lower panels) global MHD codes, for the cases when the IMF B_z is 5 nT northward (left) or 5 nT southward (right). The black lines represent flow streamlines.

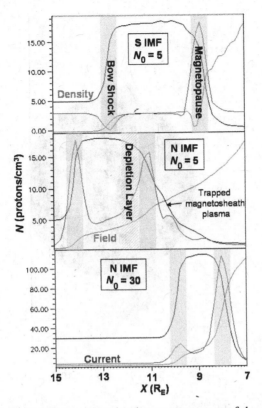

Plate 5 (Fig. 11.4). Plots of number density, y component of the electric current, and magnetic field strength (respectively the top, middle, and bottom curves at the left-hand side of each panel) along the stagnation streamline for (IMF, N_0) = (S, 5), (N, 5), and (N, 30), respectively, where N and S stand for northward and southward and N_0 stands for the upstream density (cm^{-3}). The solar wind speed is 400 km/s, the temperature 232 100 K in each case. The solid shadings identify the bow shock and the magnetopause and the hatched shading indicates the depletion layer. (OpenGGCM.)

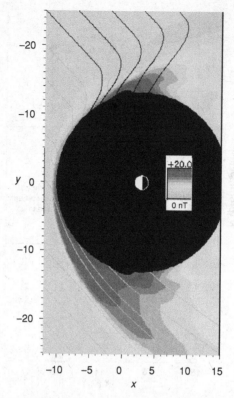

Plate 6 (Fig. 11.5). Field strength and field lines in the equatorial magnetosheath for a Parker-spiral interplanetary magnetic field (BATS-R-US) seen from above the Earth's north pole with the Sun to the left.

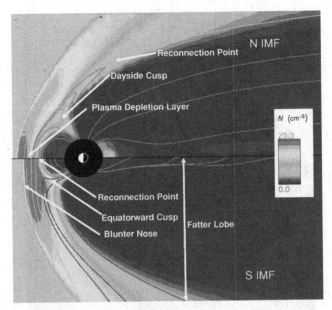

Plate 7 (Fig. 11.6). The basic structural elements of the magnetosphere as depicted in global (OpenGGCM) MHD simulations can be seen here outlined in density contours and magnetic field lines. The figure identifies contrasts between the magnetosphere as shaped under northward (upper) and southward (lower) IMF conditions. The solar wind conditions are the same in both halves of the figure, except for the IMF direction.

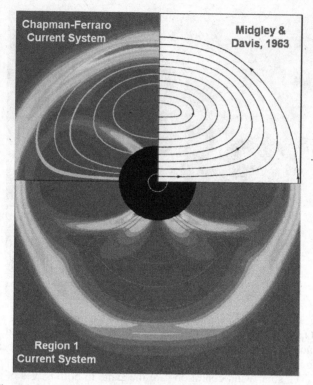

Plate 8 (Fig. 11.7). Magnetohydrodynamic simulations (BATS-R-US) showing contours in the $x = 0$ plane (which contains the terminator) of current density and current streamlines of the Chapman–Ferraro current system (zero IMF) (upper half) and the region-1 current system (5 nT southward IMF) (lower half). Chapman–Ferraro current streamlines from the analytical calculation of Midgley and Davis (1963) are shown for comparison. Solar wind speed 400 km/s, density 5 cm^{-3}, temperature 232 100 K.

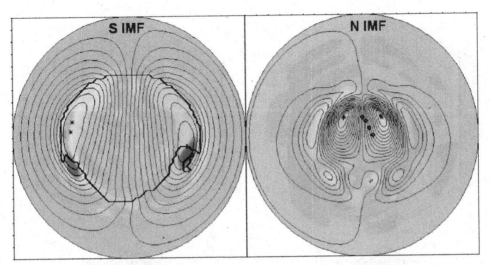

Plate 9 (Fig. 11.9). Contours of electrical potential from the OpenGGCM model on northern-hemisphere polar plots (noon towards the top). The background images show parallel (region-1) current for 5 nT southward IMF (left) and 5 nT northward IMF (right). Other solar wind parameters are as before. The trans-polar voltages are 150 kV (left) and 26 kV (right). The range on the contour scales is the same in both figures, $\pm 1.6\,\mu A/m^2$. The latitude range extends from $50°$ (outer edge) to $90°$ (center). The black closed contour in the left-hand panel marks the border between open and closed field lines. Compare with the schematic diagram in Fig. 10.5, which shows the corresponding plasma flow (right) and electric field lines (left; the corresponding perspective in this figure is from the upper-left corner) for the case with southward IMF.

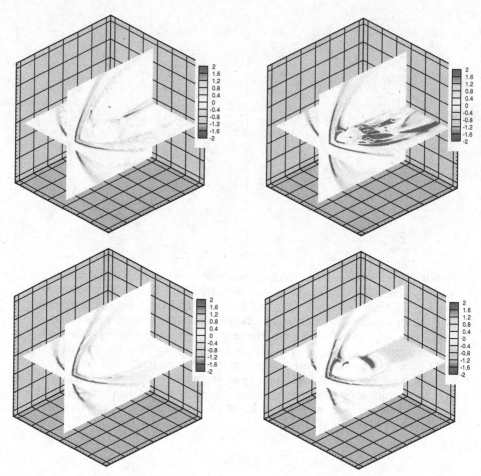

Plate 10 (Fig. 11.10). Values of $\mathbf{j} \cdot \mathbf{E}$ (the scales are in pW/m^3) in the noon–midnight meridian (left) and equatorial plane (right), computed by the OPENG-GCM (upper) and BATS-R-US (lower) MHD codes. In the left-hand panels $B_z = 5\,\text{nT}$ and in the right-hand panels $B_z = -6\,\text{nT}$.

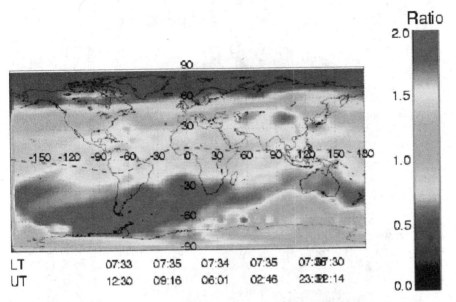

Plate 11 (Fig. 12.3). Global distribution for 14 April 2002 of the column-integrated ratio of atomic oxygen and molecular nitrogen across the middle and upper thermosphere from observations of the GUVI instrument on the TIMED satellite (Paxton *et al.*, 1999). For this April season the global circulation is already more solstice-like, which produces low values in the northern-summer high latitudes and a maximum in the southern-winter hemisphere. The shift of the peak in the southern hemisphere from polar to winter mid latitudes is due to the competition between the solar-driven circulation and that driven by the high-latitude magnetospheric source of heat. (Courtesy of L. Paxton.)

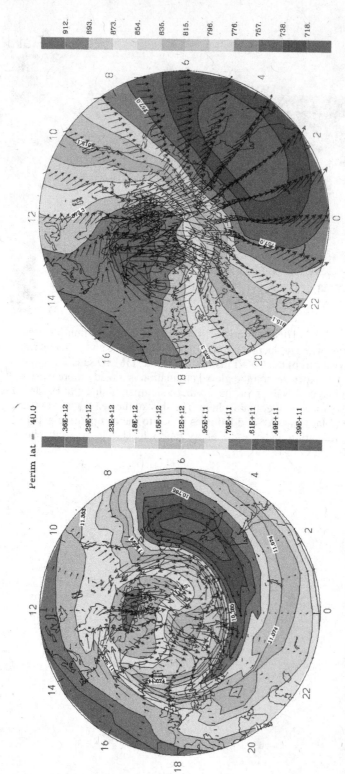

Plate 12 (Fig. 12.6). (Left) Simulation of plasma drift and electron density at the altitude of the *F*-region peak in the upper thermosphere for moderate geomagnetic activity (for a global planetary K_p index ∼ 3). The region poleward of 40° latitude is shown at 18 UT. (Right) Neutral wind and temperature over the same region at 300 km altitude in response to plasma drift. The maximum plasma drift velocity is ∼ 700 m/s and the peak wind speed is ∼ 380 m/s. (From Fuller-Rowell *et al.*, 2008.)

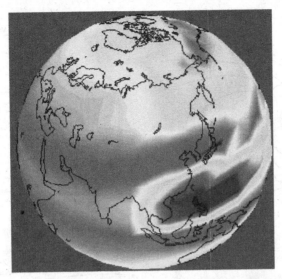

Plate 13 (Fig. 12.10). Global structure of the F-region plasma density showing the equatorial ionization (or Appleton) anomaly at about 15° latitude on either side of the magnetic equator, from a numerical simulation using a coupled thermosphere, ionosphere, plasmasphere model with self-consistent mid- and low-latitude electrodynamics.

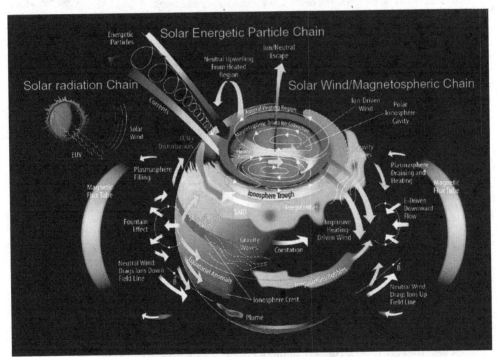

Plate 14 (Fig. 12.12). Indication of the complexity of the ionosphere–thermosphere–mesosphere (ITM) system and the range of physical processes operating. Chapter 12 touches briefly on just a few of the physical profiles behind this complexity. (Courtesy of J. Grebowski, NASA GSFC.)

URANUS

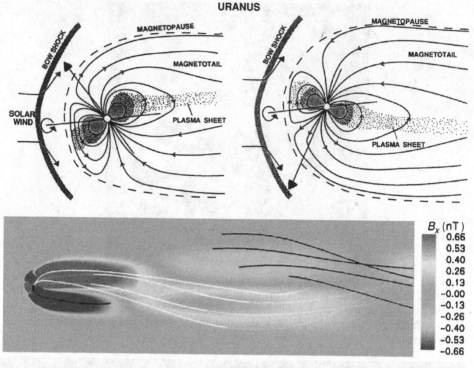

Plate 15 (Fig. 13.9). The magnetosphere of Uranus at solstice (the time of the Voyager 2 flyby). The upper left and right panels show the configuration at different phases of the planet's 18-hour spin period (Bagenal, 1992). The lower panel shows a numerical simulation of the helical magnetotail (Tóth *et al.*, 2004).

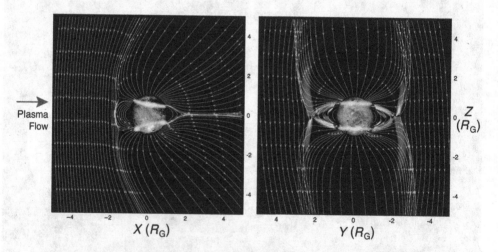

Plate 16 (Fig. 13.11). Numerical model of the magnetosphere of Ganymede, with the satellite and the location of the auroral emissions superimposed (based on Jia *et al.*, 2008). (Left) The view looking at the anti-Jupiter side of Ganymede. (Right) The view looking in the direction of the plasma flow at the upstream side (orbital trailing side) of Ganymede, with Jupiter to the left. The shaded areas show the regions of currents parallel to the magnetic field.

References

Abbett, W. P.: 2007, ApJ 665, 1469

Acuna, M. H., Connerney, J. E. P., & Wasilewski, P.: 2001, JGR 106, 23403

Alfvén, H.: 1941, Arkiv för Matematik, Astronomi och Fysik 27A(25), 1

Alfvén, H.: 1943, Arkiv för Matematik, Astronomi och Fysik 29(2), 1

Alfvén, H.: 1947, MNRAS 107(2), 211

Alfvén, H.: 1977, http://ads.abs.harvard.edu/abs/1977eccp.book . . . . . A

Allen, C. W.: 1973, *Astrophysical Quantities*, Athlone Press

Altschuler, M. D. & Newkirk, G.: 1969, SPh 9, 131

Aly, J. J. & Amari, T.: 1989, A&A 221, 287

Amari, T., Luciani, J. F., Mikic, Z., & Linker, J.: 2000, ApJL 529, L49

Antiochos, S. K.: 1987, ApJ 312, 886

Antiochos, S. K., DeVore, C. R., & Klimchuk, J. A.: 1999, ApJ 510, 485

Antonia, R. A., Ould-Rouis, M., Anselmet, F., & Zhu, Y.: 1997, J. Fluid Mech. 332, 395

Arber, T. D., Haynes, M., & Leake, J. E.: 2007, ApJ 666, 541

Arge, C. N. & Mullan, D. J.: 1998, SPh 182, 293

Armstrong, J. W., Cordes, J. M., & Rickett, B. J.: 1981, Nature 291, 561

Armstrong, J. W., Rickett, B. J., & Spangler, S. R.: 1995, ApJ 443, 209

Arnold, V. I.: 1973, *Ordinary Differential Equations*, MIT Press

Arya, S. & Freeman, J. W.: 1991, JGR 96, 14 183

Athay, R. G.: 1982, ApJ 263, 982, doi:10.1086/160565

Axford, W. I.: 1968, JGR 73, 6855

Axford, W. I.: 1984, in E. W. Hones Jr (ed.), *Magnetic Reconnection in Space and Laboratory Plasmas*, AGU, pp. 1–8

Axford, W. I. & Hines, C. O: 1961, Can. J. Phys. 39, 1433

Axford, W. I. & McKenzie, J. F.: 1992, in E. Marsch & R. Schwenn (eds.), *Solar Wind Seven Colloquium*, p. 1

Backus, G.: 1958, Ann. Phys. 4, 372

Bagenal, F.: 1992, Ann. Rev. Earth Planet. Sci. 20, 289

Bagenal, F.: 2007, J. Atmos. Terr. Phys. 69, 387, doi:10.1016/j.jastp.2006.08.012

Bagenal, F., Dowling, T., & McKinnon, W. (eds.): 2004, *Jupiter: Planet, Satellites, Magnetosphere*, Cambridge University Press

Barnes, A.: 1979, in C. F. Kennel, E. N. Parker, & L. J. Lanzerotti (eds.), *Solar System Plasma Physics*, North-Holland, p. 249

Bartley, W. C., Burkata, R. P., McCracken, K. G., & Rao, U. R.: 1966, JGR 71, 3297

Batchelor, G. K.: 1953, *The Theory of Homogeneous Turbulence*, Cambridge University Press

Baumjohann, W. & Treumann, R. A.: 1997, *Basic Space Plasma Physics*, Imperial College Press

Belcher, J. W. & Davis, L.: 1971, JGR 76, 3534

Belcher, J. W., Lazarus, A. J., McNutt, R. L., Jr, & Gordon, G. S., Jr: 1993, JGR 98, 15 177

Benz, A. O.: 1993, *Plasma Astrophysics*, Kluwer

Berchem, J. & Russell, C. T.: 1984, JGR 89, 6689

Berger, M. A.: 1988, A&A 201, 355

Berger, M. A. & Field, G. B.: 1984, J. Fluid Mech. 147, 133

Berger, T. E. & Title, A. M.: 1996, ApJ 463, 365

Berger, T. E., Löfdahl, M. G., Shine, R. S., & Title, A. M.: 1998, ApJ 495, 973

Beveridge, C. & Longcope, D. W.: 2005, SPh 227, 193

Bhardwaj, A., & Gladstone, G. R.: 2000, Rev. Geophys. 38, 295

Bieber, J. W., Wanner, W., & Matthaeus, W. H.: 1996, JGR 101, 2511

Biermann, L.: 1946, Naturwissenshaften 33, 118

Biermann, L.: 1948, Z. Astrophys. 25, 161

Biermann, L.: 1951, Z. Astrophys. 29, 274

Biermann, L.: 1957, The Observatory 77, 109

Bigg, E. K.: 1964, Nature 203, 1008

Birkeland, K.: 1908, *The Norwegian Aurora Polaris Expedition 1902–1903*, Vol. I, H. Aschehoug and Co., Christiania

Birkeland, K.: 1913, *The Norwegian Aurora Polaris Expedition 1902–1903*, Vol. II, H. Aschehoug and Co.

Birn, J. & Hesse, M.: 2001, JGR 106, 3737, doi:10.1029/1999JA001001

Birn, J., Drake, J. F., Shay, M. A., *et al.*: 2001, JGR (Space Physics) 106(A3), 3715

Biskamp, D.: 1986, Phys. Fluids 29, 1520

Biskamp, D.: 2000, *Magnetic Reconnection in Plasmas*, Cambridge University Press, Cambridge

Biskamp, D.: 2003, *Magnetohydrodynamic Turbulence*, Cambridge University Press

Biskamp, D. & Müller, W. C.: 2000, Phys. Plasmas 7, 4889

Blanchard, G. T., Lyons, L. R., de la Beaujardiére, O., Doe, R. A., & Mendillo, M.: 1996, JGR 101, 15265

Bogdan, T. J., Carlsson, M., Hansteen, V. H., *et al.*: 2003, ApJ 599, 626

Boldyrev, S.: 2005, ApJ 626, L37

Boldyrev, S.: 2006, Phys. Rev. Lett. 96, 115002

Bolton, S. J., Thorne, R. M., Bourdarie, S., De Pater, I., & Mauk, B.: 2004, Jupiter's inner radiation belts, in *Jupiter: The Planet, Satellites and Magnetosphere*, pp. 671–688

Bougher, S. W. & Roble, R. G.: 1991, JGR 96, 11045

Bougher, S. W., Roble, R. G., & Fuller-Rowell, T.: 2002, Simulations of the upper atmospheres of the terrestrial planets, in *Atmospheres in the Solar System: Comparative Aeronomy*, p. 261

Boyle, C. B., Reiff, P. H., & Hairston, M. R.: 1997, JGR 102, 111, doi:10.1029/96JA01742

Braginskii, S.I.: 1964, Sov. Phys. 20, 1462

Braginskii, S. I.: 1965, in M. A. Leontovich (ed.), *Reviews of Plasma Physics*, 205–311, Consultants' Bureau, New York

Brain, D. A., Bagenal, F., Acuña, M. H., & Connerney, J. E. P.: 2003, JGR 108(A12), 1424, doi:10.1029/2002JA009482

Brandenburg, A. & Sandin, C.: 2004, A&A 427, 13

Brandenburg, A. & Subramanian, K.: 2005, Phys. Rep. 417, 1, doi:10.1016/j.physrep.2005.06.005

Breech, B., Matthaeus, W. H., Minnie, J., *et al.*: 2007, JGR submitted

Brice, N. M. & Ioannidis, G. A.: 1970, Icarus 13, 173

Broadfoot, A. L., Belton, M. J., Takacs, P. Z., *et al.*: 1979, Science 204, 979

Brown, D. S., Nightingale, R. W., Alexander, D., *et al.*: 2003, SPh 216, 79

Brown, R. A.: 1976, ApJL 206, 179

Brown, R. A. & Chaffee, F. H., Jr: 1974, ApJL 187, 125

Browning, M. K., Miesch, M. S., Brun, A. S., & Toomre, J.: 2006, ApJL 648, L157

Brun, A. S., Miesch, M. S., & Toomre, J.: 2004, ApJ 614, 1073

Bruno, R. & Carbone, V.: 2005, Living Rev. Solar Phys. 2, 4

Bruno, R., Bavassano, B., Pietropaolo, E., Carbone, V., & Veltri, P.: 1999, GRLe 26, 3185

Bullard, E. C. & Gellman, H.: 1954, Phil. Trans. R. Soc. A 247, 213

Buneman, O.: 1992, IEEE Trans. Plasma Sci., 20, 672

Bunce, E. J., Cowley, S. W. H., & Yeoman, T. K.: 2004, JGR 109, A09S13

Burch, J. L.: 2005, Rev. Geophys. Space Phys. 3, 43

Burke, W. J., Weimer, D. R., & Maynard, N. C.: 1999, JGR 104, 9989, doi:10.1029/1999JA900031

Burlaga, L. F.: 1988, JGR 93, 7217

Burlaga, L. F.: 1991, Geophys. Res. Lett. 18, 79

Burlaga, L. F. & Ness, N. F.: 1969, SPh 9, 467

Burlaga, L. F., Klein, L., Sheeley, N. R., *et al.*: 1982, GRLe 9, 1317

Burton, R. K., McPherron, R. L., & Russell, C. T.: 1975, JGR 80, 4204

Bushby, P. J.: 2005, Astron. Nachr. 326, 218

Cane, H. V. & Richardson, I. G.: 2003, JGR 108, (A8) 1156, doi:10.1029/2002JA009817

Carlsson, M. & Stein, R. F.: 1995, ApJL 440, L29

Carlsson, M., Stein, R. F., Nordlund, Å., & Scharmer, G. B.: 2004, ApJL 610, L137

Cattaneo, F.: 1999, ApJL 515, L39

Cattaneo, F. & Hughes, D. W.: 1996, Phys. Rev. E 54, 4532

Cattaneo, F. & Hughes, D. W.: 2001, Astronomy and Geophysics 42, 18

Caunt, S. E. & Korpi, M. J.: 2001, A&A 369, 706

Chae, J.: 2001, ApJ 560, L95

Chapman, S.: 1954, ApJ 120, 151

Chapman, S.: 1957, Smithson. Contrib. Astrophys. 2, 1

Chapman, S. & Cowling, T. G.: 1960, *The Mathematical Theory of Non-Uniform Gases*, Cambridge University Press

Chapman, S. & Ferraro, V. C. A.: 1929, MNRAS 89, 470

Chappell, C. R.: 1988, Rev. of Geophys. 26, 229

Charbonneau, P.: 2005, Living Rev. Solar Phys. 2, 2

Cheung, M. C. M., Schüssler, M., & Moreno-Insertis, F.: 2007, A&A 461, 1163

Choudhuri, A. R., Schüssler, M., & Dikpati, M.: 1995, A&A 303, L29

Clarke, J. T., Grodent, D., Cowley, S. W. H., *et al.*: 2004, in W. B. McKinnon, F. Bagenal, & T. E. Dowling (eds.), *Jupiter: The Planet, Satellites and Magnetosphere*, Cambridge University Press

Clarke, J. T., Gérard, J.-C., Grodent, D., *et al.*: 2005, Nature 433, 717

Cliver, E. W., Feynman, J., & Garrett, H. B.: 1990, JGR (Space Physics) 95, 17 103

Cole, K. D.: 1964, JGR 69, 3595

Cole, T. W. & Slee, O. B.: 1980, Nature 285, 93

Coleman, P. J.: 1966, Phys. Rev. Lett. 17, 207

Coleman, P. J.: 1968, ApJ 153, 371

Connerney, J. E. P.: 1981, JGR 86, 76

Connerney, J. E. P.: 1993, JGR 98, 18 659

Connerney, J. E. P. & Ness, N. F.: 1988, in M. S. Matthew, F. Vilas, & C. R. Chapman (eds.), *Mercury*, University of Arizona Press

Connerney, J. E. P., Acuna, M. H., Ness, N. F., Spohn, T., & Schubert, G.: 2004, Space Sci. Rev. 111, 1

Coroniti, F. V. & Kennel, C. F.: 1979, in J. Arons, C. Max, & C. McKee (eds.), *Particle Acceleration in Planetary Magnetospheres*, American Institute of Physics, p. 169

Covas, E., Tavakol, R., Moss, D., & Tworkowski, A.: 2000, A&A 360, L21

Cowley, S. C., Longcope, D. W., & Sudan, R. N.: 1997, Phys. Rep. 283, 227

Cowley, S. W. H., Nichols, J. D., & Bunce, E. J.: 2002, Planet. Space Sci. 50, 717

Cowley, S. W. H., Bunce, E. J., & Nichols, J. D.: 2003a, JGR 108, (A4) 8002, doi:10.1029/2002JA009329

Cowley, S. W. H., Bunce, E. J., Stallard, T. S., & Miller, S.: 2003b, GRLe 30, 1220

Cowling, T. G.: 1933, MNRAS 94, 39

Craig, I. J. D. & Sneyd, A. D.: 2005, SPh 232, 41

Craig, I. J. D., Fabling, R. B., Henton, S. M., & Rickard, G. J.: 1995, ApJ 455, L197

Cranmer, S. R. & Van Ballegooijen, A. A.: 2005, ApJSS 156, 265

Crary, F. J., Clarke, J. T., Dougherty, M. K., *et al.*: 2005, Nature 433, 720

Cravens, T. E.: 1997, *Physics of the Solar System*, Cambridge University Press

Cravens, T. E. & Gombosi, T. I.: 2004, Adv. Space Res. 33, 1968, doi:10.1016/j.asr.2003.07.053

Crooker, N. U., Eastman, T. E., & Stiles, G. S.: 1979, JGR (Space Physics) 84, 869

Crooker, N. U., Siscoe, G. L., Shodhan, S., Webb, D. F., Gosling, J. T., & Smith, E. J.: 1993, JGR 98, 9371

Crooker, N. U., Huang, C.-L., Lamassa, S. M., Larson, D. E., Kahler, S. W., & Spence, H. E.: 2004, JGR (Space Physics) 109(A18), 3107, doi:10.1029/2003JA010170

Dasso, S., Milano, L. J., Matthaeus, W. H., & Smith, C. W.: 2005, ApJL 635, L181

Davidson, P. A.: 2001, *An Introduction to Magnetohydrodynamics*, Cambridge University Press

Davidson, P. A.: 2004, *Turbulence*, Oxford University Press

De Keyser, J., Roth, M., & Söding, A.: 1998, GRLe 25, 2649, doi:10.1029/98GL51938

De Pontieu, B., Martens, P. C. H., & Hudson, H. S.: 2001, ApJ 558, 859, doi:10.1086/322408

De Pontieu, B., McIntosh, S. W., Carlsson, M., *et al.*: 2007, Science 318, 1574

De Zeeuw, D. L., Sazykin, S., Wolf, R. A., Gombosi, T. I., Ridley, A. J., & Toth, G.: 2004, JGR (Space Physics) 109(A12), A12 219

Delamere, P. A. & Bagenal, F.: 2003, JGR (Space Physics) 108, 1276, doi:10.1029/2002JA009706

Delamere, P. A. & Bagenal, F.: 2008, GRLe 35

Delamere, P. A., Bagenal, F., Ergun, R., & Su, Y.-J.: 2003, JGR (Space Physics) 108, 1241, doi:10.1029/2002JA009530

Delamere, P. A., Steffl, A., & Bagenal, F.: 2004, JGR 109, A10216, doi:10.1029/2003JA010354

Delamere, P. A., Bagenal, F., Dols, V., & Ray, L. C.: 2007, GRLe 34, 9105, doi:10.1029/2007GL029437

Demars, H. G. & Schunk, R. W.: 1994, JGR 99, 2215

Démoulin, P., Henoux, J. C., Priest, E. R., & Mandrini, C. H.: 1996, A&A 308, 643

Démoulin, P., Bagala, L. G., Mandrini, C. H., Hénoux, J. C., & Rovira, M. G.: 1997, A&A 325, 305

Démoulin, P., Mandrini, C. H., van Driel-Gesztelyi, L., *et al.*: 2002, A&A 382, 650

Denskat, K. U., Beinroth, H. J., & Neubauer, F. M.: 1983, J. Geophys. 54, 60

Dessler, A. J. & Fejer, J. A.: 1963, Planet. Space Sci. 11, 505

Dessler, A. J. & Hill, T. W.: 1975, GRLe 2, 567

Dessler, A. J., Sandel, B. R., & Atreya, S. K.: 1981, Planet. Space Sci. 29, 215, doi:10.1016/0032-0633(81)90035-0

DeVore, C. R.: 2000, ApJ 539, 944

Dikpati, M. & Charbonneau, P.: 1999, ApJ 518, 508

Dorelli, J. C., Bhattacharjee, A., & Raeder, J.: 2007, JGR 112((A2) A02202), doi 10.1029/2006JA011877

Draine, B. T.: 1986, MNRAS 220, 133

Dungey, J. W.: 1953a, Phil. Mag. 44, 725

Dungey, J. W.: 1953b, MNRAS 113, 679

Dungey, J. W.: 1958, *Electrodynamical Phenomena in Cosmical Physics*, Cambridge University Press

Dungey, J. W.: 1961, Phys. Rev. Lett. 6, 47

Dungey, J. W.: 1994, JGR 99, 19189

Edlén, B.: 1942, Arkiv för Matematik, Astronomi, och Fysik 28B(1), 1

Einaudi, G. & Velli, M.: 1999, Phys. of Plasmas 6, 4146

Elliott, J. A.: 1993, in R. Dendy (ed.), *Plasma Physics*, Cambridge University Press, p. 29

Elmegreen, B. G. & Scalo, J.: 2004, ARA&A 42, 211

Elsässer, W. M.: 1946, Phys. Rev. 69, 106, doi:10.1103/PhysRev.69.106

Elsässer, W. M.: 1950, Phys. Rev. 79, 183

Ergun, R. E., Su, Y.-J., Andersson, L., *et al.*: 2006, JGR (Space Physics) 111(A10), 6212, doi:10.1029/2005JA011253

Fairfield, D. H., Otto, A., Mukai, T., *et al.*: 2000, JGR 105, 21,159

Fan, Y., Fisher, G. H., & McClymont, A. N.: 1994, ApJ 436, 907

Farrugia, C. J., Elphic, R. C., Southwood, D. J., & Cowley, S. W. H.: 1987, Planet. Space Sci. 35, 227

Farrugia, C. J., Southwood, D. J., & Cowley, S. W. H.: 1988, Adv. Space Res. 8, 249, doi:10.1016/0273-1177(88)90138-X

Farrugia, C. J., Erkaev, N. V., Biernat, H. K., Burlaga, L. F., Lepping, R. P., & Osherovich, V. A.: 1997, JGR (Space Physics) 102, 7087

Fejer, B. G. & Scherliess, L.: 1997, JGR 102, 24047, doi:10.1029/97JA02164

Fejer, J. A.: 1964, JGR 69, 123

Feldman, W. C., Ashbridge, J. R., Bame, S. J., Montgomery, M. D., & Gary, S. P.: 1975, JGR 80, 4181

Feldman, W. C., Ashbridge, J. R., Bame, S. J., & Gosling, J. T.: 1977, in O. R. White (ed.), *The Solar Output and its Variation*, Colorado Associated University Press, p. 351

Ferrière, K. M., Zimmer, C., & Blanc, M.: 2001, JGR 106, 327, doi:10.1029/2000JA000133

Feynman, J. & Ruzmaikin, A.: 1994, JGR 99, 17,645

Finn, J. & Antonsen, Jr, T. M.: 1985, Comments Plasma Phys. Controlled Fusion 9, 111

Fleck, B., Domingo, V., & Poland, A.: 1995, *The SOHO mission*, Kluwer

Fontenla, J. M.: 2005, A&A 442, 1099

Forbes, J. M.: 1995, *The Upper Mesosphere and Lower Thermosphere: A Review of Experiment and Theory*, Vol. 87 of *Geophysical Monograph Series*, American Geophysical Union, p. 67

Forbes, T. G.: 2001, in M. Hoshino, R. L. Stenzel, & K. Shibata (eds.), *Magnetic Reconnection in Space and Laboratory Plasmas*, Terra Scientific Publishing, p. 423

Forbes, T. G.: 2006, in J. Birn & E. R. Priest (eds.), *Reconnection of Magnetic Fields: Magnetohydrodynamics and Collisionless Theory and Observations*, Cambridge University Press, in press

Forbes, T. G. & Isenberg, P. A.: 1991, ApJ 373, 294

Forbes, T. G., Linker, J. A., Chen, J., *et al.*: 2006, Space Sci. Rev. 123, 251, Report of Working Group D

Forbush, S. E., Gill, P. S., & Vallarta, M. S.: 1949, Rev. Mod. Phys. 21, 44

Formisano, V. & Kennel, C. F.: 1969, J. Plasma Phys. 3, 55

Fossum, A. & Carlsson, M.: 2005, Nature 435, 919

Frank, L. A., Paterson, W. R., & Khurana, K. K.: 2002, JGR 107(A1), 1003, doi:10.1029/2001JA000077

Freeman, J. W., Totten, T., & Arya, S.: 1992, EOS Trans. AGU 73, 238

Frisch, U.: 1995, *Turbulence*, Cambridge University Press

Fuller-Rowell, T. J., Richmond, A. D., & Maruyama, N.: 2008, in P. M. Kintner, A. J. Coster, T. J. Fuller-Rowell, A. J. Mannucci, M. Mendillo, & R. Heelis (eds.), *Midlatitude Ionospheric Dynamics and Disturbances*, Vol. 181 of *Geophysical Monograph Series*, American Geophysical Union, p. 187

Furth, H. P., Killeen, J., & Rosenbluth, M. N.: 1963, Phys. Fluids 6, 459

Fyfe, D., Joyce, G., & Montgomery, D. C.: 1977, J. Plasma Phys. 17, 369

Gabriel, A. H.: 1976, Phil. Trans. R. Soc. 281, 399

Gailitis, A.: 1970, Magnetohydrodynamics 6, 14

Galeev, A. A.: 1979, Space Sci. Rev. 23, 411

Galsgaard, K. & Nordlund, Å.: 1996, JGR 101(10), 13445, doi:10.1029/96JA00428

Gary, S. P., Skoug, R. M., Steinberg, J. T., & Smith, C. W.: 2001, GRLe 28, 2759

Gary, S. P., Saito, S., & Li, H.: 2008, GRLe 35, 2104, doi:10.1029/2007GL032327

Gazis, P. R., Barnes, A., Mihalov, J. D., & Lazarus, A. J.: 1994, JGR 99, 6561

Ghosh, S. & Goldstein, M. L.: 1997, J. Plasma Phys. 57, 129

Ghosh, S., Matthaeus, W. H., Roberts, D. A., & Goldstein, M. L.: 1998, JGR 103, 23705

Gilbert, H., Kilper, G., & Alexander, D.: 2007, ApJ 671, 978

Gilman, P. A.: 1983, ApJSS 53, 243

Giovanelli, R. G.: 1946, Nature 158, 81

Gladstone, G. R., Waite, J. H., Jr., Lewis, D., *et al.*: 2002, Nature 415, 1000

Glatzmaier, G. A.: 1984, J. Computational Physics 55, 461

Glatzmaier, G. A.: 2002, Ann. Rev. Earth Planet. Sci. 30, 237, doi:10.1146/30.091201.140817

Glatzmaier, G. A. & Gilman, P. A.: 1982, ApJ 256, 316

Glatzmaier, G. A. & Roberts, P. H.: 1995, Nature 377, 203

Gold, T.: 1959, JGR 64, 1219

Goldreich, P. & Farmer, A. J.: 2007, JGR 112(A05225), 5225, doi:10.1029/2006JA012163

Goldreich, P. & Sridhar, S.: 1995, ApJ 438, 763

Goldstein, B. E. & Siscoe, G. L.: 1972, in P. J. Coleman, C. P. Sonnett & J. M. Wilcox (eds.), *Solar Wind 5*, NASA Spec. Publ. NASA SP-308, p. 506

Goldstein, M. L. & Wong, H. K.: 1987, JGR 92, 4695

Goldstein, M. L., Burlaga, L. F., & Matthaeus, W. H.: 1984, JGR 89, 3747

Goldstein, M. L., Roberts, D. A., & Fitch, C. A.: 1994, JGR 99, 11,519

Goldstein, M. L., P., Escoubet C., & Laakso, H.: 2006, Adv. Space Res. 38

Gombosi, T. I., Powell, K. G., De Zeeuw, D. L., *et al.*: 2004, Computing in Science and Engineering 6(2), 14

Gombosi, T. I., Powell, K. G., & van Leer, B.: 2000, JGR (Space Physics) 105(A6), 13141

Goodman, M. L.: 2000, ApJ 533, 501

Gopalswamy, N.: 2007, in K.-L. Klein & A. L. MacKinnon (eds.), *Lecture Notes in Physics*, Vol. 725 Springer Verlag, p. 139

Gosling, J. T.: 1990, in *Physics of Magnetic Flux Ropes*, Vol. 58 of *Geophysical Monograph Series*, American Geophysical Union, pp. 343–364

Gosling, J. T.: 1993, JGR 98, 18 937

Gosling, J. T.: 2007, in T. V. Johnson, L. McFadden, & P. Weissman (eds.), *Encyclopedia of the Solar System*, p. 99

Gosling, J. T., Asbridge, J. R., Bame, S. J., Feldman, W. C., Borrini, G., & Hansen, R. T.: 1981, JGR 86, 5438

Gosling, J. T., Baker, D. N., Bame, S. J., Feldman, W. C., Zwickl, R. D., & Smith, E. J.: 1987, JGR 92, 8519

Gosling, J. T., Birn, J., & Hesse, M.: 1995, GRLe 22, 869

Gosling, J. T., Skoug, R. M., McComas, D. J., & Smith, C. W.: 2005, JGR (Space Physics) 110(A9), 1107, doi:10.1029/2004JA010809

Gosling, J. T., Eriksson, S., Blush, L. M., *et al.*: 2007, GRLe 34, 20108, doi:10.1029/2007GL031492

Grant, H. L., Stewart, R. W., & Moilliet, A.: 1962, J. Fluid Mech. 12, 241

Grappin, R., Pouquet, A., & Léorat, J.: 1983, A&A 126, 51

Green, R. M.: 1965, in R. Lust (ed.), *Stellar and Solar Magnetic Fields. Proc. IAU Symp.* 22, North-Holland, p. 398

Greene, J. M.: 1988, JGR 93(A8), 8583

Greene, J. M.: 1992, J. Comput. Phys. 98, 194

Greene, J. M.: 1993, Phys. Fluids, B 5, 2355

Griffiths, D. J.: 1999, *Introduction to Electrodynamics*, Prentice Hall

Grodent, D., Clarke, J. T., Waite, J. H., Jr, Cowley, S. W. H., Gerard, J.-C., , & Kim, J.: 2003a, JGR 108, 1366

Grodent, D., Gerard, J.-C., Clarke, J. T., Gladstone, G. R., & Waite, J. H., Jr: 2003b, JGR 108(A10), 1366, doi:10.1029/2003JA010017

Grotrian, W.: 1933, Z. Astrophys. 7, 26

Grotrian, W.: 1939, Naturwissenschaften 27, 214

Guckenheimer, J. & Holmes, P.: 1983, *Nonlinear Oscillations, Dynamical Systems, and Bifurcations of Vector Fields*, Vol. 42, Springer-Verlag

Gudiksen, B. V. & Nordlund, Å.: 2002, ApJL 572, L113, doi:10.1086/341600

Guild, T., Spence, H., Kepko, L., *et al.*: 2004, J. Atmos. Solar–Terrestrial Phys. 66(15–16), 1351

Gurnett, D. A., Persoon, A. M., Kurth, W. S., *et al.*: 2007, Science 316, 442

Haerendel, G.: 1992, in J. J. Hunt (ed.), *Study of the Solar–Terrestrial System*, Vol. 346 of *ESA Special Publications*, p. 23

Hale, G. E., Ellerman, F., Nicholson, S. B., & Joy, A. H.: 1919, ApJ 49, 153

Hamilton, K., Smith, C. W., Vasquez, B. J., & Leamon, R. J.: 2008, JGR (Space Physics) 113(A12), 1106, doi:10.1029/2007JA012559

Handy, B. N., Acton, L. W., Kankelborg, C. C., *et al.*: 1999, SPh 187, 229

Hansen, C. J., Esposito, L., Stewart, A. I. F., *et al.*: 2006, Science 311, 1422

Hansteen, V.: 1993, ApJ 402, 741

Hansteen, V. H. & Leer, E.: 1995, JGR 100(9), 21 577, doi:10.1029/95JA02300

Hansteen, V. H., Leer, E., & Holzer, T. E.: 1997a, ApJ 482, 498

Hansteen, V. H., Leer, E., & Holzer, T. E.: 1997b, ApJ 482, 498

Harris, E. G.: 1962, Nuovo Cim. 23, 115

Haverkorn, M., Gaensler, B. M., McClure-Griffiths, N. M., Dickey, J. M., & Green, A. J.: 2004, ApJ 609, 776

Haynes, A. L., Parnell, C. E., Galsgaard, K., & Priest, E. R.: 2007, Proc. R. Soc. A 463, 1097

Heelis, R. A.: 2004, J. Atm. Solar–Terr. Phys. 66, 825, doi:10.1016/j.jastp.2004.01.034

Heerikhuisen, J. & Craig, I. J. D.: 2004, in V. Florinski, N. V. Pogorelov, & G. P. Zank (eds.), *Physics of the Outer Heliosphere*, Vol. 719, American Institute of Physics, p. 305

Hellinger, P., Trávníček, P., Kasper, J. C., & Lazarus, A. J.: 2006, GRLe 33, L09101, doi:10.1029/2006GL025925

Hendrix, D. L., van Hoven, G., Mikic, Z., & Schnack, D. D.: 1996, ApJ 470, 1192, doi:10.1086/177942

Herring, J. R. & Tennekes, H.: 1989, *Lecture Notes on Turbulence*, World Scientific

Herzenberg, A.: 1958, Phil. Trans. R. Soc. A 250, 543

Hesse, M.: 1991, http://ccmc.gsfc.nasa.gov/metrics/metrics report GEM.pdf

Hesse, M. & Schindler, K.: 1988, JGR 93(A6), 5559

Hesse, M., Birn, J., & Kuznetsova, M.: 2001, JGR 106, 3721, doi:10.1029/1999JA001002

Hesse, M., Forbes, T. G., & Birn, J.: 2005, ApJ 631, 1227

Heuer, M. & Marsch, E.: 2007, JGR 112, doi:10.1029/2006JA011979

Heyvaerts, J., Priest, E. R., & Rust, D. M.: 1977, ApJ 216, 123

Hiei, E., Hundhausen, A. J., & Sime, D. G.: 1993, GRLe 20, 2785

Higdon, J. C.: 1984, ApJ 285, 109

Hill, R. J.: 1997, J. Fluid Mech. 353, 67

Hill, T. W.: 1979, JGR 84, 6554

Hill, T. W.: 1980, Science 207, 301

Hill, T. W.: 1983, in R. L. Carovillano & J. M. Forbes (eds.), *Solar–Terrestrial Physics*, Reidel, p. 261

Hill, T. W.: 1994, in J. L. Burch & J. H. Waite (eds.), *Solar System Plasmas in Space and Time*, p. 199

Hill, T. W., Dessler, A. J., & Michel, F. C.: 1974, GRLe 1, 3

Hill, T. W., Dessler, A. J., & Wolf, R. A.: 1976, GRLe 3, 429

Hill, T. W., Dessler, A. J., & Maher, L. J.: 1981, JGR 86, 9020

Hinze, J. O.: 1975, *Turbulence*, McGraw-Hill

Hollweg, J. V.: 1973, ApJ 181, 547

Hollweg, J. V. & Isenberg, P. A.: 2002, JGR 107, (A7) 1147, doi:10.1029/2001JA000270

Holton, J. R.: 1972, *An Introduction to Dynamic Meteorology; International Geophysics Series*, Vol. 16, Academic Press

Holzer, T. E.: 2005, in *Solar Wind 11/SOHO 16, Connecting Sun and Heliosphere*, Vol. 592 of *ESA Special Publications*, p. 115

Hones Jr, E.W.: 1976, in D. J. Williams (ed.), *Physics of Solar Planetary Environments*, American Geophysical Union, p. 558

Horbury, T. S., Burgess, D., Fränz, M., & Owen, C. J.: 2001, GRLe 28, 677, doi:10.1029/2000GL000121

Hornig, G. & Schindler, K.: 1996, Phys. Plasmas 3, 781

Howard, R. F.: 1991, SPh 136, 251

Howes, G. G., Cowley, S. C., Dorland, W., Hammett, G. W., Quataert, E., & Schekochihin, A. A.: 2007, in D. Shaikh & G. P. Zank (eds.), *Proc. Sixth Annual International Astrophysical Conf.*, American Institute of Physics, p. 3

Hoyle, F.: 1949, *Some Recent Researches in Solar Physics*, Cambridge University Press

Hu, Y. Q. & Low, B. C.: 1982, SPh 81, 107

Huba, J. D. & Lyon, J. G.: 1999, J. Plasma Phys. 61, 391

Hughes, W. J. & Sibeck, D. G.: 1987, GRLe 14, 636

Hundhausen, A. J.: 1972, Physics and Chemistry in Space 5

Ichimaru, S.: 1975, ApJ 202, 528

Ieda, A., Machida, S., Mukai, T., *et al.*: 1998, JGR 103, 4453

Iijima, T. & Potemra, T. A.: 1976, JGR (Space Physics) 81, 2165

Imshennik, V. S. & Syrovatskii, S. I.: 1967, Soviet Phys. JETP Lett. (English trans.) 15, 656

Ioannidis, G. & Brice, N.: 1971, Icarus 14, 360

Iroshnikov, P. S.: 1964, Sov. Astron. 7, 566

Isenberg, P. A. & Matthaeus, W. H.: 2003, ApJ 592, 564

Janse, Å. M., Lie-Svendsen, Ø., & Leer, E.: 2007, A&A 474, 997

Jensen, T. H.: 1989, ApJ 343, 507

Jia, X., Walker, R. J., Kivelson, M. G., Khuranaand, J. A., & Linker, K. K.: 2008, JGR

Jian, L., Russell, C. T., Luhmann, J. G., & Skoug, R. M.: 2006, SPh 239, 393

Jokipii, J. R. & Parker, E. N: 1968, Phys. Rev. Lett. 21, 44

Jokipii, J. R. & Thomas, B.: 1981, ApJ 243, 1115

Jokipii, J. R., Giacalone, J., & Kóta, J.: 2007, Planet. Space Sci. 55, 2267, doi:10.1016/j.pss.2007.05.007

Jurac, S., & Richardson, J. D.: JGR (Space Physics) 110(A9), 9220

Kabin, K. & Papitashvili, V. O.: 1998, Earth Planets Space 50, 87

Kallenrode, M.-B.: 2001, *Space Physics*, Springer-Verlag

Käpylä, P. J., Korpi, M. J., Ossendrijver, M., & Stix, M.: 2006, A&A 455, 401

Kasper, J. C., Lazarus, A. J., & Gary, S. P.: 2002, GRLe 29, 1839, doi:10.1029/2002 GL015128

Kelley, M. C.: 1989, *The Earth's Ionosphere*, Academic Press

Kennel, C. F.: 1995, *Convection and Substorms. Paradigms of Magnetospheric Phenomenology*, Oxford University Press

Khurana, K. K.: 2001, JGR 106, 25 999

Khurana, K. K. & Schwarzl, H. K.: 2005, JGR 110, A07227, doi:10.1029/2004 JA010757

Khurana, K. K., Kivelson, M. G., Vasyliūnas, V. M., *et al.*: 2004, in W. McKinnon, F. Bagenal, & T. Dowling (eds.), *Jupiter: Planet, Satellites, Magnetosphere*, Cambridge University Press

Killie, M. A., Janse, Å. M., Lie-Svendsen, Ø., & Leer, E.: 2004, ApJ 604, 842

Kingsep, A. S., Chukbar, K. V., & Yan'kov, V. V.: 1990, in B. Kadomtsev (ed.), *Rev. Plasma Phys.*, Vol. 16, Consultants Bureau, p. 243

Kivelson, M. G.: 2007, in A. C.-L. Chian Kamide, Y. (ed.), *Handbook of Solar–Terrestrial Environment*, Springer Verlag

Kivelson, M. G. & Bagenal, F.: 2007, in T. V. Johnson, L. McFadden, & P. Weissman (eds.), *Encyclopedia of the Solar System*, p. 519

Kivelson, M. G. & Russell, C. T.: 1995, *Introduction to Space Physics*, Cambridge University Press

Kivelson, M. G. & Southwood, D. J.: 2005, JGR 110, A12209, doi:10.1029/2005 JA011176

Kivelson, M. G., Bagenal, F., Kurth, W. S., Neubauer, F. M., Paranicas, C., & Saur, J.: 2004, in W. McKinnon, F. Bagenal, & T. Dowling (eds.), *Jupiter: The Planet, Satellites and Magnetosphere*, Cambridge University Press, p. 513

Klein, L. W. & Burlaga, L. F.: 1982, JGR 87, 613

Knetter, T., Neubauer, F. M., Horbury, T., & Balogh, A.: 2004, JGR (Space Physics) 109(A18), 6102, doi:10.1029/2003JA010099

Kohl, J. L., Noci, G., Cranmer, S. R., & Raymond, J. C.: 2006, A&A Rev. 13, 31

Kolmogorov, A. N.: 1941a, Dokl. Akad. Nauk SSSR 30, 301. (Reprinted in Proc. R. Soc. A, 434, 9–13, 1991.)

Kolmogorov, A. N.: 1941b, Dokl. Akad. Nauk SSSR 32, 16. (Reprinted in Proc. R. Soc. A, 434, 15–17, 1991.)

Kopp, R. A. & Pneuman, G. W.: 1976, SPh 50, 85

Kraichnan, R. H.: 1965, Phys. Fluids 8, 1385

Kraichnan, R. H. & Montgomery, D. C.: 1980, Rep. Prog. Phys. 43, 547

Krimigis, S. M., Carbary, J. F., Keath, E. P., *et al.*: 1981, JGR 86, 8227

Kronberg, E. A., Woch, J., Krupp, N., Lagg, A., Khurana, K. K., & Glassmeier, K.-H.: 2005, JGR 110, doi:10.1029/2004JA010777

Krupp, N., Lagg, A., Livi, S., *et al.*: 2001, JGR 106, 26017

Krupp, N., Vasyliūnas, V. M., Woch, J., *et al.*: 2004, in F. Bagenal, T. Dowling, & W. McKinnon (eds.), *Jupiter: Planet, Satellites, Magnetosphere*, Cambridge University Press

Kuijpers, J.: 1997, ApJ 489, L201

Kuperus, M.: 1996, SPh 169, 349

Kupo, I., Mekler, Y., & Eviatar, A.: 1976, ApJL 205, L51

Kurth, W. S., Gurnett, D. A., Clarke, J. T., *et al.*: 2005, Nature 433, 722

Kurth, W. S., Lecacheux, A., Averkamp, T. F., Groene, J. B., & Gurnett, D. A.: 2007, GRLe 34, 2201

Kusano, K., Maeshiro, T., Yokoyama, T., & Sakurai, T.: 2002, ApJ 577, 501

Laming, J. M.: 2004, ApJ 614, 1063

Landau, L. D. & Lifshitz, E. M.: 1959, *Fluid Mechanics*, Pergamon Press

Lankalapalli, S., Flaherty, J. E., Shephard, M. S., & Strauss, H.: 2007, J. Comput. Phys. 225(1), 363

Larmor, J.: 1919, Br. Assoc. Adv. Sci. Rep. 159–160

Lau, Y.-T. & Finn, J. M.: 1990, ApJ 350, 672

Lazarian, A. & Vishniac, E. T.: 1999, ApJ 517, 700

Le, G., Zheng, Y., Russell, C. T., *et al.*: 2008, JGR 113, A01205, doi:10.1029/2007JA012377

Leake, J. E., Arber, T. D., & Khodachenko, M. L.: 2005, A&A 442, 1091, doi:10.1051/0004-6361:20053427

Leamon, R. J., Matthaeus, W. H., Smith, C. W., & Wong, H. K.: 1998a, ApJL 507, L181

Leamon, R. J., Smith, C. W., Ness, N. F., Matthaeus, W. H., & Wong, H. K.: 1998b, JGR 103, 4775

Lee, M. A.: 1984, Adv. Space Res. 4, 295

Lee,.M. A. & Ip, W.-H.: 1987, JGR 92, 11 041

Leenaarts, J., Carlsson, M., Hansteen, V., & Rutten, R. J.: 2007, A&A 473, 625

Leer, E. & Holzer, T. E.: 1980, JGR 85, 4681

Leer, E., Holzer, T. E., & Shoub, E. C.: 1992, JGR 97, 8183

Lees, L: 1964, AIAA J. 1576–1582

Leka, K. D., Canfield, R. C., McClymont, A. N., & Van Driel Gesztelyi, L.: 1996, ApJ 462, 547

Lepping, R. P. & Behannon, K. W.: 1986, JGR 91, 8725

Lepping, R. P., Jones, J. A., & Burlaga, L. F.: 1990, JGR 95, 11 957

Lesieur, M.: 1990, *Turbulence in Fluids*, Nijhoff

Lie-Svendsen, Ø. & Rees, M. H.: 1996, JGR 101, 2415, doi:10.1029/95JA02690

Lie-Svendsen, Ø., Leer, E., & Hansteen, V. H.: 2001, JGR 106, 8217, doi:10.1029/2000JA000409

Lielausis, O.: 1975, Atomic Energy Rev. 13, 527

Liewer, P. C., Neugebauer, M., & Zurbuchen, T.: 2004, SPh 223, 209

Lin, J. & Van Ballegooijen, A. A.: 1995, ApJ 629, 582

Lin, J. & Forbes, T. G.: 2000, JGR 105, 2375

Lin, J., Ko, Y.-K., Sui, L., *et al.*: 2005, ApJ 622, 1251

Litvinenko, Y. E. & Craig, I. J. D.: 1999, SPh 189, 315

Litvinenko, Y. E., Forbes, T. G., & Priest, E. R.: 1996, SPh 167, 445

Liu, Y., Richardson, J. D., Belcher, J. W., Kasper, J. C, & Skoug, R. M.: 2006, JGR (Space Physics) 111, A09108

Longcope, D. W.: 1996, SPh 169, 91

Longcope, D. W.: 2005, Living Rev. Solar Phys. 2, Online Article

Longcope, D. W. & Cowley, S. C.: 1996, Phys. Plasmas 3(8), 2885

Longcope, D. W. & Klapper, I.: 2002, ApJ 579, 468

Longcope, D. W. & Strauss, H. R.: 1994, ApJ 437, 851

Longcope, D. W., Ravindra, B., & Barnes, G.: 2007, ApJ 668, 571

Low, B. C.: 2006a, ApJ 646, 1288

Low, B. C.: 2006b, ApJ 649, 1064

Luhmann, J. G: 1995a, JGR (Space Physics) 100, 12,035

Luhmann, J. G.: 1995b, in *Introduction to Space Physics*, Cambridge University Press

Luhmann, J. G. & Cravens, T. E.: 1991, Space Sci. Rev. 55, 201

Lyons, L. R. & Speiser, T. W.: 1985, JGR 90, 8543

MacBride, B. T., Smith, C. W., & Forman, M. A.: 2008, ApJ submitted

MacKay, D. H., Gaizauskas, V., Rickard, G. J., & Priest, E. R.: 1997, ApJ 486, 534

Maclean, R. C., Parnell, C. E., de Moortel, I., *et al.*: 2006, in *SOHO-17. 10 Years of SOHO and Beyond*, Vol. 617 of *ESA Special Publications*

Malkus, W. V. R. & Proctor, M. R. E.: 1975, J. Fluid Mech. 67, 417

Manchester, W. I.: 2007, ApJ 666, 532

Mariska, J. T.: 1993, *The Solar Transition Region*, Cambridge University Press

Marsch, E. & Tu, C.-Y.: 1994, Ann. Geophys. 12, 1127

Marsch, E., Ao, X.-Z., & Tu, C.-Y.: 2004, JGR 109, A04102, doi:10.1029/2003 JA010330

Martens, P. C. & Zwaan, C.: 2001, ApJ 558, 872

Matsushita, S.: 1967, in S. Matsushita & W. H. Campbell (eds.), *Physics of Geomagnetic Phenomena*, Academic Press, p. 302

Matthaeus, J. V., Shebalin, W. H., & Montgomery, D.: 1983, J. Plasma Phys. 29, 525

Matthaeus, W. H. & Zhou, Y.: 1989, Phys. Fluids B 1, 1929

Matthaeus, W. H., Goldstein, M. L., & Roberts, D. A.: 1990, JGR 95, 20 673

Matthaeus, W. H., Bieber, J. W., & Zank, G. P.: 1995, US Natl. Rep. Int. Union Geod. Geophys. 1991–1994, Rev. Geophys. 33, 609

Matthaeus, W. H., Ghosh, S., Oughton, S., & Roberts, D. A.: 1996a, JGR 101, 7619

Matthaeus, W. H., Zank, G. P., & Oughton, S.: 1996b, J. Plasma Phys. 56, 659

Matthaeus, W. H., Zank, G. P., Smith, C. W., & Oughton, S.: 1999, Phys. Rev. Lett. 82, 3444

Matthaeus, W. H., Minnie, J., Breech, B., Parhi, S., Bieber, J. W., & Oughton, S.: 2004, GRLe 31, L12 803

McComas, D. J. & Bagenal, F.: 2007, GRLe 34, 20 106, doi:10.1029/2007GL031078

McComas, D. J., Phillips, K. L., Bame, S. J., Gosling, J. T., Goldstein, B. E., & Neugebauer, M.: 1995, Space Sci. Rev. 72, 93

McComas, D. J., Barraclough, B. L., Funsten, H. O., *et al.*: 2000, JGR 105, 10 419

McComas, D. J., Elliott, H. A., Schwadron, N. A., Gosling, J. T., Skoug, R. M., & Goldstein, B. E.: 2003, GRLe 30, 24, doi:10.1029/2003GL017136

McComas, D. J., Allegrini, F., Bagenal, F., *et al.*: 2007a, Science 318, 217, doi:10.1126/science.1147393

McComas, D. J., Velli, M., Lewis, W. S., *et al.*: 2007b, Rev. Geophys. 45, RG1004, doi:10.1029/2006RG000195

McCracken, K. G. & Ness, N. F.: 1966, JGR 71, 3315

McDonough, T. R. & Brice, N. M.: 1973, Nature 242, 513

McIntosh, S. W. & Judge, P. G.: 2001, ApJ 561, 420

McIntosh, S. W., Bogdan, T. J., Cally, P. S., *et al.*: 2001, ApJL 548, L237

McKenzie, D. E.: 2000, SPh 195, 381

McNutt, R. L., Belcher, J. W., Sullivan, J. D., Bagenal, F., & Bridge, H. S.: 1979, Nature 280, 803

McNutt, R. L., Haggerty, D. K., Hill, M. E., *et al.*: 2007, Science 318, 220

Mead, G. D.: 1964, JGR (Space Physics) 1181–1195

Mendillo, M., Baumgardner, J., Flynn, B., & Hughes, W. J.: 1990, Nature 348, 312

Merrill, R. T., McElhinney, M. W., & McFadden, P. L.: 1996, *The Magnetic Field of the Earth*, Academic Press

Michel, F. C.: 1982, Rev. Mod. Phys. 54, 1

Michel, F. C. & Sturrock, P. A.: 1974, Planet. Space Sci. 22, 1501

Midgley, J. E. & Davis, L.: 1963, JGR (Space Physics) 68(18), 5111

Miesch, M. S.: 2005, Living Rev. Solar Phys. 2, 1

Miley, G.: 1980, ARA&A 18, 165

Mininni, P. D., Montgomery, D. C., & Pouquet, A.: 2005, Phys. Rev. E 71, 046304

Minter, A. H. & Balser, D. S.: 1997, ApJ 484, L133

Minter, A. H. & Spangler, S. R.: 1996, ApJ 458, 194

Minter, A. H. & Spangler, S. R.: 1997, ApJ 485, 182

Moffatt, H. K.: 1978, *Magnetic Field Generation in Electrically Conducting Fluids*, Cambridge University Press

Moffatt, H. K. & Ricca, R. L.: 1992, Proc. R. Soc. A 439, 411

Moldwin, M. B. & Hughes, W. J.: 1991, JGR 96, 14051

Moldwin, M. B. & Hughes, W. J.: 1992a, GRLe 19, 1081

Moldwin, M. B. & Hughes, W. J.: 1992b, JGR 97, 19252

Moldwin, M. B. & Hughes, W. J.: 1993, JGR 98, 81

Moldwin, M. B. & Hughes, W. J.: 1994, JGR 99, 184

Moldwin, M. B., Ford, S., Lepping, R., Slavin, J., & Szabo, A.: 2000, GRLe 27, 57

Monin, A. S.: 1959, Dokl. Akad. Nauk. SSR 125, 515

Monin, A. S. & Yaglom, A. M.: 1971, *Statistical Fluid Mechanics: Mechanics of Turbulence*, Vol. 1, MIT Press

Monin, A. S. & Yaglom, A. M.: 1975, *Statistical Fluid mechanics: Mechanics of Turbulence*, Vol. 2, MIT Press

Montgomery, D. C.: 1982, Phys. Scr. T 2(1), 83

Montgomery, D. C. & Turner, L.: 1981, Phys. Fluids 24, 825

Montgomery, D. C., Brown, M. R., & Matthaeus, W. H.: 1987, JGR 92, 282

Moore, T. E. & Horwitz, J. L.: 2007, Rev. Geophys. 45

Morabito, L. A., Synnott, S. P., Kupferman, P. N., & Collins, S. A.: 1979, Science 204, 972

Muller, R.: 1983, SPh 85, 113

Mulligan, T. & Russell, C. T.: 2001, JGR 106, 10581

Murphy, N., Smith, E. J., Tsurutani, B. T., Balogh, A., & Southwood, D. J.: 1995, Space Sci. Rev. 72 (1–2), 447

Nagy, A. F., Winterhalter, D., Sauer, K., *et al.*: 2004, Space Sci. Rev. 111, 33

Nakada, M. P.: 1969, SPh 7, 303

Nakamura, M. S., Matsumoto, H., & Fujimoto, M.: 2002, GRLe 29(8), 1247, doi:10.1029/2001GL013780

Nandy, D., Hahn, M., Canfield, R. C., & Longcope, D. W.: 2003, ApJ 597, L73

Ness, N. F.: 1969, Rev. Geophys. Space Phys. 7, 97

Ness, N. F., Scearce, C. S., & Cantarano, S.: 1966, JGR 71, 3305

Neugebauer, M.: 2006, JGR (Space Physics) 111(A10), 4103, doi:10.1029/2005 JA011497

Neugebauer, M. & Snyder, C. W.: 1962, Science 138, 1095

Neugebauer, M., Clay, D. R., Goldstein, B. E., Tsurutani, B. T., & Zwickl, R. D.: 1984, JGR 89, 5395

Neugebauer, M., Forsyth, R. J., Galvin, A. B., *et al.*: 1998, JGR 103, 14587, doi:10.1029/98JA00798

Newcomb, W. A.: 1958, Ann. Phys. (NY) 3, 347

Nichols, J. D. & Cowley, S. W. H.: 2004, Ann. Geophys. 22, 1799

Nichols, J. D. & Cowley, S. W. H.: 2005, Ann. Geophys. 23, 799

Nindos, A., Zhang, J., & Zhang, H.: 2003, ApJ 594, 1033

Nisenson, P., Van Ballegooijen, A. A., de Wijn, A. G., & Sütterlin, P.: 2003, ApJ 587, 458

Nordlund, Å.: 1982, A&A 107, 1

November, L. J. & Simon, G. W.: 1988, ApJ 333, 427

Ogilvie, K. W., Scudder, J. D., Vasyliūnas, V. M., Hartle, R. E., & Siscoe, G. L.: 1977, JGR 82, 1807

Ohtani, S. & Raeder, J.: 2004, JGR (Space Physics) 109(A1), A01207

Øieroset, M., Mitchell, D. L., Phan, T. D., Lin, R. P., Crider, D. H., & Acuña, M. H.: 2006, Space Sci. Rev. 111, 185

Olsen, E. L. & Leer, E.: 1999, JGR 104, 9963

Olson, P., Christensen, U., & Glatzmaier, G. A.: 1999, JGR 104, 10383, doi:10.1029/1999 JB900013

Omidi, N. & Sibeck, D. G.: 2007, GRLe 34, L04106, doi:10.1029/2006GL028698

Ossendrijver, M.: 2003, Astron. Astrophys. Rev. 11, 287

Ossendrijver, M., Stix, M., & Brandenburg, A.: 2001, A&A 376, 713

Ossendrijver, M., Stix, M., Brandenburg, A., & Rüdiger, G.: 2002, A&A 394, 735

Otto, A.: 2001, JGR 106, 3751, doi:10.1029/1999JA001005

Otto, A. & Fairfield, D. H.: 2000, JGR 105, 21 175

Oughton, S., Priest, E. R., & Matthaeus, W. H.: 1994, J. Fluid Mech. 280, 95

Pallier, L. & Prange, R.: 2004, GRLe 31, L06701

Papadopoulos, K., Goodrich, C., Wiltberger, M., Lopez, R., & Lyon, J. G.: 1999, Phys. Chem. Earth C Solar–Terrestial and Planetary Science 24(1–3), 189

Pariat, E., Aulanier, G., Schmieder, B., Georgoulis, M. K., Rust, D. M., & Bernasconi, P. N.: 2004, ApJ 614, 1099

Pariat, E., Démoulin, P., & Berger, M. A.: 2005, A&A 439, 1191

Parker, E. N.: 1955, ApJ 122, 293

Parker, E. N.: 1957, JGR 62, 509

Parker, E. N.: 1958, ApJ 128, 664

Parker, E. N.: 1963, *Interplanetary Dynamical Processes*, Wiley-Interscience

Parker, E. N.: 1965, Space Sci. Rev. 4, 666

Parker, E. N.: 1972, ApJ 174, 499

Parker, E. N.: 1973, J. Plasma Phys. 9, 49

Parker, E. N.: 1979, *Cosmical Magnetic Fields, Their Origin and Their Activity*, Clarendon Press

Parker, E. N.: 1994, *Spontaneous Current Sheets in Magnetic Fields*, Oxford University Press

Parker, E. N.: 1996, JGR 101, 10 587

Parker, E. N.: 2000, in *Magnetospheric Current Systems*, Vol. 118 of *Geophysical Monograph Series*, American Geophysical Union, p. 1

Parker, E. N.: 2004, Phys. Plasmas 11, 2328

Parker, E. N.: 2007, *Conversations on Electric and Magnetic Fields in the Cosmos*, Princeton University Press

Parnell, C. E., Smith, J. M., Neukirch, T., & Priest, E. R.: 1996, Phys. Plasmas 3(3), 759

Paschmann, G., Sckopke, N., Haerendel, G., et al.: 1978, Space Sci. Rev. 22, 717

Paxton, L. J., Christensen, A. B., Humm, D. C., et al.: 1999, in A. M. Larar (ed.), *Proc. SPIE*, Vol. 3756, pp. 265–276

Peale, S. J., Cassen, P., & Reynolds, R. T.: 1979, Science 203, 892

Pedlosky, J.: 1987, *Geophysical Fluid Dynamics*, Springer-Verlag, New York

Perrault, P. & Akasofu, S.-I.: 1978, Geophys. J. R. Astron. Soc. 54, 547

Peter, H., Gudiksen, B. V., & Nordlund, Å.: 2004, ApJL 617, L85, doi:10.1086/427168

Petschek, H. E.: 1964, in W. N. Hess (ed.), *The Physics of Solar Flares*, NASA SP-50, Washington DC, p. 425

Pevtsov, A. A., Canfield, R. C., & Metcalf, T. R.: 1994, ApJ 425, L117

Pevtsov, A. A., Canfield, R. C., & Metcalf, T. R.: 1995, ApJ 440, L109

Pevtsov, A. A., Maleev, V. M., & Longcope, D. W.: 2003, ApJ 593, 1217

Pneuman, G. W.: 1984, SPh 94, 387

Podesta, J. J.: 2007, J. Fluid Mech, submitted

Podesta, J. J., Roberts, D. A., & Goldstein, M. L.: 2006, JGR 111, A10109, doi:10.1029/2006JA011834

Podesta, J. J., Roberts D. A., & Goldstein, M. L.: 2007, ApJ 664, 543

Podesta, J. J., Forman, M. A., & Smith, C. W.: 2008, J. Plasma Phys. in press

Politano, H. & Pouquet, A.: 1998a, GRLe 25, 273

Politano, H. & Pouquet, A.: 1998b, Phys. Rev. E 57, R21

Pond, S., Stewart, R. W., & Burling, R. W.: 1963, J. Atmos. Sci. 20, 319

Pontius, Jr, D. H. & Hill, T. W.: 1989, JGR 94, 15 041

Pontius, Jr, D. H., Wolf, R. A., Hill, T. W., Spiro, R. W., Yang, Y. S., & Smyth, W. H.: 1998, JGR 103, 19 935

Pope, S. B.: 2000, *Turbulent Flows*, Cambridge University Press

Porco, C. C., Helfenstein, P., Thomas, P. C., *et al.*: 2006, Science 311, 1393

Pouquet, A., Frisch, U., & Leorat, J.: 1976, J. Fluid Mech. 77, 321

Prandtl, L. & Tietjens, O. G.: 1934, *Applied Hydro and Aeromechanics*, McGraw-Hill

Priest, E. R.: 1982, *Solar Magnetohydrodynamics*, Reidel

Priest, E. R.: 1996, in K. Tsingansos (ed.), *Solar and Astrophysical MHD Flows*, Kluwer, p. 151

Priest, E. R. & Démoulin, P.: 1995, JGR 100, 23 443

Priest, E. R. & Forbes, T. G.: 1986, JGR 91, 5579

Priest, E. R. & Forbes, T. G.: 2000, *Magnetic Reconnection – MHD Theory and Applications*, Cambridge University Press

Priest, E. R. & Raadu, M. A.: 1975, SPh 43, 177

Priest, E. R. & Titov, V. S.: 1996, Phil. Trans. R. Soc. A 354, 2951

Priest, E. R., Ballegooijen, A. A. Van, & MacKay, D. H.: 1996, ApJ 460, 530

Priest, E. R., Bungey, T. N., & Titov, V. S.: 1997, Geophys. Astrophys. Fluid Dyn. 84, 127

Priest, E. R., Hornig, G., & Pontin, D. I.: 2003, JGR 108(A7) 1285, doi:10.1029/2002JA009812

Proctor, M. R. E. & Gilbert, A. D.: 1995, *Lectures on Solar and Planetary Dynamos*, Cambridge University Press

Rädler, K.-H.: 1980, Astron. Nachr. 301, 101

Rädler, K.-H. & Rheinhardt, M.: 2006, ArXiv Astrophysics e-prints

Rädler, K.-H. & Stepanov, R.: 2006, Phys. Rev. E 73(5), 056311, doi:10.1103/PhysRevE.73.056311

Raeder, J.: 2000, JGR (Space Physics) 105(A6), 13 149

Raeder, J.: 2006, Ann. Geophys. 24, 381

Raeder, J. & Maynard, N. C.: 2001, JGR 106, 345, doi:10.1029/2000JA000600

Raeder, J., Berchem, J., & Ashour-Abdalla, M.: 1998, JGR (Space Physics) 103(A7), 14 787

Raeder, J., McPherron, R. L., Frank, L. A., *et al.*: 2001a, JGR 106, 381, doi:10.1029/2000JA000605

Raeder, J., Wang, Y. L., Fuller-Rowell, T. J., & Singer, H. J.: 2001b, SPh 204(1–2), 325

Raouafi, N.-E. & Solanki, S. K.: 2006, A&A 445, 735

Rastatter, L., Hesse, M., Kuznetsova, M., Sigwarth, J. B., Raeder, J., & Gombosi, T. I.: 2005, JGR (Space Physics) 110(A7), A07212

Rempel, M.: 2006, ApJ 647, 662

Richardson, I. G.: 2004, Space Sci. Rev. 111, 267, doi:10.1023/B:SPAC.0000032689.52830.3e

Richardson, I. G. & Cane, H. V.: 1993, JGR 98, 15 295

Richardson, J. D.: 2002, Planet. Space Sci. 50, 503

Richardson, J. D. & Smith, C. W.: 2003, GRLe 30(5), 1206, doi:10.1029/2002 GL016551

Richardson, J. D., Paularena, K. I., Lazarus, A. J., & Belcher, J. W.: 1995a, GRLe 22, 1469

Richardson, J. D., Paularena, K. I., Lazarus, A. J., & Belcher, J. W.: 1995b, GRLe 22, 325

Richmond, A. D.: 1995, J. Atm. Solar–Terr. Phys. 57, 1103

Ridley, A. J. & Kihn, E. A.: 2004, GRLe 31(7), L07801

Riley, P., Schatzman, C., Cane, H. V., Richardson, I. G., & Gopalswamy, N.: 2006, Astrophys. J. 647, 648

Rishbeth, H. & Garriott, O. K.: 1969, *Introduction to Ionospheric Physics*, Vol. 14 of *International Geophysics Series*, Academic Press

Roberts, D. A., Klein, L. W., Goldstein, M. L., & Matthaeus, W. H.: 1987, JGR 92, 11 021

Rosenbauer, H., Grunwaldt, H., Montgomery, M. D., Paschmann, G., & Sckopke, N.: 1975, JGR (Space Physics) 80, 2723

Rosenbluth, M. N., Monticello, D. A., Strauss, H. R., & White, R. B.: 1976, Phys. Fluids 19, 1987

Rosenthal, C. S., Bogdan, T. J., Carlsson, M., *et al.*: 2002, ApJ 564, 508

Rosner, R., Tucker, W. H., & Vaiana, G. S.: 1978, ApJ 220, 643

Rossi, B. & Olbert, S.: 1970, *Introduction to the Physics of Space*, McGraw-Hill

Rüdiger, G. & Hollerbach, R.: 2004, *The Magnetic Universe: Geophysical and Astrophysical Dynamo Theory*, Wiley-VCH

Russell, C. T.: 1993, JGR 98, 18 681

Russell, C. T.: 2004, Adv. Space Res. 33, 2004

Russell, C. T.: 2006, Adv. Space Res. 37, 1467

Russell, C. T. & Elphic, R. C.: 1978, Space Sci. Rev. 22, 681

Russell, C. T. & Elphic, R. C.: 1979, Nature 279, 616

Russell, C. T. & Mulligan, T.: 2002, Planetary and Space Science 50, 527

Russell, C. T. & Walker, R. J.: 1985, JGR 90, 11 067

Russell, C. T., Khurana, K. K., Kivelson, M. G., & Huddleston, D. E.: 2000, Adv. Space Res. 26, 1499

Rust, D. M.: 2001, JGR 106, 25 075

Rust, D. M. & Kumar, A.: 1994, SPh 155, 69

Ruzmaikin, A. A., Feynman, J., Goldstein, B. E., Smith, E. J., & Balogh, A.: 1995, JGR 100, 3395

Sanchez, E., Summers, D., & Siscoe, G. L.: 1990, JGR (Space Physics) 95, 20 743

Saunders, M. A. & Russell, C. T.: 1986, JGR 91, 5589

Saur, J., Neubauer, F. M., Connerney, J. E. P., Zarka, P., & Kivelson, M. G.: 2004, Plasma interaction of Io with its plasma torus, in *Jupiter: The Planet, Satellites and Magnetosphere*, pp. 537–560

Scalo, J. & Elmegreen, B. G.: 2004, ARA&A 42, 275

Scharmer, G. B., Bjelksjö, K., Korhonen, T. K., Lindberg, B., & Petterson, B.: 2003, in St. L. Keil & S. V. Avakyan (eds.), *Innovative Telescopes and Instrumentation for Solar Astrophysics, Proc, SPIE*, Vol. 4853, p. 341

Schatten, K. H., Wilcox, J. M., & Ness, N. F.: 1969, SPh 6, 442

Scherrer, P. H., Bogart, R. S., Bush, R. I., *et al.*: 1995, SPh 162, 129

Schindler, K., Hesse, M., & Birn, J.: 1988, JGR 93, 5547

Schlegel, K.: 2007, in *Space Weather: Physics and Effects*, Springer

Schmidt, G.: 1966, *Physics of High Temperature Plasmas*, Academic Press

Schmitt, J. H. M. M. & Liefke, C.: 2004, A&A 417, 651

Schneider, N. M. & Bagenal, F.: 2007, in R. Lopes (ed.), *Io After Galileo*, Praxis

Scholer, M.: 1989, JGR 94, 8805

Schrijver, C. J.: 2002, Astron. Nachr. 323, 157

Schrijver, C. J.: 2005, in *Solar Wind 11/SOHO 16, Connecting Sun and Heliosphere*, Vol. 592 of *ESA Special Publications*, p. 213

Schrijver, C. J. & Zwaan, C.: 2000, *Solar and Stellar Magnetic Activity*, Cambridge University Press

Schrijver, C. J., Title, A. M., Berger, T. E., *et al.*: 1999, SPh 187, 261

Schrijver, C. J., Derosa, M. L., Metcalf, T. R., *et al.*: 2006, SPh 235, 161

Schubert, G., Solomatov, V. S., Tackley, P. J., & Turcotte, D. L.: 1988, in R. J. Phillips, S. W. Bougher, & D. M. Hunten (eds.), *Venus II*, University of Arizona Press, p. 429

Schulz, M. & Lanzerotti, L. J.: 1974, *Particle Diffusion in the Radiation Belts*, Springer-Verlag

Schunk, R. W.: 1977, Rev. Geophys. Space Phys. 15, 429

Schunk, R. W. & Nagy, A. F.: 2000, *Ionospheres – Physics, Plasma Physics, and Chemistry*, Cambridge University Press

Schwadron, N. A., Fisk, L. A., & Zurbuchen, T. H.: 1999, ApJ 521, 859

Schwenn, R. & Marsch, E.: 1991, in L. J. Lanzerotti, M. C. E. Huber, & D. Stöffler (eds.), *Physics and Chemistry in Space*, Vol. 21, Springer-Verlag

Seehafer, N.: 1978, SPh 58, 215

Seehafer, N.: 1986, SPh 105, 223

Shapiro, A. H: 1961, *Shape and Flow*, Anchor Books

Sheeley, N. R., Wang, Y.-M., Hawley, S. H., *et al.*: 1997, ApJ 484, 472

Shu, F. H.: 1992, *Physics of Astrophysics*, Vol. II, University Science Books

Siscoe, G. L.: 1966, Planet. Space Sci. 14, 947

Siscoe, G. L.: 1983, Solar system magnetohydrodynamics, in R. L. Carovillano & J. M. Forbes (eds.), *Solar–Terrestrial Physics*, Reidel, pp. 11–100

Siscoe, G. L.: 1979, in C. F. Kennel, L. J. Lanzerotti, and E. N. Parker (eds.), *Solar System Plasma Physics*, Vol. II, North-Holland

Siscoe, G. L.: 1988, in T. Chang, G. B. Crew, & J. R. Jasperse (eds.), *Physics of Space Plasmas (1987)*, Scientific Publishers, p. 3

Siscoe, G. L., Raeder, J., & Ridley, A. J.: 2004, JGR (Space Physics) 109(A9), A09203

Siscoe, G. L.: 2006, J. Atmos. Solar–Terr. Phys. 68, 2119

Siscoe, G. L. & Siebert, K. D.: 2006, J. Atmos. Solar–Terrestrial Phys. 68(8), 911

Siscoe, G. L. & Summers, D.: 1981, JGR 86, 8471

Siscoe, G. L., Davis, L., Jr Coleman, P. J., Jr, Smith, E. J., & Jones, D. E.: 1968, JGR 73, 61

Siscoe, G. L., Frank, L. A., Ackerson, K. L., & Paterson, W. R.: 1994, GRLe 21, 2975

Siscoe, G. L., Erickson, G. M., Sonnerup, B. U. Ö., *et al.*: 2001, JGR 106, 13015, doi:10.1029/2000JA000062

Siscoe, G. L., Crooker, N. U., Erickson, G. M., *et al.*: 2002a, Planet. Space Sci. 50, 461

Siscoe, G. L., Crooker, N. U., & Siebert, K. D.: 2002b, JGR (Space Physics) 107(A10), 1321

Siscoe, G. L., Erickson, G. M., Sonnerup, B. U. Ö., *et al.*: 2002c, JGR 107, 1075, doi:10.1029/2001JA000109

Siscoe, G. L., Crooker, N. U., & Elliott, H. A.: 2006, SPh 239, 293

Skartlien, R.: 2000, ApJ 536, 465

Slavin, J. A.: 1998, in A. Nishida, D. N. Baker, & S. W. H. Cowley (eds.), *New Perspectives on the Earth's Magnetotail*, Vol. 102 of *Geophysical Monograph Series*, American Geophysical Union, p. 225

Slavin, J. A.: 2004, Adv. Space Res. 33, 1859

Slavin, J. A., Smith, E. J., Tsurutani, B. T., *et al.*: 1984, GRLe 11, 657

Slinker, S. P., Fedder, J. A., Emery, B. A., *et al.*: 1999, JGR (Space Physics) 104(A12), 28 379

Slinker, S. P., Fedder, J. A., Ruohoniemi, J. M., & Lyon, J. G.: 2001, JGR 106, 361, doi:10.1029/2000JA000603

Smith, C. W., Matthaeus, W. H., Zank, G. P., Ness, N. F., Oughton, S., & Richardson, J. D.: 2001a, JGR 106, 8253

Smith, C. W., Mullan, D. J., Ness, N. F., Skoug, R. M., & Steinberg, J.: 2001b, JGR (Space Physics) 106(A9), 18625

Smith, C. W., Hamilton, K., Vasquez, B. J., & Leamon, R. J.: 2006a, ApJL 645, L85

Smith, C. W., Isenberg, P. A., Matthaeus, W. H., & Richardson, J. D.: 2006b, ApJ 638, 508

Smith, E. J.: 1973, JGR 78, 2054

Smith, N. S., Padhye, C. W., & Matthaeus, W. H.: 2001, JGR 106, 18 635

Smith, P. A. Isenberg C. W., & Matthaeus, W. H.: 2003, ApJ 592, 564

Smrekar, S. E., Elkins-Tanton, L., Leitner, J. J., *et al.*: 2007, in L. W. Esposito (ed.), *Venus as a Terrestrial Planet*, American Geophysical Union

Söding, A. F., Neubauer, M., Tsurutani, B. T., Ness, N. F., & Lepping, R. P.: 2001, Ann. Geophys. 19, 667

Sokolov, I. V., Powell, K. G., Gombosi, T. I., & Roussev, I. I.: 2006, J. Comput. Phys. 220(1), 1

Somov, B.: 1992, *Physical Processes in Solar Flares*, Kluwer

Song, P., Gombosi, T. I., & Ridley, A. J.: 2001, JGR 106, 8149, doi:10.1029/2000JA000423

Song, P. & Russell, C. T.: 1992, JGR (Space Physics) 97, 1411

Sonnerup, B. U. Ö.: 1970, J. Plasma Phys. 4, 161

Sonnerup, B. U. Ö. & Cahill, L. J.: 1967, JGR 72, 171

Sonnerup, B. U. Ö. & Priest, E. R.: 1975, J. Plasma Phys. 14, 283

Sorisso-Valvo, L., Carbone, V., Veltri, P., Consolini, G., & Bruno, R.: 1999, GRLe 26, 1801

Sorriso-Valvo, L., Marino, R., Carbone, V., *et al.*: 2007, Phys. Rev. Lett. 99, 115 001

Southwood, D. J. & Kivelson, M. G.: 1989, JGR 94, 299

Southwood, D. J. & Kivelson, M. G.: 2007, JGR 112, A12222, doi:10.1029/2007JA012254

Spangler, S. R.: 1991, ApJ 376, 540

Spangler, S. R.: 1999, ApJ 522, 879

Spangler, S. R.: 2007, in D. Shaikh & G. P. Zank (eds.), *Turbulence and Nonlinear Processes in Astrophysical Plasmas, Proc. Sixth Annual International Astrophysics Conf.*, Vol. 932, American Institute of Physics, p. 85

Spitzer, L. & Härm, R.: 1953, Phys. Rev. 89, 977, doi:10.1103/PhysRev. 89.977

Spreiter, J. R. & Stahara, S. S.: 1980, JGR (Space Physics), 85(NA12), 6769

Spreiter, J. R., Summers, A. L., & Alksne, A. Y.: 1966, Planet. Space Sci. 14, 223

Spreiter, J. R., Alksne, A. Y., & Summers, A. L.: 1968, in R. L. Carovillano, J. F. McClay, and H. R. Radoski (eds.), *Physics of the Magnetosphere*, Reidel, p. 301

Spreiter, J. R., Summers, A. L., & Rizzi, A. W.: 1970, Planet. Space Sci. 18, 1281

Sreenivasan, K. R.: 1995, Phys. Fluids 7, 2778

St-Maurice, J.-P. & Schunk, R. W.: 1977, Planet. Space Sci. 25, 907, doi:10.1016/0032-0633(77)90003-4

Stahara, S. S., Rachiele, R. R., Spreiter, J. R., & Slavin, J. A.: 1989, JGR (Space Physics) 94, 13 353

Stawicki, O., Gary, S. P., & Li, H.: 2001, JGR 106, 8273

Steenbeck, M., Krause, F., & Rädler, K. H.: 1966, Z. Naturforschung Teil A 21, 369

Steffl, A. J., Bagenal, F., & Stewart, A. I. F.: 2004, Icarus 172, 91

Steffl, A. J., Delamere, P. A., & Bagenal, F.: 2006, Icarus 180, 124

Stein, R. F. & Nordlund, A.: 1998, ApJ 499, 914

Steinolfson, R. S. & Van Hoven, G.: 1984, Phys. Fluids 27, 1207

Stenflo, J. O.: 1994, *Solar Magnetic Fields. Polarized Radiation Diagnostics*, Vol. 189 of *Astrophysics and Space Sciences Library*, Kluwer

Sterling, A. C. & Hollweg, J. V.: 1988, ApJ 327, 950

Stern, D. P.: 1966, Space Sci. Rev. 6, 147

Stevenson, D. J.: 1981, Science 214, 611

Stevenson, D. J.: 2001, Nature 412, 214

Stevenson, D. J.: 2003, Earth Plan. Sci. Lett. 208, 1

Stevenson, D. J., Spohn, T., & Schubert, G.: 1983, Icarus 54, 466

Stix, M.: 2004, *The Sun: An Introduction*, Springer-Verlag

Stix, T. H.: 1992, *Waves in Plasmas*, American Institute of Physics

Strachan, N. & Priest, E. R.: 1994, Geophys. Astrophys. Fluid Dyn. 74, 245

Strauss, H. R.: 1976, Phys. Fluids 19, 134

Strauss, H. R.: 1977, Phys. Fluids 20, 1354

Strauss, H. R.: 1986, Phys. Fluids 29, 3668

Sturrock, P. A.: 1994, *Plasma Physics: An Introduction to the Theory of Astrophysical, Geophysical & Laboratory Plasmas*, Cambridge University Press

Su, Y.-J., Ergun, R. E., Bagenal, F., & Delamere, P. A.: 2003, JGR 108(A2), 1094, doi:10.1029/2002JA009247

Su, Y.-J., Jones, S. T., Ergun, R. E., *et al.*: 2006, JGR 111(A10), 6211, doi:10.1029/2005JA011252

Sweet, P. A.: 1950, MNRAS 110, 69

Sweet, P. A.: 1958, in B. Lehnert (ed.), *Electromagnetic Phenomenon in Cosmical Physics*, Cambridge University Press, p. 123

Syrovatskii, S. I.: 1971, Sov. Phys. JETP 33(5), 933

Taktakishvili, A., Kuznetsova, M. M., Hesse, M., *et al.*: 2007, JGR (Space Physics) 112(A9), A09203

Tennekes, H. & Lumley, J. L.: 1972, *A First Course in Turbulence*, MIT Press

Terada, N., Machida, S., & Shinagawa, H.: 2002, JGR 107, (A12) 1471, doi:10.1029/2001JA009224

Thomas, N., Bagenal, F., Hill, T. W., & Wilson, J. K.: 2004, The Io neutral clouds and plasma torus, in *Jupiter: The Planet, Satellites and Magnetosphere*, pp. 561–591

Thomas, V. A. & Winske, D.: 1991, GRLe 18, 1943

Thompson, S. M., Kivelson, M. G., Khurana, K. K., *et al.*: 2005, JGR (Space Physics) 110(A9), 2212, doi:10.1029/2004JA010714

Titov, V. S.: 1992, SPh 139, 401

Titov, V. S., Priest, E. R., & Démoulin, P.: 1993, A&A 276, 564

Titov, V. S., Hornig, G., & Démoulin, P.: 2002, JGR 107(A8), 1164, doi:10.1029/2001JA000278

Tokar, R. L., Johnson, R. E., Hill, T. W., *et al.*: 2006, Science 311, 1409

Tomczyk, S., McIntosh, S. W., Keil, S. L., *et al.*: 2007, Science 317, 1192

Tóth, G.: 2000, J. Comput. Phys. 161(2), 605

Tóth, G., Kovács, D., Hansen, K. C., & Gombosi, T. I.: 2004, JGR 109(A18), 11 210, doi:10.1029/2004JA010406

Tóth, G., De Zeeuw, D. L., Gombosi, T. I., & Powell, K. G.: 2006, J. Comput. Phys. 217(2), 722

Totten, T. L., Freeman, J. W., & Arya, S.: 1995, JGR 100, 13

Trafton, L., Parkinson, T., & Macy, Jr, W.: 1974, ApJL 190, 85

Trang, F. S. & Rickett, B. J.: 2007, ApJ 661, 1064

Tranquille, C., Sanderson, T. R., Marsden, R. G., Wenzel, K.-P., & Smith, E. J.: 1987, JGR 92, 6

Tsinober, A. B.: 1975, Magnitnaya Gidrodinamika 1, 7

Tsurutani, B. T.: 1991, in A. D. Johnstone (ed.), *Cometary Plasma Processes*, American Geophysical Union, p. 189

Tu, C.-Y. & Marsch, E.: 2002, JGR 107(A09), 1249

Tur, T. J. & Priest, E. R.: 1976, SPh 48, 89

Turkmani, R., Vlahos, L., Galsgaard, K., Cargill, P. J., & Isliker, H.: 2005, ApJL 620, L59

Ugai, M.: 1988, Computer Phys. Commun. 49, 185

Unti, T. W. J. & Neugebauer, M.: 1968, Phys. Fluids 11(3), 563

Vainshtein, S. I. & Cattaneo, F.: 1992, ApJ 393, 165

Van Allen, J. A. & Bagenal, F.: 1998, in J. K. Beatty, C. C. Petersen, & A. Chaikin (eds.), *The New Solar System*, Cambridge University Press and Sky Publishing

Van Ballegooijen, A. A.: 1985, ApJ 298, 421

Van Ballegooijen, A. A.: 1988, Geophys. Astrophys. Fluid Dynam. 41, 181

Van Ballegooijen, A. A. & Martens, P. C. H.: 1989, ApJ 343, 971

Van Ballegooijen, A. A., DeLuca, E. E., Squires, K., & Mackay, D. H.: 2007, J. Atmos. Solar–Terrestrial Phys. 69, 24, doi:10.1016/j.jastp.2006.06.007

van Driel-Gesztelyi, L., Démoulin, P., & Mandrini, C. H.: 2003, Adv. Space Res. 32, 1855, doi:10.1016/S0273-1177(03)90619-3

Vasquez, B. J., Abramenko, V. I., Haggerty, D. K., & Smith, C. W.: 2007a, JGR (Space Physics) 112(A11), 11 102, doi:10.1029/2007JA012504

Vasquez, B. J., Smith, C. W., Hamilton, K., MacBride, B. T., & Leamon, R. J.: 2007b, JGR 112, A07101

Vasyliūnas, V. M.: 1970, in B. M. McCormack (ed.), *Particles and Fields in the Magnetosphere*, Reidel, p. 60

Vasyliūnas, V. M.: 1975a, in V. Formisano (ed.), *The Magnetospheres of the Earth and Jupiter*, Reidel, p. 179

Vasyliūnas, V. M.: 1975b, Rev. Geophys. 13, 303

Vasyliūnas, V. M.: 1979, Space Sci. Rev. 24, 609

Vasyliūnas, V. M.: 1983, in A. J. Dessler (ed.), *Physics of the Jovian Magnetosphere*, Cambridge University Press, p. 395

Vasyliūnas, V. M.: 1988, in *Proc. Joint Varenna–Abastumani International School and Workshop on Plasma Astrophysics, ESA SP-285*, Vol. 1

Vasyliūnas, V. M.: 1989a, in T. D. Guyenne and J. J. Hunt (eds.), *Plasma Astrophysics*, ESA SP-285, Vol. I, Noordwijk, The Netherlands, p. 31

Vasyliūnas, V. M.: 1989b, GRLe 16, 1465

Vasyliūnas, V. M.: 2001, GRLe 28, 2177, doi:10.1029/2001GL013014

Vasyliūnas, V. M.: 2004, Adv. Space Res. 33, 2113, doi:10.1016/j.asr.2003.04.051

Vasyliūnas, V. M: 2005a, Ann. Geophys. 23, 1347

Vasyliūnas, V. M: 2005b, Ann. Geophys. 23, 2589

Vasyliūnas, V. M: 2007, Ann. Geophys. 25, 255

Vasyliūnas, V. M.: 2008, Ann. Geophys. 26, 1341

Vasyliūnas, V. M. & Pontius, D. H.: 2007, JGR (Space Physics) 112, A10204, doi:10.1029/2007JA012457

Vasyliūnas, V. M. & Song, P.: 2005, JGR (Space Physics) 110, A02301, doi:10.1029/2004JA010615

Vasyliūnas, V. M., Kan, J. R., Siscoe, G. L., & Akasofu, S.-I.: 1982, Planet. Space Sci. 30, 359, doi:10.1016/0032-0633(82)90041-1

Verdini, A. & Velli, M.: 2007, ApJ 662, 669

Verma, M. K., Roberts, D. A., & Goldstein, M. L.: 1995, JGR 100, 19 839

Vernazza, J. E., Avrett, E. H., & Loeser, R.: 1981, ApJSS 45, 635

Violante, L., Cattaneo, M., Bavassano, B., Moreno, G., & Richardson, J. D.: 1995, JGR (Space Physics) 100, 12 047

Vögler, A. & Schüssler, M.: 2007, A&A 465, L43

Vogt, J. & Glassmeier, K.-H.: 2001, Adv. Space Res. 28, 863

Volish, A. D. & Koliesnikov, Y. B.: 1976, Dokl. Akad. Nauk. SSSR 229, 573

von Steiger, R. & Murdin, P.: 2000, *Encyclopedia of Astronomy and Astrophysics*, doi:10.1888/0333750888/2265

Waite, J. H., Combi, M. R., Ip, W.-H., *et al.*: 2006, Science 311, 1419

Waite Jr, J. H., Bagenal, F., Seward, F., *et al.*: 1994, JGR 99, 14799

Walker, R. J. & Russell, C. T.: 1991, JGR 90, 7397

Walker, R. J. & Russell, C. T.: 1995, in M. G. Kivelson & C. T. Russell (eds.), *Introduction to Space Physics*, Cambridge University Press

Wang, Y.-M., Sheeley, Jr, N. R., Walters, J. H., *et al.*: 1998, ApJL 498, L165

Wang, Y. L., Raeder, J., & Russell, C. T.: 2004, Ann. Geophys. 22, 1001

Wesson, J. A.: 1987, *Tokamaks*, Oxford University Press

Whang, Y. C.: 1977, SPh 53, 507

White, W. W., Siscoe, G. L., Erickson, G. M., *et al.*: 1998, GRLe 25, 1605

Wikstøl, Ø., Hansteen, V. H., Carlsson, M., & Judge, P. G.: 2000, ApJ 531, 1150, doi:10.1086/308475

Wiltberger, M., Lyon, J. G., & Goodrich, C. C.: 2003, J. Atmos. Solar–Terrestrial Phys. 65(11–13), 1213

Wiltberger, M., Wang, W., Burns, A. G., Solomon, S. C., Lyon, J. G., & Goodrich, C. C.: 2004, J. Atmos. Solar–Terrestrial Phys. 66(15–16), 1411

Woch, J., Krupp, N., & Lagg, A.: 2002, GRLe 29, 1138

Wolf, R. A.: 1970, JGR 75, 4677

Wolf, R. A.: 1983, in R. L. Carovillano & J. M. Forbes (eds.), *Solar–Terrestrial Physics*, Reidel, 303

Wolf, R. A.: 1995, Class notes, Houston, TX

Wolf, R. A., Spiro, R. W., Sazykin, S., & Toffoletto, F. R.: 2007, J. Atmos. Solar–Terrestrial Phys. 69(3), 288

Wolff, R. S., Goldstein, B. E., & Yeates, C. M.: 1980, JGR 85, 7697

Wood, B. E., Müller, H.-R., Zank, G. P., Linsky, J. L., & Redfield, S.: 2005, ApJL 628, L143

Wu, H., Hill, T. W., Wolf, R. A., & Spiro, R. W.: 2007, JGR 112(A11), 2206, doi:10.1029/2006JA012032

Yamada, M.: 2007, Phys. Plasmas 14(5), 058102

Yan, M., Lee, L. C., & Priest, E. R.: 1992, JGR 97, 8277

Yang, Y. S., Wolf, R. A., Spiro, R. W., Hill, T. W., & Dessler, A. J.: 1994, JGR 99, 8755

Yeh, T.: 1976, JGR 81(13), 2140

Yokoyama, T. & Shibata, K.: 1994, ApJL 436, L197

Yoshimura, H.: 1975, ApJSS 29, 467

Young, D. T.: 1997, in J. T. Shirley & R. W. Fairbridge (eds.), *Encyclopedia of Planetary Sciences*, Van Nostrand Reinhold

Young, D. T.: 1998, in D. T. Young, R. F. Pfaff, & J. E. Borovsky (eds.), *Measurement Techniques in Space Plasmas: Particles*, American Geophysical Union

Zank, G. P. & Matthaeus, W. H.: 1992, J. Plasma Phys. 48, 85

Zank, G. P. & Matthaeus, W. H.: 1993, Phys. Fluids A 5, 257

Zank, G. P., Matthaeus, W. H., & Smith, C. W.: 1996, JGR 101, 17 093

Zarka, P., Lamy, L., Cecconi, B., Prange, R., & Rucker, H. O.: 2007, Nature 450, 265

Zeleny, L. M. & Kuznetsova, M. M.: 1988, Astron. Zhurn. 65, 626, in Russian

Zhang, J. C., Liemohn, M. W., De Zeeuw, D. L., *et al.*: 2007, JGR (Space Physics) 112(A4), A04208

Zieger, B., Vogt, J., Glassmeier, K.-H., & Gombosi, T. I.: 2004, JGR (Space Physics) 109(A18), 7205, doi:10.1029/2004JA010434

Zirker, J. B., Martin, S. F., Harvey, K., & Gaizauskas, V.: 1997, SPh 175, 27

Zwan, B. J. & Wolf, R. A.: 1976, JGR (Space Physics) 81, 1636

Zwickl, R. D., Asbridge, J. R., Bame, S. J., Feldman, W. C., Gosling, J. T., & Smith, E. J.: 1983, in *Solar Wind Five*, p. 711

Index

Printed in the United States
By Bookmasters